Climate Change Impact and Adaptation in Agricultural Systems

CABI CLIMATE CHANGE SERIES

Climate change is a major environmental challenge to the world today, with significant threats to ecosystems, food security, water resources and economic stability overall. In order to understand and research ways to alleviate the effects of climate change, scientists need access to information that not only provides an overview of and background to the field, but also keeps them up to date with the latest research findings.

This series addresses many topics relating to climate change, including strategies to develop sustainable systems that minimize impact on climate and/or mitigate the effects of human activity on climate change. Coverage will encompass all areas of environmental and agricultural sciences. Aimed at researchers, upper level students and policy makers, titles in the series provide international coverage of topics related to climate change, including both a synthesis of facts and discussions of future research perspectives and possible solutions.

Titles Available

1. Climate Change and Crop Production
   Edited by Matthew P. Reynolds
2. Crop Stress Management and Global Climate Change
   Edited by José L. Araus and Gustavo A. Slafer
3. Temperature Adaptation in a Changing Climate: Nature at Risk
   Edited by Kenneth B. Storey and Karen K. Tanino
4. Plant Genetic Resources and Climate Change
   Edited by Michael Jackson, Brian Ford-Lloyd and Martin Parry

Climate Change Impact and Adaptation in Agricultural Systems

Edited by

Jürg Fuhrer

Agroscope, Zurich, and Oeschger Centre for Climate Change Research, University of Bern, Switzerland

Peter J. Gregory

East Malling Research, Kent, and School of Agriculture, Policy and Development, University of Reading, UK

CABI is a trading name of CAB International

CABI	CABI
Nosworthy Way	745 Atlantic Avenue
Wallingford	8th Floor
Oxfordshire OX10 8DE	Boston, MA 02111
UK	USA
Tel: +44 (0)1491 832111	Tel: +1 (617)682-9015
Fax: +44 (0)1491 833508	
E-mail: info@cabi.org	E-mail: cabi-nao@cabi.org
Website: www.cabi.org	

© CAB International 2014. All rights reserved. No part of this publication may be reproduced in any form or by any means, electronically, mechanically, by photocopying, recording or otherwise, without the prior permission of the copyright owners.

A catalogue record for this book is available from the British Library, London, UK.

The Library of Congress has cataloged the hardcover edition as follows:

Climate change impact and adaptation in agricultural systems / edited by Jürg Fuhrer, Peter J. Gregory.
   pages cm. -- (CABI climate change series ; 5)
  ISBN 978-1-78064-289-5 (hbk : alk. paper) 1. Crops and climate. 2. Climatic changes. 3. Agricultural systems. I. Fuhrer, Jürg. II. Gregory, P. J. III. C.A.B. International. IV. Series: CABI climate change series ; 5.
  S600.7.C54C659 2014
  630.2'515--dc23
                       2014011559

ISBN-13: 978 1 78639 535 1 (PB)

Commissioning editor: Vicki Bonham
Editorial assistant: Emma McCann
Production editor: Simon Hill

Typeset by Columns XML Ltd, Reading
Printed and bound by CPI Group (UK) Ltd, Croydon, CR0 4YY
Book Services Ltd, Didcot, Oxon
First printed in hardback in 2014. Transferred to POD paperback in 2019.

Contents

Contributors		vii
Foreword		xi
	Climate Change Impact and Adaptation in Agricultural Systems – Introduction *Jürg Fuhrer and Peter J. Gregory*	1
1	**Climate Projections for 2050** *Markku Rummukainen*	7
2	**Rainfed Intensive Crop Systems** *Jørgen E. Olesen*	17
3	**Climate Sensitivity of Intensive Rice–Wheat Systems in Tropical Asia: Focus on the Indo-Gangetic Plains** *Anil Kumar Singh and Himanshu Pathak*	31
4	**Climate Change Challenges for Low-Input Cropping and Grazing Systems – Australia** *Steven Crimp, Mark Howden, Chris Stokes, Serena Schroeter and Brian Keating*	47
5	**Diversity in Organic and Agroecological Farming Systems for Mitigation of Climate Change Impact, with Examples from Latin America** *Walter A.H. Rossing, Pablo Modernel and Pablo A. Tittonell*	69
6	**UK Fruit and Vegetable Production – Impacts of Climate Change and Opportunities for Adaptation** *Rosemary Collier and Mark A. Else*	88
7	**Intensive Livestock Systems for Dairy Cows** *Robert J. Collier, Laun W. Hall and John F. Smith*	110
8	**Climate Change and Integrated Crop–Livestock Systems in Temperate-Humid Regions of North and South America: Mitigation and Adaptation** *Alan J. Franzluebbers*	124

9	**Land Managed for Multiple Services** *Richard Aspinall*	140
10	**Adaptation of Mixed Crop–Livestock Systems in Asia** *Fujiang Hou*	155
11	**Enhancing Climate Resilience of Cropping Systems** *Heidi Webber, Helena Kahiluoto, Reimund Rötter and Frank Ewert*	167
12	**Shaping Sustainable Intensive Production Systems: Improved Crops and Cropping Systems in the Developing World** *Clare Stirling, Jon Hellin, Jill Cairns, Elan Silverblatt-Buser, Tadele Tefera, Henry Ngugi, Sika Gbegbelegbe, Kindie Tesfaye, Uran Chung, Kai Sonder, Rachael A. Cox, Nele Verhulst, Bram Govaerts, Phillip Alderman and Matthew Reynolds*	186
13	**The Role of Modelling in Adapting and Building the Climate Resilience of Cropping Systems** *Helena Kahiluoto, Reimund Rötter, Heidi Webber and Frank Ewert*	204
14	**Agroforestry Solutions for Buffering Climate Variability and Adapting to Change** *Meine van Noordwijk, Jules Bayala, Kurniatun Hairiah, Betha Lusiana, Catherine Muthuri, Ni'matul Khasanah and Rachmat Mulia*	216
15	**Channelling the Future? The Use of Seasonal Climate Forecasts in Climate Adaptation** *Lauren Rickards, Mark Howden and Steve Crimp*	233
16	**Agricultural Adaptation to Climate Change: New Approaches to Knowledge and Learning** *Julie Ingram*	253
17	**What are the Factors that Dictate the Choice of Coping Strategies for Extreme Climate Events? The Case of Farmers in the Nile Basin of Ethiopia** *Temesgen Tadesse Deressa*	271
Index		281

Contributors

Phillip Alderman, International Maize and Wheat Improvement Center (CIMMYT), Apdo. Postal 6-641, 06600 Mexico, DF, Mexico. E-mail: p.alderman@cigar.org

Richard Aspinall, James Hutton Institute, Craigiebuckler, Aberdeen, AB15 8QH, UK. E-mail: Richard.Aspinall@hutton.ac.uk

Jules Bayala, World Agroforestry Centre (ICRAF), JL. CIFOR, Situ Gede, PO Box 161, Bogor 16001, Indonesia. E-mail: J.Bayala@cgiar.org

Jill Cairns, International Maize and Wheat Improvement Center (CIMMYT), PO Box MP 163, Mount Pleasant, Harare, Zimbabwe. E-mail: j.cairns@cigar.org

Uran Chung, International Maize and Wheat Improvement Center (CIMMYT), Apdo. Postal 6-641, 06600 Mexico, DF, Mexico. E-mail: u.chung@cigar.org

Robert J. Collier, University of Arizona, Department of Animal Science, Shantz Building 236, PO Box 210038, Tucson, AZ 85721-0038, USA. E-mail: rcollier@ag.arizona.edu

Rosemary Collier, Warwick Crop Centre, School of Life Sciences, University of Warwick, Wellesbourne, Warwick CV35 9EF, UK. E-mail: Rosemary.Collier@warwick.ac.uk

Rachael A. Cox, International Maize and Wheat Improvement Center (CIMMYT), Apdo. Postal 6-641, 06600 Mexico, DF, Mexico. E-mail: r.cox@cigar.org

Steven Crimp, CSIRO Climate Adaptation Flagship and Ecosystem Sciences, Canberra, ACT 2601, Australia. E-mail: steven.crimp@csiro.au

Temesgen Tadesse Deressa, Guest Scholar, Africa Growth Initiative, Global Economy and Development, Brookings Institute, 1775 Massachusetts Ave., NW, Washington, DC 20036, USA. E-mail: tderessa@brookings.edu

Mark A. Else, East Malling Research, New Road, East Malling, Kent ME19 6BJ, UK. E-mail: mark.else@emr.ac.uk

Frank Ewert, University of Bonn, Institute of Crop Science and Resource Conservation (INRES), Crop Science Group, Katzenburgweg 5, D-53115 Bonn, Germany. E-mail: frank.ewert@uni-bonn.de

Alan J. Franzluebbers, USDA – Agricultural Research Service, 3218 Williams Hall, NCSU Campus, Box 7619, Raleigh, NC 27695, USA. E-mail: Alan.Franzluebbers@ars.usda.gov

Jürg Fuhrer, Climate Group/Air Pollution, Agroscope, Reckenholzstrasse 191, CH-8046 Zurich, Switzerland, and Oeschger Centre for Climate Change Research, University of Bern, Zähringerstrasse 25, CH-3012 Bern, Switzerland. E-mail: juerg.fuhrer@agroscope.admin.ch

Sika Gbegbelegbe, International Maize and Wheat Improvement Center (CIMMYT), CRAF House, United Nations Avenue, Gigiri PO Box 1041, Village Market-00621, Nairobi, Kenya. E-mail: s.gbegbelegbe@cigar.org

Bram Govaerts, International Maize and Wheat Improvement Center (CIMMYT), Apdo. Postal 6-641, 06600 Mexico, DF, Mexico. E-mail: b.govaerts@cigar.org

Peter J. Gregory, East Malling Research, New Road, East Malling, Kent, ME19 6BJ, and School of Agriculture, Policy and Development, University of Reading, Reading, RG6 6AR, UK. E-mail: Peter.Gregory@emr.ac.uk

Kurniatun Hairiah, Brawijaya University, Jl. Veteran, Malang 65145, Indonesia. E-mail: kurniatunhairiah@gmail.com

Laun W. Hall, University of Arizona, Department of Animal Science, Shantz Building 236, PO Box 210038, Tucson, AZ 85721-0038, USA. E-mail: lwhall@email.arizona.edu

Jon Hellin, International Maize and Wheat Improvement Center (CIMMYT), Apdo. Postal 6-641, 06600 Mexico, DF, Mexico. E-mail: j.hellin@cigar.org

Fujiang Hou, State Key Laboratory of Grassland Agro-Ecosystem, China College of Pastoral Agriculture Science and Technology, Lanzhou University, Lanzhou, 730020, Gansu, China. E-mail: cyhoufj@lzu.edu.cn

Mark Howden, CSIRO Climate Adaptation Flagship and Ecosystem Sciences, Canberra, ACT 2601, Australia. E-mail: Mark.Howden@csiro.au

Julie Ingram, Countryside and Community Research Institute, University of Gloucestershire, Gloucester, Gloucestershire, GL2 9HW, UK. E-mail: jingram@glos.ac.uk

Helena Kahiluoto, MTT Agrifood Research Finland, Plant Production Research, Lönnrotinkatu 5, 50100 Mikkeli, Finland. E-mail: helena.kahiluoto@mtt.fi

Brian Keating, CSIRO Sustainable Agriculture Flagship, Ecosciences Precinct, Brisbane QLD 4001, Australia. E-mail: brian.keating@csiro.au

Ni'matul Khasanah, World Agroforestry Centre (ICRAF), JL. CIFOR, Situ Gede, PO Box 161, Bogor 16001, Indonesia. E-mail: n.khasanah@cgiar.org

Betha Lusiana, World Agroforestry Centre (ICRAF), JL. CIFOR, Situ Gede, PO Box 161, Bogor 16001, Indonesia. E-mail: B.Lusiana@cgiar.org

Pablo Modernel, Farming Systems Ecology, Wageningen University, 6700 AN Wageningen, The Netherlands, and Facultad Agronomía, Universidad de la República, CP 12900, Montevideo, Uruguay. E-mail: pablo.modernelhristoff@wur.nl

Rachmat Mulia, World Agroforestry Centre (ICRAF), JL. CIFOR, Situ Gede, PO Box 161, Bogor 16001, Indonesia. E-mail: r.mulia@cgiar.org

Catherine Muthuri, World Agroforestry Centre (ICRAF), JL. CIFOR, Situ Gede, PO Box 161, Bogor 16001, Indonesia. E-mail: C.Muthuri@cgiar.org

Henry Ngugi, International Maize and Wheat Improvement Center (CIMMYT), Apdo. Postal 6-641, 06600 Mexico, DF, Mexico. E-mail: h.ngugi@cigar.org

Jørgen E. Olesen, Aarhus University, Department of Agroecology, DK-8830 Tjele, Denmark. E-mail: jorgene.olesen@agrsci.dk

Himanshu Pathak, Centre for Environment Science and Climate Resilient Agriculture, Indian Agricultural Research Institute, New Delhi 110 012, India. E-mail: hpathak.iari@gmail.com

Matthew Reynolds, International Maize and Wheat Improvement Center (CIMMYT), Apdo. Postal 6-641, 06600 Mexico, DF, Mexico. E-mail: m.reynolds@cigar.org

Lauren Rickards, Melbourne Sustainable Society Institute, University of Melbourne, Southern Annex, Ground Floor Alice Hoy Building (Blg 162) Monash Road, Parkville, Victoria 3010, Australia. E-mail: lauren.rickards@unimelb.edu.au

Walter A.H. Rossing, Farming Systems Ecology, Wageningen University, 6700 AN Wageningen, The Netherlands. E-mail: Walter.Rossing@wur.nl

Reimund Rötter, MTT Agrifood Research Finland, Plant Production Research, Lönnrotinkatu 5, 50100 Mikkeli, Finland. E-mail: reimund.rotter@mtt.fi

Markku Rummukainen, Centre for Environmental and Climate Research, Lund University, Sölvegatan 37, SE-223 63 Lund, Sweden. E-mail: Markku.Rummukainen@nateko.lu.se

Serena Schroeter, CSIRO Climate Adaptation Flagship, Highett VIC 3190, Australia. E-mail: Serena.schroeter@csiro.au

Elan Silverblatt-Buser, International Maize and Wheat Improvement Center (CIMMYT), Apdo. Postal 6-641, 06600 Mexico, DF, Mexico. E-mail: esilverblatt@gmail.com

Anil Kumar Singh, RVS Agricultural University, Gwalior 474 002, India. E-mail: aksingh.icar@gmail.com

John F. Smith (deceased), University of Arizona, Department of Animal Science, Shantz Building 236, PO Box 210038, Tucson, AZ 85721-0038, USA.

Kai Sonder, International Maize and Wheat Improvement Center (CIMMYT), Apdo. Postal 6-641, 06600 Mexico, DF, Mexico. E-mail: k.sonder@cigar.org

Clare Stirling, International Maize and Wheat Improvement Center (CIMMYT), Ynys Mon, Wales, LL74 8NS, UK, E-mail: c.stirling@cgiar.org

Chris Stokes, CSIRO Climate Adaptation Flagship and Ecosystem Sciences, Aitkenvale QLD 4814, Australia. E-mail: Chris.Stokes@csiro.au

Tadele Tefera, International Maize and Wheat Improvement Center (CIMMYT), CRAF House, United Nations Avenue, Gigiri PO Box 1041, Village Market-00621, Nairobi, Kenya. E-mail: t.tefera@cgiar.org

Kindie Tesfaye, International Maize and Wheat Improvement Centre (CIMMYT), PO Box 5689, Addis Ababa, Ethiopia. E-mail: k.tesfaye@cigar.org

Pablo A. Tittonell, Farming Systems Ecology, Wageningen University, 6700 AN Wageningen, The Netherlands. E-mail: pablo.tittonell@wur.nl

Meine van Noordwijk, World Agroforestry Centre (ICRAF), JL. CIFOR, Situ Gede, PO Box 161, Bogor 16001, Indonesia. E-mail: m.vannoordwijk@cgiar.org

Nele Verhulst, International Maize and Wheat Improvement Center (CIMMYT), Apdo. Postal 6-641, 06600 Mexico, DF, Mexico. E-mail: n.verhulst@cigar.org

Heidi Webber, University of Bonn, Institute of Crop Science and Resource Conservation (INRES), Crop Science Group, Katzenburgweg 5, D-53115 Bonn, Germany. E-mail: hwebber@uni-bonn.de

Foreword

Climate change impact and adaptation in agricultural systems

Climate is changing. We see the climatic changes as predicted by models from the 1990s coming true today, and these changes will continue into the future according to recent predictions. The latest 5th Assessment Report of the Intergovernmental Panel on Climate Change (IPCC) states that cumulative emissions of CO_2 largely determine global mean surface warming by the late 21st century and beyond, and that most aspects of climate change will persist for many centuries even if emissions of CO_2 are stopped. This will continue to impact on many, if not all, aspects of human life, including the provision of sufficient and safe food. We face changes that require us to adapt our agricultural systems to higher temperatures and more extreme weather conditions. We need to do this in an era where the world population will increase to 9–10 billion in 2050, with major increases in developing countries. These enormous challenges require new resilient agri-food systems that will provide us with enough food while preserving our natural resources and protecting the environment. They need to produce more than they do today, as we face an increased demand of 70% as a result of a growing population and changing diets by 2050. To meet the challenges, we must explore options for adapted systems.

Starting from a summary of projections of global and regional climate change, this book discusses in detail the need for the adaptation of different agricultural systems in different regions in the world, with each of these systems having its own characteristics, climate sensitivities and possibilities for change. Systems addressed range from intensive rice–wheat systems in the Indo-Gangetic Plains to integrated livestock systems in the Americas and Asia, and to organic systems in Latin America; from intensive rainfed cropping systems in Europe to low-input systems in dry areas of Australia; from fruit and vegetable production systems in the UK to agroforestry systems, and to intensive livestock systems in the USA. The book discusses the need for and constraints on transferring knowledge from science to practice, using the example of farmers in the Nile Basin in Ethiopia, and the support that can be expected from using seasonal weather forecasts in farm management. It addresses the need for and the role of models; this is particularly important, as models allow us to explore the resilience of cropping systems under variable and future climatic conditions. The book also pays attention to the other functionalities of production systems, the so-called ecosystems services, and the role of multifunctional landscapes.

This book is a timely production, with a wealth of up-to-date knowledge on the impact of climatic change on agricultural systems and the options for adaptation. It illustrates how we can realize the triple win of climate-smart agriculture: increasing food production while mitigating climate change in systems that are adapted to changing conditions. The book shows that we do have options. We need to explore and exploit these options carefully, and by acting today, I am sure that we will have sophisticated new systems by the time we need them most. International organizations such as the Consortium of International Agricultural Research Centres (CGIAR), with its 15 agricultural research centres focusing on the developing world, together with the agricultural research organizations and universities such as Wageningen UR in the Netherlands and many others, and the private sector also, will all play a key role in this endeavour.

Prof Dr Martin J. Kropff
Rector Magnificus Wageningen University
Member of the Consortium Board of the CGIAR

Climate Change Impact and Adaptation in Agricultural Systems – Introduction

Jürg Fuhrer[1] and Peter J. Gregory[2]

[1]*Climate Group/Air Pollution, Agroscope, Zurich, and Oeschger Centre for Climate Change Research, University of Bern, Switzerland;* [2]*East Malling Research, Kent, and School of Agriculture, Policy and Development, University of Reading, UK*

Background

The world's population has doubled since the 1970s to over 7 billion in 2013, and it is expected to grow to about 9 billion by 2050. The largest increase is projected for the developing world, with countries experiencing a growth rate higher than 3% year^{-1} predominately located in Africa and the Middle East (UN, 2013). This population increase is accompanied by a growing demand for food, and changes in dietary requirements towards more livestock products, fat and sugar, as observed recently in the emerging economies of the Middle East, Latin America and Asia (FAO, WFP and IFAD, 2012). In the near future, these dietary changes may override population growth as a major driver behind increasing land requirement for the production of food (Kastner *et al.*, 2012).

World food production has increased by 18% over the past 20 years, but to satisfy the future food demand of all people, further increases in agricultural productivity will be required, together with reductions in postharvest losses and food waste (Beddington *et al.*, 2012). But, current yield growth rates in major staple crops are insufficient to reach the expected necessary doubling of crop production by 2050 (Ray *et al.*, 2013). Current productivity in rainfed systems is reaching, on average, little more than half of its potential (FAO, 2012). To close the yield gap in these systems requires improved agricultural practices that conserve soil, water and air quality, and biodiversity (Sawyer and Cassmann, 2013). These improved practices will continue to depend on inputs of fertilizers, pesticides, water and energy, although at levels that preserve environmental integrity (Tilman *et al.*, 2011). It will be one of the great challenges for agriculture to increase the productivity of land-based production systems on the basis of limited resources such as fertile soil and water while minimizing environmental impacts and maintaining, regulating and supporting ecosystem services to society (Bommarco *et al.*, 2012). Sustainable – or ecological – intensification to increase productivity per unit of land area becomes even more challenging when considering climate change, which will rapidly alter the conditions under which agricultural systems are managed.

The Importance of Climate Change

Increasing global mean temperature is the fundamental characteristic of climate change, and global climate models project a gradual increase of several degrees by the end of the 21st century (Rummukainen, Chapter 1, this volume). It is important to note that the projected future pace and scale of change may exceed any historical trend. According to the IPCC (Intergovernmental Panel on Climate Change; 2007), average northern hemisphere temperatures during the second half of the 20th century were 'very likely' higher than during any other 50-year period in the past 500 years and 'likely' the highest in at least the past 1300 years. Furthermore, projections from models indicate that decadal average warming by 2030 is 'very likely' to be at least twice as large as the corresponding model-estimated natural variability during the 20th century.

The consequences of global climate change will differ regionally as warming will not be uniform, with some regions warming more than others and some regions getting wetter and others getting drier. Using a statistical multi-parameter approach, Diffenbaugh and Giorgi (2012) identified the Amazon, the Sahel and tropical West Africa, Indonesia and the Tibetan Plateau as persistent hotspots in the CMIP5 (Coupled Model Intercomparison Project Phase 5; http://cmip-pcmdi.llnl.gov/cmip5/) global model ensemble, and areas of southern Africa, the Mediterranean, the Arctic and Central America/western North America as additional hotspot regions under a strong climate-forcing pathway. Several of these hotspot regions are important for agricultural production. In a study concerning tropical regions, Ericksen et al. (2011), using different climate change thresholds, identified southern Africa and selected areas in north-east Brazil, Mexico, Pakistan, India and Afghanistan as being highly exposed to climate change.

As a consequence of the different spatial patterns of change, implications for agriculture need to be assessed at the scale at which decisions are taken. This requires using climate scenarios downscaled from global climate model outputs to estimate anomalies relative to the current climate at local and regional scales. For instance, while Knox et al. (2012) provided evidence of consistent yield losses by 2050 for the major regional and local crops across Africa (wheat, maize, sorghum and millet) and South Asia (maize and sorghum) caused mainly by higher temperatures, Sultan et al. (2013) found contrasting results for the Sudanian and Sahelian regions, with crop yields in the former region (sorghum and millet) strongly sensitive to high temperatures and those in the latter region primarily responsive to changes in rainfall. Singh and Pathak (Chapter 3, this volume) discuss more specifically the generally negative implications of warming in intensive rice–wheat systems in tropical Asia, a region with a particularly high population density. In temperate, more humid regions, climate warming and increasing precipitation may have both positive and negative effects, as discussed by Olesen (Chapter 2, this volume). Hence, coping with the projected changes in climatic conditions will require the development of spatially differentiated strategies to mitigate negative trends and to take advantage of positive trends in crop and livestock productivity.

As the climate continues to change, it is expected that increasing variability with more frequent droughts, heatwaves, storms or hail will cause greater uncertainty in production. Climate extremes, i.e. extreme weather or extreme climate events, are defined according to the IPCC (2012) as 'the occurrence of a value of a weather or climate variable above (or below) a threshold value near the upper (or lower) ends of the range of observed values of the variable'. However, projections of the frequency of climate extremes remain uncertain. In the short term, the signal is small relative to natural variability, while in the longer term the signal is larger, but projections remain uncertain due to both uncertain greenhouse gas emissions and climate model outputs (Hawkins and Sutton, 2009). Uncertainties in climate model outputs propagate to impact models used to investigate the implications of climate change and to evaluate possible adaptation strategies (Kahiluoto et al., Chapter 13, this volume). From an analysis of a number of CMIP3 and CMIP5 model outputs, Ramirez-Villegas et al. (2013) concluded that at least an additional 5–30 years of work was still required to improve regional temperature simulations, and at least 30–50 years for precipitation simulations, for these to be inputted directly into impact models. On top of that, simulated climate change impacts vary among the impact models, as shown recently by Asseng et al. (2013) for wheat, which necessitates using multi-model ensembles.

Need for Adaptation

Without adapting the structure and management of the various types of agricultural systems, the negative impacts of rapid changes in temperature and rainfall would lower agricultural outputs in most regions. However, adaptation may be critical in vulnerable regions, with 'vulnerability' defined as the degree to which a system is susceptible to, or unable to cope with, the adverse effects of climate change, including climate variability and extremes. Vulnerability is a function of the character, magnitude and rate of climate change and variation to which a system is exposed, its sensitivity and its adaptive capacity (IPCC, 2012). Regions considered most vulnerable are located predominantly in sub-Saharan Africa (Malone and Brenkert, 2009).

Many generic options for adaptation exist, as summarized by Vermeulen et al. (2012), and agricultural systems are intrinsically dynamic and the adoption of new practices is not new. As an example, crop and livestock production systems in the USA have expanded across a diversity of growing conditions over the past 150 years, and responded to variations in climate and other natural resources (Walthall et al., 2012). Also, smallholder, subsistence and pastoral systems located in marginal environments, areas of high variability of rainfall or high risk of natural hazards, are often characterized by livelihood strategies aiming to reduce overall vulnerability to climate shocks and to manage their impacts ex post (Morton, 2007).

Options for adaptation in intensive cropping systems are discussed by Webber et al. (Chapter 11, this volume), by Crimp et al. (Chapter 4, this volume) specifically for low-input and grazing systems in Australia and by Stirling et al. (Chapter 12, this volume) for cropping systems in the developing world. Collier et al. (Chapter 7, this volume) present a range of measures to cope with increasing heat stress in intensive livestock systems in the USA, and Collier and Else (Chapter 6, this volume) summarize climate-related risks and adaptation options in horticultural systems in the UK. However, due to significant uncertainties in regional climate projections discussed above, decision making is challenging (Hallegatte, 2009). The overall uncertainty attached to projections of climate change impacts the requirements to be considered when planning adaptation for agriculture. Vermeulen et al. (2013) propose a framework for prioritizing adaptation approaches in different time frames. Moreover, decision makers and the farming industry need to search for anticipatory and robust adaptation strategies, i.e. no-regret strategies that are not specific to individual climate impact projections and which lead to higher climate resilience.

In addition to different timescales, spatial scales at which climate adaptations are developed and assessed are of major importance. Responses at different decision levels should be considered, and to support decision making, trade-offs among the following factors are typically involved (van Delden et al., 2011): (i) the scale at which end-users or policy makers require information; (ii) the scale at which processes take place and the representation of those processes in a single model; (iii) the way to integrate model components representing processes occurring at different scales; and (iv) the limitations posed by practical restrictions, such as data limitations and computation speed. The majority of the studies on the impacts of and adaptation to climate change in agriculture have been conducted at the field or farm level. Since effects and responses depend on local conditions and interactions with soils, climate and cropping systems, there is considerable need for an increased number of integrated regional studies (Olesen et al., 2011). In this volume, several contributions deal with regional approaches to adaptation. Rossing et al. (Chapter 5, this volume) describe how diversity in organic and agroecological farming may help to mitigate climate change impacts in Latin America. Franzluebbers (Chapter 8, this volume) discusses adaptation options for mixed crop–livestock systems of North and South America, and Hou (Chapter 10) presents some ideas of how integrated crop–livestock systems in Asia may be adapted while also mitigating climate change.

Types of Adaptation

According to Vermeulen *et al.* (2013), adaptation of agricultural systems involves both better management of agricultural risks and adaptation to progressive climate change. This should result in higher climate resilience of the production systems. As reviewed by Webber *et al.* (Chapter 11, this volume), the climate resilience of a system reflects its capacity to persist in its essential functions in the face of climatic shocks. Transformation towards higher resilience requires, in many circumstances, higher adaptive capacity which is the ability of the system to adjust to climate variability and extremes. This involves, for instance, better access to new technologies and improved seed, and education (Stirling *et al.*, Chapter 12, this volume). At lower levels of climate change, i.e. in the nearer term, 'incremental adaptation' to manage known risks better is most realistic. According to Stafford-Smith *et al.* (2011), incremental adaptation implies that adjustments are aimed at enabling the decision maker to continue to meet current objectives under changed conditions. Use of seasonal forecasts could be important in assisting decision making, but as discussed by Rickards *et al.* (Chapter 15, this volume), proper use of inevitably uncertain forecasts needs further development and strong stakeholder involvement. At higher levels of climate change, 'transformative adaptation' requires fundamental changes to those objectives, as it focuses on two major types of change (Rickards and Howden, 2012): changes in the goals (e.g. change of land use or function) and/or changes in location. The costs of transformational adaptation are expected to be high, but may become profitable as the severity of climate change increases. As an example, van Noordwijk *et al.* (Chapter 14, this volume) discuss how agroforestry could contribute to adapting food production systems through the physical buffering effects of trees on the micro- and mesoclimate, multifunctional landscapes and rural livelihood systems, provided that the advice not to oversimplify and overspecialize their farms and landscapes is taken up by the farmers. Another example of alternative land use is the various forms of organic farming, as discussed by Rossing *et al.* (Chapter 5, this volume).

Barriers to Adaptation

Several of the contributions present a range of options for adaptation that are specific to different types of production systems; however, implementation of these options is not always easy. As discussed by Deressa (Chapter 17, this volume) for the case of farmers in the Nile Basin of Ethiopia, to sustain livelihoods during extreme climate events requires policy measures targeting specific socio-economic and environmental factors such as education of the head of the household, farm income, livestock ownership, farmer-to-farmer extension, ownership of a radio and better-quality homes. Stafford-Smith *et al.* (2011) presented a number of barriers to adaptation, such as psychological and social barriers, cognitive response to uncertainty and governance structure and institutional barriers. According to Marshall *et al.* (2012), four elements of transformational capacity are essential: (i) how risks and uncertainty are managed; (ii) the extent of skills in planning, learning and reorganizing; (iii) the level of financial and psychological flexibility to undertake change; and (iv) the willingness to undertake change. To increase adaptive – or transformational – capacity, Ingram (Chapter 16, this volume) suggests a combination of scientific and human development solutions based on the promotion of adaptation technologies, facilitation of farmer learning, co-production of knowledge and provision of climate information (communicating risk, weather forecasting, etc.). However, Rickards *et al.* (Chapter 15, this volume) discuss the difficulties in using seasonal climate forecasts, and the potential disadvantages of producers relying too strongly on these forecasts.

Finally, land use for food production has a major influence on the environment and on many agroecosystem services other than food production. Aspinall (Chapter 9, this volume) argues that human activity and land management need to be considered in conjunction with environmental system processes in order to produce multiple benefits across the landscape. This complex interaction of ecosystems, land use and land management presents a major challenge in reaching sustainable and climate-resilient agricultural production systems (Sayer et al., 2013). Using scenario analysis and optimization on different spatial scales may help to identify important trade-offs between land use and ecosystem services (Seppelt et al., 2013), but further developments are necessary to make this approach acceptable to stakeholders for their decision-making process.

Scope of the Book

This book provides insights into sensitivities and possible adaptive measures to climate change across a number of different production systems, ranging from intensive cropping and horticultural systems to mixed crop–livestock systems, to intensive livestock systems, and across regions ranging from humid temperate to dryland areas. It also provides a discussion of the opportunities for and the limitations to adaptation.

References

Asseng, S., Ewert, F., Rosenzweig, C., Jones, J.W., Hatfield, J.L., Ruane, A.C., et al. (2013) Uncertainty in simulating wheat yields under climate change. *Nature Climate Change* 3, 827–832.

Beddington, J., Asaduzzaman, M., Clark, M., Fernández, A., Guillou, M., Jahn, M., et al. (2012) *Achieving Food Security in the Face of Climate Change*. Final Report from the Commission on Sustainable Agriculture and Climate Change. CGIAR Research Program on Climate Change, Agriculture and Food Security (CCAFS), Copenhagen (www.ccafs.cgiar.org/commission, accessed 15 November 2013).

Bommarco, R., Kleijn, D. and Potts, S.G. (2012) Ecological intensification: harnessing ecosystem services for food security. *Trends in Ecology and Evolution* 28, 230–238.

Diffenbaugh, N.S. and Giorgi, F. (2012) Climate change hotspots in the CMIP5 global climate model ensemble. *Climatic Change* 114, 813–822.

Ericksen, P., Thornton, P., Notenbaert, A., Cramer, L., Jones, P. and Herrero, M. (2011) *Mapping Hotspots of Climate Change and Food Insecurity in the Global Tropics*. CCAFS Report No 5. CGIAR Research Program on Climate Change, Agriculture and Food Security (CCAFS), Copenhagen (www.ccafs.cgiar.org, accessed 15 November 2013).

FAO (2012) *Statistical Year Book 2012. World Food and Agriculture*. FAO, Rome (http://www.fao.org/docrep/015/i2490e/i2490e00.htm, accessed 15 November 2013).

FAO, WFP and IFAD (2012) *The State of Food Insecurity in the World 2012. Economic Growth is Necessary but not Sufficient to Accelerate Reduction of Hunger and Malnutrition*. FAO, Rome.

Hallegatte, S. (2009) Strategies to adapt to an uncertain climate change. *Global Environmental Change – Human Policy Dimension* 19, 240–247.

Hawkins, E. and Sutton, R. (2009) The potential to narrow uncertainty in regional climate predictions. *Bulletin of the American Meteorological Society* 90, 1095–1107.

IPCC (2007) *Climate Change 2007 – The Physical Science Basis*. Contribution of Working Group I to the Fourth Assessment Report of the Intergovernmental Panel on Climate Change. Solomon, S., Qin, D., Manning, M., Chen, Z., Marquis, M., Averyt, K.B., et al. (eds). Cambridge University Press, Cambridge, UK, and New York.

IPCC (2012) *Managing the Risks of Extreme Events and Disasters to Advance Climate Change Adaptation*. A Special Report of Working Groups I and II of the Intergovernmental Panel on Climate Change. Field, C.B., Barros, V., Stocker, T.F., Qin, D., Dokken, D.J., Ebi, K.L., et al. (eds). Cambridge University Press, Cambridge, UK, and New York, 582 pp.

Kastner, T., Ibarrola Rivas, M.J., Koch, W. and Nonhebel, S. (2012) Global changes in diets and the consequences for land requirements for food. *Proceedings of the National Academy of Sciences* 109, 6868–6872.

Knox, J., Hess, T., Daccache, A. and Wheeler, T. (2012) Climate change impacts on crop productivity in Africa and South Asia. *Environmental Research Letters* 7, 034032.

Malone, E.L. and Brenkert, A. (2009) Vulnerability, sensitivity, and coping/adaptive capacity worldwide. In: Ruth, M. and Ibarraran, M. (eds) *The Distributional Effects of Climate Change: Social and Economic Implications*. Elsevier Science, Dordrecht, the Netherlands, pp. 8–45.

Marshall, N., Park, S., Adger, W., Brown, K. and Howden, S. (2012) Transformational capacity and the influence of place and identity. *Environmental Research Letters* 7, 034022.

Morton, J.F. (2007) The impact of climate change on smallholder and subsistence agriculture. *Proceedings of the National Academy of Sciences* 104, 19680–19685.

Olesen, J.E., Trnka, M., Kersebaum, K.C., Skjelvag, A.O., Seguin, B., Peltonen-Sainio, P., et al. (2011) Impacts and adaptation of European crop production systems to climate change. *European Journal of Agronomy* 34, 96–112.

Ramirez-Villegas, J., Challinor, A.J., Thornton, P.K. and Jarvis, A. (2013) Implications of regional improvement in global climate models for agricultural impact research. *Environmental Research Letters* 8, 024018.

Ray, D.K., Mueller, N.D., West, P.C. and Foley, J.A. (2013) Yield trends are insufficient to double global crop production by 2050. *PLoS ONE* 8, e66428.

Rickards, L. and Howden, M. (2012) Transformational adaptation: agriculture and climate change. *Crop Pasture Science* 63, 240–250.

Sawyer, J. and Cassmann, K.G. (2013) Agricultural innovation to protect the environment, *Proceedings of the National Academy of Sciences* 110, 8345–8348.

Sayer, J., Sunderland, T., Ghazoul, J., Pfund, J.-L., Sheil, D., Meijaard, E., et al. (2013) Ten principles for a landscape approach to reconciling agriculture, conservation, and other competing land uses. *Proceedings of the National Academy of Sciences* 110, 8349–8356.

Seppelt, R., Lautenbach, S. and Volk, M. (2013) Identifying trade-offs between ecosystem services, land use, and biodiversity: a plea for combining scenario analysis and optimization on different spatial scales. *Current Opinion in Environmental Sustainability* 5, 1–6.

Stafford Smith, M., Horrocks, L., Harvey, A. and Hamilton, C. (2011) Rethinking adaptation for a 4°C world. *Philosophical Transactions of the Royal Society* A 369, 196–216.

Sultan, B., Roudier, P., Quirion, P., Alhassane, A., Muller, B., Dingkuhn, M., et al. (2013) Assessing climate change impacts on sorghum and millet yields in the Sudanian and Sahelian savannas of West Africa. *Environmental Research Letters* 8, 014040.

Tilman, D., Balzer, C., Hill, J. and Befort, B.L. (2011) Global food demand and the sustainable intensification of agriculture. *Proceedings of the National Academy of Sciences* 108, 20260–20264.

UN (2013) *World Population Prospects: The 2012 Revision, Highlights and Advance Tables*. ESA/P/WP.228, United Nations, Department of Economic and Social Affairs, Population Division. United Nations, New York.

van Delden, H., van Vliet, J., Rutledge, D.T. and Kirkby, M.J. (2011) Comparison of scale and scaling issues in integrated land-use models for policy support. *Agriculture, Ecosystems and Environment* 142, 18–28.

Vermeulen, S.J., Aggarwal, P.K., Ainslie, A., Angelone, C., Campbell, B.M., Challinor, A.J., et al. (2012) Options for support to agriculture and food security under climate change. *Environmental Science and Policy* 15, 136–144.

Vermeulen, S.J., Challinor, A.J., Thornton, P.K., Campbell, B.M., Eriyagama, N., Vervoort, J., et al. (2013) Addressing uncertainty in adaptation planning for agriculture. *Proceedings of the National Academy of Sciences* 110, 8357–8362.

Walthall, C.L., Hatfield, J., Backlund, P., Lengnick, L., Marshall, E., Walsh, M., et al. (2012) *Climate Change and Agriculture in the United States: Effects and Adaptation*. USDA Technical Bulletin 1935. US Department of Agriculture (USDA), Washington, DC.

1 Climate Projections for 2050

Markku Rummukainen
Centre for Environmental and Climate Research, Lund University, Sweden

1.1 Introduction

In order to assess the implications of climate change in terms of impacts and adaptation needs, projections of the future climate are needed. Climate models are the primary means of such simulations. The results are often coined 'climate scenarios' but should really be called projections, as they are built on alternative scenarios of future land-use changes and greenhouse gas emissions. The basis for climate projections is discussed in this chapter, together with a selection of general results that are of key relevance for agriculture, which stem from state-of-the-art climate projections. This chapter provides the background for the subsequent chapters in this book, and discusses climate projections for the next few decades. While the focus is on the period until 2050, it should be noted that climate change will very likely continue well beyond the middle of the 21st century. Indeed, the long-term prospects are about not only a changed climate but also a climate that is changing over time, i.e. it is about continuous change over a long time. The same is thus also true for our knowledge requirements regarding climate change impacts, as well as the motivation and need for climate change adaptation; however, these may take form.

1.2 Basis for Climate Change Projections

1.2.1 General

While we observe and experience our contemporary climate and intrinsically may expect its past behaviour to also give us a good picture of things to come, future conditions are innately unknown to us. This is especially true for the consequences of the use of fossil fuels and land-use change, which increasingly adds greenhouse gases to the atmosphere, not least carbon dioxide but also methane, nitrous oxide, etc. There are emissions that affect the tropospheric ozone, which has an effect on the climate, in addition to impacts on health and vegetation. Human activities also affect the amount of sulfate particles and soot in the atmosphere, which further compounds our impact on the climate. Land-use change, in addition to affecting carbon sources and sinks, affects the physical properties of the land surface, which further adds to the forces that the climate now responds to on global, regional and local scales (Pitman *et al.*, 2011).

That we force the climate and that the climate responds is certain (IPCC, 2007). Climate change projections for the future have, however, uncertainties. This should not be confused with the view that they are left wanting; evaluation of climate models suggests that they perform well in many respects, and as they are based on physical principles, their results do have considerable credibility.

Model shortcomings are one source of uncertainty. Scenario uncertainty concerning underlying future emissions and land-use pathways is another. In addition, the climate system exhibits internal variability that arises from the complex interplay between the atmosphere, the ocean and the other climate system components. The relative importance of these sources of uncertainty is well established (e.g. Hawkins and Sutton, 2009).

Climate projection uncertainty is smaller at the global scale compared to the regional scale – and even more so, compared to the local scale. This is due largely to the ubiquitous internal variability that can simultaneously affect different regions in contrasting ways but which is largely cancelled out in the global mean. The relative importance of internal variability for climate projection uncertainty declines over time, as the forced climate change signals become greater. At the same time, the uncertainty linked to the emissions and land-use change scenarios grows. The uncertainty attributed to climate models has a more constant presence compared to the other two factors. These sources of uncertainty are discussed below.

1.2.2 Climate-forcing scenarios: fossil fuels and land-use change

Underlying climate model projections, i.e. forward-looking simulations of the evolution of the climate system, are scenarios of climate-forcing factors. In terms of the climate over the next few decades and beyond, this concerns anthropogenic emissions of greenhouse gases, particles and their precursors and the indirect greenhouse gases (see above), as well as land-use change. While today's energy systems, consumption patterns, food and fibre production do lock us on to a path of continued climate change in the short and medium term, the longer-term situation is less certain. Thus, scenario assumptions of emissions and land-use change are an important part of uncertainty in climate projections.

Knowledge of both the underlying climate-forcing scenario (emissions, land-use change) and of the climate model (cf. 'climate sensitivity', see below) is paramount when considering a specific climate change projection; for example, in terms of temperature change. Climate models, emissions and land-use scenarios have evolved over time. Early on, more or less idealized scenarios were used, which were followed by more versatile ones. Over the past 10 years or so, most global and regional climate projections have been based on the IPCC Special Report on Emissions Scenarios (SRES; Nakićenović and Swart, 2000), which span a range of possible future emissions pathways. The most recent climate projections are based on the RCP scenarios (representative concentration pathway; Moss et al., 2010). These are coined RCP2.6, RCP4.5, RCP6.0 and RCP8.5. The SRES and the RCP scenarios are set up in different ways. The former provides greenhouse gas emission and land-use change pathways, based on underlying assumptions regarding socio-economic drivers such as population and economic and technical development. The atmospheric concentrations of greenhouse gases are then derived from the emissions scenarios, for use in climate models. The RCP scenarios provide radiative forcing/greenhouse gas concentration scenarios for the 21st century. The number attached to each scenario designates the radiative forcing in W m^{-2} by 2100. There is an accompanying effort with RCPs for the generation of corresponding greenhouse gas emissions, land use and socio-economic developments.

There is no one-to-one comparability of the RCPs and the SRES, but they span much of the same range of alternative future climate forcing. The RCPs, however, also include a scenario (RCP2.6) that corresponds to considerably lower emissions than any of the SRES scenarios, and as such is aligned with a considerable mitigation effort. Still, neither the SRES nor the RCPs are recommendations for policy, or forecasts. There are no probabilities affixed to them.

Overall, when interpreting climate model results, information is needed about the underlying emissions and land-use scenarios, as climate change projections largely scale with the emissions scenario. How much so depends, however, on the climate models themselves.

1.2.3 Climate sensitivity

Alongside assumptions regarding the emissions pathways and future land-use change, a second key uncertainty in climate change projections is the sensitivity of the climate system. Simply put, this is a measure of how much the climate changes when it is forced.[1] Climate sensitivity is the net measure of the direct effect of the forcing and the feedback that arises within the climate system. An example of key feedback is that a warmer atmosphere can hold more moisture; as water vapour is a greenhouse gas, this enhances the initial warming. Possible changes in clouds are another key feedback. The sign of the overall climate sensitivity is robustly known (positive, i.e. feedback enhances the change due to some initiating factor, such as emissions), but its magnitude is generally only known within a range of values. The range of climate sensitivity in climate models overlaps with the body of estimates based on historical and contemporary climate variations, which provides confidence in the models and their results.

1.2.4 Climate models

Climate models are sophisticated simulation models that build on the physical, chemical and biological understanding of the climate system, written in computer code. There are today quite a few global and regional climate models in the world, constantly under further development, evaluation and in use in research. The majority of climate models have interacting components for the atmosphere, the ocean and the land surface, but there are also models with interactive carbon cycle and vegetation components, as well as some with yet additional climate system components. The latter are today known as 'Earth System Models'. In the case of models that do not carry a vegetation component, relevant properties are prescribed.

While climate models do exhibit various biases, they also perform well in many respects (Randall et al., 2007), including the overall global and regional climate characteristics and the reproduction of observed changes over time.

Global climate models are fundamental when considering the response of the climate system to forcing. Solving the equations in climate models requires, however, extensive computational power, not least as simulations span from decades to centuries and often need to be repeated with several variations. This constrains the resolution of the models. Even today, many global climate models have a resolution ('grid size') of a few hundred kilometres. This is insufficient for resolving variable landforms and other physiographical details that have a significant effect on the near-surface climate in many regions. Global model results are therefore applied to regions by various downscaling techniques (Rummukainen, 2010) such as statistical models and regional climate models. The latter is also known as dynamic downscaling.

The global climate modelling community has a long tradition of organizing co-ordinated simulation experiments that span many climate models and different sets of simulations. Many of the results in the Fourth Assessment Report of the IPCC (Intergovernmental Panel on Climate Change; Meehl et al., 2007) came from the so-called CMIP3 coordinated study, which was followed by CMIP5 (Taylor et al., 2012). Coordinated regional climate model studies are fewer but have, during the past few years, emerged for several regions, not least Europe (Christensen and Christensen, 2007; Kjellström et al., 2013), the Americas (Menendez et al., 2010; Mearns et al., 2012) and Africa (Paeth et al., 2011). Coordinated studies of course provide more information for the characterization of model-related uncertainties, either by co-consideration of all results or by allowing a specific scenario to be tested in a wider context, including how it compares with other scenarios.

Advanced climate models are based on fundamental physical laws. This enables their use in projections of the future beyond the observed period. There are limitations on model resolution, as mentioned above, meaning that small-scale processes that cannot be resolved have to be parameterized (i.e. represented with approximate descriptions). For example, cloud formation cannot be simulated explicitly in climate models as it ultimately involves very detailed mechanisms. Rather, it may be parameterized in terms of relevant large-scale ambient conditions in the models. Parameterization is, however, also based on the physical understanding of the involved processes.

Parameterizations are formulated in somewhat different ways in different climate models. This explains why climate models as a whole exhibit a range of climate sensitivities. This range overlaps observational estimates.

1.2.5 Internal variability

Finally, the climate system is non-linear. This manifests itself in ubiquitous internal variability within the climate system, resulting in inter-annual variability and also variability at the decadal scale. The global mean temperature, for example, exhibits some inter-annual variability in concert with the large-scale interaction between the ocean and the atmosphere in the Pacific, known as El Niño–Southern Oscillation (ENSO). ENSO also has various strong regional signals around the world of anomalous warmth and coolness, as well as unusually wet and dry conditions. Different variability patterns characterize yet other world regions, including the Arctic and North Atlantic Oscillations (AO, NAO; Thompson and Wallace, 1998), the Pacific-North American Pattern (PNA), as well as the more regular monsoon circulations.

The presence of significant regional-scale climate variability implies that, to begin with, while climate change is indisputably discernible at the large scale, it still may remain within regional-scale variability, meaning that it may be more difficult to identify conclusively at this scale. The same applies to climate projections and, consequently, the emergence of statistically significant change occurs later in many regions than in the global mean (Giorgi and Bi, 2009; Kjellström et al., 2013). Mahlstein et al. (2012) find, for example, that statistically significant regional precipitation changes emerge only once global mean warming climbs above 1.4°C, which is roughly a doubling of the warming until the beginning of the 2000s. There is not, however, an absence of ongoing regional changes before clear signals emerge; rather, regional climates undergo transitions that may manifest themselves earlier as changes in, for example, the likelihood of extreme events (Stott et al., 2004; Jaeger et al., 2008), before the mean climate shows a significant response.

1.3 Projections

1.3.1 Temperature

Temperature change is a fundamental characteristic of climate change ('global warming' is often used synonymously with the present-day 'climate change'). The observed global mean change since the pre-industrial era is large compared to variability over comparable timescales, and now amounts to c.0.8–0.9°C. To keep the global mean temperature rise under 2°C has been agreed as the international target under the UN Framework Convention on Climate Change (UNFCCC). However, the present evolution of emissions is not aligned with emissions pathways that might provide a likely chance of meeting the two-degree goal (e.g. Peters et al., 2013), suggesting that global warming may well come to exceed this UNFCCC target. The majority of climate change projections to date build on scenarios that do not include specific new climate policy measures and, consequently, result in a larger warming than the two-degree goal. The IPCC (2007) Fourth Assessment Report contained projected global warming results that ranged from around 1°C to more than 6°C for the period between the late 20th century and the late 21st century, with consideration of different emissions scenarios, climate models and information on climate change impacts on the carbon cycle. When additionally considering the observed warming since the pre-industrial until the late 20th century, the same projected change in temperature increases to c.1.5–7°C.

The climate system response to forcing is not uniform. While the overall pattern due to emissions is one of warming, some regions will warm more (or less) than others, and thus more (or less) than the global mean change (see Plate 1). For example, a 2°C global mean warming would imply temperature increases larger than 2°C over land regions.

Changes in the average temperature emerge over time in a relatively gradual manner. Changes in variability, and not least in extremes, can, however, manifest themselves in more complicated ways. Intuitively, and what is also evident in climate projections, is that warm extremes become more commonplace, whereas cold extremes less so (e.g. Zwiers et al., 2011; Orlowski and Seneviratne, 2012; Rummukainen, 2012). It is also characteristic that in areas in which there is a reduction in seasonal snow cover, such as the high northern latitudes, the reduction of cold extremes exceeds the wintertime mean temperature change. Correspondingly, in the relatively dry subtropical areas that experience increasing dryness, changes in warm extremes exceed the average regional temperature change (e.g. Kharin et al., 2013).

As extremes manifest themselves in a more or less sporadic fashion, changes in them are more difficult to pinpoint than those of climate means (Trenberth, 2012). Extremes can also change in terms of their return period or likelihood of occurrence, magnitude, geographical distribution, and so on. When posing the question of whether extreme events will change in ways that impact a specific sector or region, the vulnerability of the activity or the location needs to be specified. The use of indices may be helpful (Sillman and Roeckner, 2008; Zhang et al., 2011).

1.3.2 Precipitation

Global precipitation increases with global warming. Results suggest that the increase in the global mean of precipitation is around 2% for each 1°C rise in temperature. The projected changes are non-uniform over the globe, as evident from Plate 1 (see also, for example, Solomon et al., 2009). Regional changes are often larger or smaller than the global mean change. There is a distinct large-scale pattern that is coined 'wet gets wetter' and 'dry gets drier' (cf. Held and Soden, 2006). Although there are exceptions to this, it by and large summarizes the big picture well. Consequently, precipitation is projected to increase at high and middle latitudes, decrease in the subtropical regions and increase in parts of the tropics. In the transition zones between these divergent patterns, the projected change is very small or of insignificant magnitude. The exact location of the transition regions varies to some extent between models and projections. Thus, in the affected regions, while some projections may suggest an increase, others can show a decrease.

A general increase in the occurrence of heavy precipitation is, however, a typical result both for regions in which precipitation on average increases and for regions in which precipitation on average decreases. There are many measures for extreme precipitation. For example, such heavy precipitation events that at the end of the 20th century had a return period of 20 years are projected to become 1-in-15- to 1-in-10-year events by around 2050 across most of the global land area (IPCC, 2012); that is, with a rate of increase that is twice or more the rate of the global mean precipitation increase (Kharin et al., 2013). The uncertainty due to climate model quality is larger for the tropics than for many other regions. Climate projections tend to exhibit large increases in extreme precipitation compared to changes in average precipitation (Rummukainen, 2012; Kharin et al., 2013; Sillman et al., 2013).

Precipitation is a basic measure of hydrological conditions, but does not wholly describe issues relating to water availability; information is also needed on evapotranspiration, soil moisture, drought risks, runoff, etc. For example, there are studies that indicate an increasing risk of drought in subtropical regions in the Americas, southern Europe, northern and southern

Africa, South-east Asia and Australia (Dai, 2011; IPCC, 2012; Orlowsky and Seneviratne, 2012).

1.3.3 Other aspects

Temperature and precipitation are two fundamental aspects of the climate we experience. There are also a variety of other variables and processes that intimately affect us, such as cloudiness, soil moisture, evapotranspiration, snow, glaciers and sea ice, wind and sea level. Characterization of the climate also involves the consideration of sequences of events and phenomena such as, for example, drought, flooding, storms and heatwaves. Likewise, characterization of the climate concerns, in addition to average conditions, also variability patterns and extremes (IPCC, 2007, 2012; Rummukainen, 2012). While climate projections can provide information on all of these aspects, a comprehensive account is beyond this chapter. Also, which aspects are pertinent to consider depends on the question in hand: for example, the kind of climate impact or region of interest. Seasonality related to agriculture, for example, follows temperature in Europe, while it follows the succession of wet and dry periods in Africa.

1.4 Regional Patterns of Change

Global climate model projections also provide information on regional-scale climate. Plate 1 gives a first impression of regional patterns of projected temperature and precipitation change, relative to the overall global mean warming amount.

However, the detail in global climate model projections is constrained by model resolution, which in most cases corresponds to a few hundreds of kilometres (Masson and Knutti, 2011; Räisänen and Ylhäisi, 2011). Information on the quality of the models in simulating large-scale variability is important for regions in which such variability plays a significant role in shaping the regional climate (e.g. van Haren et al., 2013). Complementary regional-scale climate projections are carried out with downscaling, either by means of regional climate models (Rummukainen, 2010) or statistical downscaling (Maraun et al., 2010). Downscaling attempts to capture better the influence of variable orography and land–sea distribution on the regional climate than what is feasible to achieve with global models. Consequently, for many climate change impact studies, downscaled climate projections can be a better starting point than the direct results of global climate projections.

For example, Kjellström et al. (2013) analysed results from 21 recent regional climate models for Europe. Even though these had been forced by different global models, the projections gave very similar patterns of change for both temperature and precipitation (see Plate 2). The largest wintertime changes occur in the north-east and the largest summertime ones in the south. Precipitation tends to increase in the north and decrease in the south. However, there are also models that show small to insignificant changes in large parts of Europe, and models that suggest a general wetting. These findings largely confirm earlier regional projection results for Europe (e.g. Christensen and Christensen, 2007; Déqué et al., 2012).

A similar analysis for Africa is shown in Plate 3. In North Africa, warming is greater in the summer than in the winter. There is a similar feature in the south of Africa. Warming for most of the models here ranges from around 1°C or slightly less to somewhat above 2°C. The range of precipitation changes projected by regional climate models (RCMs) is similar to the European case (Plate 2), in the sense that it spans from a general drying to a general wetting. The RCM median result suggests drying in both the north and the south of Africa, and either an increase or no change in between these areas.

1.5 Circulation Patterns and Regional Changes

Internal variability is a ubiquitous aspect of the climate system. Its specific regional manifestations can often be analysed in terms of circulation patterns or 'large-scale variability modes'. For example, different phases of ENSO lead to significant regional temperature and precipitation anomalies in many parts of the world. Inter-annual variability in the North Atlantic, Europe and the Arctic region occurs in concert with the North Atlantic Oscillation and the Arctic Oscillation, and manifests itself as, not least, inter-annual variability in general cold-season weather. In monsoon regions, such as South-east Asia and Western Africa, the seasonally changing temperature contrast between the land and the sea generates a distinct variability of regional precipitation. Under climate change, however, some of the characteristics of circulation patterns may change. This would imply regional climate change that further deviates from the global mean, in addition to the general larger warming over land than over sea, etc. (cf. Plate 1).

While individual models may project various changes in circulation patterns as a result of climate change, global climate projections to date do not collectively suggest major shifts in ENSO, NAO and some of the other comparable modes (IPCC, 2007). Precipitation associated with monsoons is projected to increase in general, as well as the overall global area that is affected by monsoons (Hsu et al., 2012), due to higher sea surface temperature and atmospheric water vapour content resulting from global warming.

More recently, it has been postulated that the retreat of sea ice in the Arctic region may affect atmospheric circulation and promote the occurrence of more persistent weather patterns, such as extreme winters and summers at the high and mid-latitudes of the northern hemisphere (Francis and Vavrus, 2012; Liu et al., 2012). Climate models suggest that such a link may exist and play some role during the 21st century (Yang and Christensen, 2012). The expected continued reduction of the Arctic sea ice cover may counteract the effect of the overall warming on the occurrence of cold winter months in Europe and concurrent mild winters across North America. This does not mean an expectation of more cold winters in Europe, however, but rather that cold winters will still occur even when overall warming proceeds.

Overall, regional-scale climate in many parts of the world is shaped not only by global-scale constraints but also by the action of circulation patterns; the latter often warrant special attention in the analysis of regional climate change projections (Deser et al., 2012).

1.6 Discussion and Conclusion

Climate change is of global and regional concern. In addition to general warming, changes are expected in precipitation patterns and other aspects, both on average and in terms of variability and extreme events. Changes that well exceed climate variability far back in time, and certainly over the history of modern society, are to be expected. This underlines the need for knowledge on how the future can unfold, in order to anticipate the impacts and to be able to prepare for them. Climate projections offer a means to do this.

A pertinent question for analyses and impact assessments is 'which climate model and projection to choose'. Unfortunately, there is no specific answer. Rather, one needs to recognize the sources of uncertainty and, as much as possible, look at many projections from many climate models, based on different emissions scenarios. In the case of it being prohibitive to account for large sets of scenarios, the evaluation of climate models may provide guidance for choosing a smaller set of models and projections. For example, one could perhaps exclude models which exhibit larger biases in variables that are especially crucial for the impact study in hand. For the early part of the 21st century, projections are relatively similar across many emissions scenarios; and one could consider focusing on one or at most a few emissions scenarios. In any case, consideration of the results from more than one model and/or projection is always recommended. While use of a subset of models and projections does not necessarily suffice for quantification of scenario uncertainty, it can still provide a useful reminder that scenarios and projections are possible unfolding futures that are subject to uncertainty, rather than definitive forecasts.

Climate projection results are often similar when it comes to the direction of the projected changes, be these of increase, decrease or no evident change. The size of projected change, however, varies more. This is due in part to differences in climate sensitivity, i.e. in the response to a given forcing represented in models (in other words, in the underlying process descriptions). Some of the variation across models can be attributed to the simulated internal variability. Nevertheless, the magnitude of change characteristically increases with fossil fuel emissions and land-use change, and thus also over time as long as these remain unabated.

One way to condense information from multiple models and projections is to consider multi-models (ensembles). Analysis of multi-model means helps to highlight results that are consistent across the models, although this occurs at the expense of suppressing outliers. For example, the latitude zones that border the higher latitude regions with a projected consistent precipitation increase, and the subtropical regions with a projected consistent precipitation decrease, have a 'no change' appearance, while specific projections may exhibit either increases or decreases. Outliers may need to be considered for impact assessments, as they may represent extreme responses and thus give at least a partial idea of 'best case' and 'worst case' scenarios. Thus, multi-model mean results need to be amended with consideration of either some individual projections, or the model spread. The pursuit of probabilistic projection analysis of global and regional projections is a more refined method addressing the same problem (e.g. Déqué and Somot, 2010; Alessandri et al., 2011).

In the end, of course, the intended use of climate projections, such as in agricultural impact assessment, needs to guide the considerations. High-resolution information may be preferred, in which case downscaled information, be it from statistical or dynamical approaches, is probably needed. The assessment and impact model at hand may pose constraints on whether individual projections or ensemble-based results can be used. Nevertheless, proper consideration of climate projections necessarily requires insights into their basis and underlying scenario assumptions.

Note

[1] The definition of climate sensitivity refers to the long-term (equilibrium) global mean temperature rise for a doubling of the atmospheric carbon dioxide content. A specific climate scenario may feature an increase in the atmospheric carbon dioxide concentration that is either smaller or larger than a doubling. The value of the climate sensitivity should thus not be confused with a specific climate change/global warming scenario. Another concept is the 'transient climate response', which refers to the warming that manifests itself around the time when the atmospheric carbon dioxide concentration doubles. The climate sensitivity is larger than the transient climate response, which is due to the slow progression of warming signals to spread in the ocean.

References

Alessandri, A., Borrelli, A., Navarra, A., Arribas, A., Déqué, M., Rogel, P., et al. (2011) Evaluation of probabilistic quality and value of the ENSEMBLES multimodel seasonal forecasts: comparison with DEMETER. *Monthly Weather Review* 139, 581–607.

Christensen, J.H.C. and Christensen, O.B.C. (2007) A summary of the PRUDENCE model projections of changes in European climate by the end of this century. *Climatic Change* 81, 7–30.

Dai, A. (2011) Drought under global warming: a review. *Wiley Interdisciplinary Reviews: Climate Change* 2, 45–65.

Déqué, M. and Somot, S. (2010) Weighted frequency distributions express modelling uncertainties in the ENSEMBLES regional climate experiments. *Climate Research* 44, 195–209.

Déqué, M., Somot, S., Sanchez-Gomez, E., Goodess, C.M., Jacob, D., Lenderink, G., et al. (2012) The spread amongst ENSEMBLES regional scenarios: regional climate models, driving general circulation models and interannual variability. *Climate Dynamics* 38, 951–964.

Deser, C., Phillips, A., Bourdette, V. and Teng, H. (2012) Uncertainty in climate change projections: the role of internal variability. *Climate Dynamics* 38, 527–546.

Francis, J.A. and Vavrus, S.J. (2012) Evidence linking Arctic amplification to extreme weather in mid-latitudes. *Geophysical Research Letters* 39, L06801.

Giorgi, F. and Bi, X. (2009) Time of emergence (TOE) of GHG-forced precipitation change hot-spots. *Geophysical Research Letters* 36, L06709.

Hawkins, E. and Sutton, R. (2009) The potential to narrow uncertainty in regional climate predictions. *Bulletin of the American Meteorological Society* 90, 1095–1107.

Held, I.M. and Soden, B.J. (2006) Robust responses of the hydrological cycle to global warming. *Journal of Climate* 19, 5686–5699.

Hsu, P.-C., Li, T., Luo, J.-J., Murakami, H., Kitoh, A. and Zhao, M. (2012) Increase of global monsoon area and precipitation under global warming: a robust signal? *Geophysical Research Letters* 39, L06701.

IPCC (2007) *Climate Change 2007: The Physical Science Basis*. Contribution of Working Group I to the Fourth Assessment Report of the Intergovernmental Panel on Climate Change. Solomon, S., Qin, D. Manning, M., Chen, Z., Marquis, M., Averyt, K.B., et al. (eds). Cambridge University Press, Cambridge, UK, and New York.

IPCC (2012) *Managing the Risks of Extreme Events and Disasters to Advance Climate Change Adaptation*. A Special Report of Working Groups I and II of the Intergovernmental Panel on Climate Change. Field, C.B., Barros, V., Stocker, T.F., Qin, Q., Dokken, D.J., Ebi, K.L., et al. (eds). Cambridge University Press, Cambridge, UK, and New York.

Jaeger, C.C., Krause, J., Haas, A., Klein, R. and Hasselmann, K. (2008) A method for computing the fraction of attributable risk related to climate damages. *Risk Analysis* 28, 815–823.

Kharin, V.V., Zwiers, F.W., Zhang, X. and Wehner, M. (2013) Changes in temperature and precipitation extremes in the CMIP5 ensemble. *Climatic Change* 119, 345–357.

Kjellström, E., Theijll, P., Rummukainen, M., Christensen, J.H., Boberg, F., Christensen, O.B., et al. (2013) Emerging regional climate change signals for Europe under varying large-scale circulation conditions. *Climate Research* 56, 103–119.

Liu, J., Curry, J.A., Wang, H., Song, M. and Horton, R.M. (2012) Impact of declining Arctic sea ice on winter snowfall. *Proceedings of the National Academy of Sciences* 109, 4074–4079.

Mahlstein, I., Portmann, W., Daniel, J.S., Solomon, S. and Knutti, R. (2012) Perceptible changes in regional precipitation in a future climate. *Geophysical Research Letters* 39, L05701.

Maraun, D., Wetterhall, F., Ireson, A.M., Chandler, R.E., Kendon, E.J., Widmann, M., et al. (2010) Precipitation downscaling under climate change: recent developments to bridge the gap between dynamical models and the end user. *Reviews of Geophysics* 48, RG3003.

Masson, D. and Knutti, R. (2011) Spatial-scale dependence of climate model performance in the CMIP3 ensemble. *Journal of Climate* 24, 2680–2692.

Mearns, L.O., Arritt, R., Biner, S., Bukovsky, M.S., McGinnis, S., Sain, S., et al. (2012) The North American Regional Climate Change Assessment Program: overview of phase I results. *Bulletin of the American Meteorological Society* 93, 1337–1362.

Meehl, G.A., Stocker, T.F., Collins, W.D., Friedlingstein, P., Gaye, A.T., Gregory, J.M., et al. (2007) Global climate projections. In: Solomon, S., Qin, D., Manning, M., Chen, Z., Marquis, M., Averyt, K.B., et al. (eds) *Climate Change 2007: The Physical Science Basis*. Contribution of Working Group I to the Fourth Assessment Report of the Intergovernmental Panel on Climate Change. Cambridge University Press, Cambridge, UK, and New York, pp. 747–845.

Menendez, C.G., de Castro, M., Sorensson, A., Boulanger, J. and participating CLARIS modelling groups (2010) CLARIS Project: towards climate downscaling in South America. *Meteorologische Zeitschrift* 19, 357–362.

Moss, R.H., Edmonds, J.A., Hibbard, K.A., Manning, M.R., Rose, S.K., van Vuuren, D.P., et al. (2010) The next generation of scenarios for climate change research and assessment. *Nature* 463, 747–756.

Nakićenović, N. and Swart, R. (eds) (2000) *Special Report on Emissions Scenarios*. A Special Report of Working Group III of the Intergovernmental Panel on Climate Change. Cambridge University Press, Cambridge, UK, and New York.

Orlowsky, B. and Seneviratne, S.I. (2012) Global changes in extreme events: regional and

seasonal dimension. *Climatic Change* 110, 669–696.
Paeth, H., Hall, M.H.J., Gaertner, M.A., Alonso, M.D., Moumouni, S., Polcher, J., *et al.* (2011) Progress in regional downscaling of West African precipitation. *Atmospheric Science Letters* 12, 75–82.
Peters, G.P., Andrew, R.M., Boden, T., Canadell, J.G., Ciais, P., Le Quéré, C., *et al.* (2013) The challenge to keep global warming below 2°C. *Nature Climate Change* 3, 4–6.
Pitman, A.J., Arneth, A. and Ganzeveld, L. (2011) Regionalizing global climate models. *International Journal of Climatology* 32, 321–337.
Räisänen, J. and Ylhäisi, J.S. (2011) How much should climate model output be smoothed in space? *Journal of Climate* 24, 867–880.
Randall, D.A., Wood, R.A., Bony, S., Colman, R., Fichefet, T., Fyfe, J., *et al.* (2007) Climate models and their evaluation. In: Solomon, S., Qin, D., Manning, M., Chen, Z., Marquis, M., Averyt, K.B., *et al.* (eds) *Climate Change 2007: The Physical Science Basis*. Contribution of Working Group I to the Fourth Assessment Report of the Intergovernmental Panel on Climate Change. Cambridge University Press, Cambridge, UK, and New York, pp. 589–662.
Rummukainen, M. (2010) State-of-the-art with regional climate models. *Wiley Interdisciplinary Reviews: Climate Change* 1, 82–96.
Rummukainen, M. (2012) Changes in climate and weather extremes in the 21st century. *Wiley Interdisciplinary Reviews: Climate Change* 3, 115–129.
Sillman, J. and Roeckner, E. (2008) Indices for extreme events in projections for anthropogenic climate change. *Climatic Change* 86, 83–104.
Sillman, J., Kharin, V.V., Zwiers, F.W., Zhang, X. and Bronaugh, D. (2013) Climate extremes indices in the CMIP5 multimodel ensemble: Part 2. Future climate projections. *Journal of Geophysical Research: Atmospheres* 118, 2473–2493.
Solomon, S., Plattner, G.-K., Knutti, R. and Friedlingstein, P. (2009) Irreversible climate change due to carbon dioxide emissions. *Proceedings of the National Academy of Sciences* 106, 1704–1709.
Stott, P.A., Stone, D.A. and Allen, M.R. (2004) Human contribution to the European heatwave of 2003. *Nature* 432, 610–614.
Taylor, K.E., Stouffer, R.J. and Meehl, G.A. (2012) An overview of CMIP5 and the experiment design. *Bulletin of the American Meteorological Society* 93, 485–498.
Thompson, D.W. and Wallace, J.M. (1998) The Arctic Oscillation signature in the wintertime geopotential height and temperature fields. *Geophysical Research Letters* 25, 1297–1300.
Trenberth, K. (2012) Framing the way to relate climate extremes to climate change. *Climatic Change* 115, 283–290.
van Haren, R., van Oldenborgh, G.J., Lenderink, G., Collins, M. and Hazeleger, W. (2013) SST and circulation trend biases cause an underestimation of European precipitation trends. *Climate Dynamics* 40, 1–20.
Yang, S. and Christensen, J.H. (2012) Arctic sea ice reduction and European cold winters in CMIP5 climate change experiments. *Geophysical Research Letters* 39, L20707.
Zhang, X., Alexander, L., Hegerl, G.C., Jones, P., Klein Tank, A., Peterson, T.C., *et al.* (2011) Indices for monitoring changes in extremes based on daily temperature and precipitation data. *Wiley Interdisciplinary Reviews: Climate Change* 2, 851–870.
Zwiers, F.W., Zhang, X. and Feng, Y. (2011) Anthropogenic influence on long return period daily temperature extremes at regional scales. *Journal of Climate* 24, 881–992.

2 Rainfed Intensive Crop Systems

Jørgen E. Olesen

Aarhus University, Department of Agroecology, Tjele, Denmark

2.1 Introduction

Intensive crop production systems are widespread in regions with sufficient rainfall during the growing season and/or with deep soils that provide sufficient water supply for crop growth. Examples of such regions are central and north-west Europe (Olesen and Bindi, 2002), the US Midwest (Rosenzweig *et al.*, 2002), parts of the North China Plain (Piao *et al.*, 2010) and areas of Brazil and Argentina (Monfreda *et al.*, 2008). The intensive rainfed crop production systems contribute considerably to world supply of food and feed, in particular through the production of cereal grains (e.g. wheat, maize and barley) and protein and oilseed crops (e.g. soybean, oilseed rape and sunflower). These crop products are used primarily for human consumption and livestock feed, in particular for monogastric livestock such as pigs and poultry but also for concentrate in intensive ruminant livestock systems, such as dairy farming and feedlot beef production systems. Recently, an increasing proportion of maize in the USA and oilseed rape in Europe has been used for the production of first-generation biofuels (Havlik *et al.*, 2011).

These intensive cropping systems are characterized by modern farm management practices with high inputs of fertilizers and pesticides, although such inputs are in many regions increasingly being regulated by government to reduce adverse environmental impacts (Oenema *et al.*, 2009). There is increasing concern that climate change will exacerbate the environmental consequences on aquatic ecosystems of nutrient loading from intensive farming systems (Jeppesen *et al.*, 2009, 2011), which may further increase the need for measures to reduce nutrient loadings. However, such measures must to be balanced against the need to meet the growing demand for high-quality foods (Mueller *et al.*, 2012).

The cultivation of crops, their productivity and quality, are directly dependent on climatic factors, in particular temperature and rainfall. Climate change is already having an impact on agriculture (Peltonen-Sainio *et al.*, 2010; Lobell *et al.*, 2011) and has been attributed as one of the factors contributing to stagnation in wheat yields in parts of Europe, despite continued progress in crop breeding (Brisson *et al.*, 2010). Climate change is expected to continue to affect agriculture in the future (Olesen *et al.*, 2011), and the effects will vary greatly in space, but they may also change over time (Trnka *et al.*, 2011). It is generally accepted that productivity will increase in cool temperate regions such as northern Europe due to a lengthened growing season and an extension of the frost-free period, as well as the benefits from enhanced photosynthesis from increasing CO_2 concentrations (Olesen and Bindi, 2002). However, year-to-year variability in yields may also be an issue expected to increase in regions with intensive crop production, due to extreme climatic events and other factors, including pests and diseases (Kristensen *et al.*, 2011; Schaap *et al.*, 2011; Semenov and Shewry, 2011).

Intensive farming systems are generally considered to have a low sensitivity to climate change, because a given change in temperature or rainfall has a modest impact (Chloupek *et al.*, 2004) and because the farmers have resources to adapt and compensate by changing management (Reidsma et al., 2010). However, there may be a considerable difference in adaptive capacity between cropping systems and farms, depending on their specialization (Reidsma et al., 2009). Intensive systems in cool climates may therefore respond favourably to a modest climatic warming (Olesen and Bindi, 2002). On the other hand, some of the rainfed cropping systems currently located in warmer and/or drier areas are expected to be affected severely by climate change, unless irrigation can be provided (Trnka et al., 2010).

2.2 Observed Changes

2.2.1 Observed climate changes

Major areas of intensive rainfed cropping systems have experienced increases in mean temperature over the past 30 years, in some cases with changes that have exceeded trends in global mean temperatures (Lobell et al., 2011). In Europe, trends have been higher in central and north-eastern Europe and in mountainous regions (up to 0.4°C per decade), while the lowest temperature trends have occurred in the Mediterranean region, and temperature has increased more in winter than in summer in northern Europe (Haylock et al., 2008). An increase in temperature variability has been observed, due primarily to an increase in warm extremes (Klein Tank and Können, 2003). In China, the largest warming has been found in north-east China (up to 0.4°C per decade). Across some of the major agricultural regions in Europe, Asia and America, there has been a marked increase in the frequency of hot extremes, and there are indications of increasing inter-annual temperature variability, making the climate less predictable (Hansen et al., 2012).

Rainfall patterns have also been changing, but with large regional and seasonal differences (New et al., 2001). Northern Europe has seen an increase in precipitation, although this increase was dominated by changes during the winter period (Haylock et al., 2008). Other regions such as northern China have seen a reduction in total rainfall (Piao et al., 2010). However, across temperate climate zones, there has been an increase in mean precipitation per wet day, even in areas getting drier (Alexander et al., 2006).

Across many of these regions with intensive rainfed agriculture, there has also been an increase in droughts, linked partly with changes in the occurrence of rainfall and due partly to greater evapotranspiration associated with higher temperatures (Dai, 2013). The droughts might be combined with extreme heatwaves, as was the case over Western and central Europe in 2003 and over Russia in 2010, when large areas were exposed to summer temperature rises of 3–5°C (Hansen et al., 2012). The heatwave over Europe in 2003 was associated with an annual precipitation deficit up to 300 mm, and the drought was a major contributor to the estimated reduction of 30% in gross primary production of terrestrial ecosystems (Ciais et al., 2005). The heatwave also led to widespread reductions in farm income (Fink et al., 2004).

2.2.2 Observed yield trends

Intensive rainfed cropping systems vary considerably between regions. Arable cropping systems in north-west Europe are dominated by cereals, in particular winter wheat (*Triticum aestivum* L.) and spring barley (*Hordeum vulgare* L.), oilseed crops such as oilseed rape (*Brassica napus* L.) and root and tuber crops such as sugarbeet (*Beta vulgaris* L.) and potato (*Solanum tuberosum* L.). Over the past decades, cultivation of high-yielding crops such as winter wheat and grain maize (*Zea mays* L.) has increased in this region (Fig. 2.1). By far the largest proportion of wheat grown here is winter wheat, which gives higher yields than spring wheat. The increase in the area of winter wheat has, to a large extent, happened at the expense of less productive spring cereal crops. The large increases in yield over the

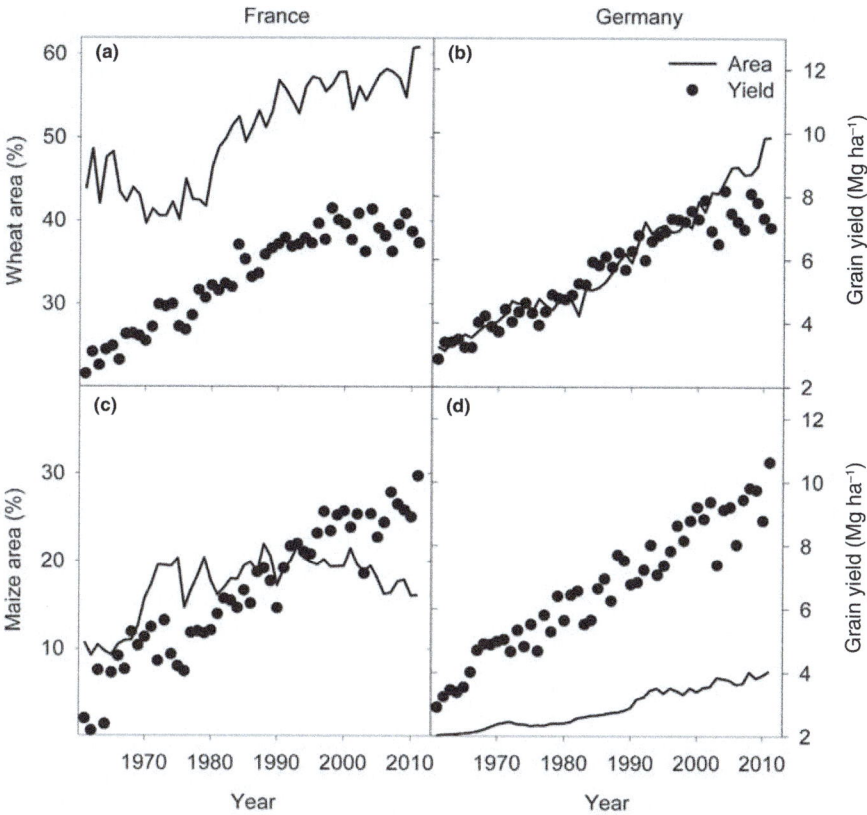

Fig. 2.1. Developments of grain yield and area (% of total cereal area) of wheat (a, b) and grain maize (c, d) for France (a, c) and Germany (b, d). (Source: FAOStat.)

50-year period shown in Fig. 2.1 has been brought about through intensified cropping by changes in genotypes and the increased use of fertilizers and pesticides.

Wheat yields in Europe have been stagnating over the past 10–20 years, as also shown in Fig. 2.1, where yields in France have not been increasing since 1994 and in Germany not since 1998. Rather, there seems to be a greater variability of wheat grain yield in the recent two decades. Stagnating wheat yields in France have been attributed to lower yields under rising temperatures (Brisson et al., 2010), but changes in management may also have played a role in some countries (Finger, 2010).

In contrast to wheat, yields of grain maize show a continued increase in both France and Germany (Fig. 2.1), so that grain maize yields now exceed those of winter wheat. The area of grain maize is growing in northern parts of Europe (Elsgaard et al., 2012), as also shown for Germany in Fig. 2.1, and this increase appears to be linked with the observed warming. As with wheat, it appears that variability of maize yields has increased in recent years, and this may be related to changes in the occurrence of climatic extremes (Peltonen-Sainio et al., 2010). Maize in southern parts of France is irrigated, but the increase in maize cultivation in northern Europe is largely rainfed.

Analyses of the effects of observed climate change on yield potential in Europe have shown positive effects for maize and sugarbeet, which have benefited from an increase in the duration of the growing season for these crops (Supit et al., 2010). Yield benefits have been greatest in the UK and other parts of northern Europe. For potato, warming may also have contributed to higher yields in these northern regions (Gregory and Marshall, 2012). In contrast, warmer and more variable climatic conditions that increase the risk of drought have decreased crop yields in parts of central Europe (Supit et al., 2010; Trnka et al., 2012).

The cropping systems on the North China Plain and in north-east China are dominated by winter wheat and maize double-cropping systems or monocultures of winter wheat or maize, all grown in intensive systems with high inputs of fertilizers and pesticides. Soybean (Glycine max L.) is also grown during summer in rotation with these crops. Wheat in the double-cropping system is normally irrigated, whereas maize is mostly rainfed. Observations have revealed a shortening of the growing period for wheat in China, although the duration of the grain-filling period may not have changed much (Xiao et al., 2013). This is because warming has occurred mainly during the vegetative (pre-flowering) stages in both wheat and maize (Liu et al., 2010). It has been estimated that the warming in this region has contributed to about 5% yield reduction in wheat (You et al., 2009; Lobell et al., 2011) and 7% yield reduction in maize (Lobell et al., 2011). There are indications that the autonomous adaptation of new crop varieties with longer pre-flowering periods has been instrumental in mitigating some of the estimated yield reductions from warming on the North China Plain (Liu et al., 2010).

The intensive rainfed cropping systems in North and South America include grain maize and soybean in the US Corn Belt and vast areas of Brazil and Argentina. Yields of maize and soybean in these regions mostly continue to increase (Ray et al., 2012). It has been estimated that temperature has reduced yields through these regions for both crops, although there are also areas experiencing yield benefits from warming, for example in northern parts of the USA (Sakurai et al., 2011). In contrast, precipitation changes have caused a more mixed picture with larger spatial differences. Lobell et al. (2013) estimated that the overall effect of observed climate change on yields of maize and soybean in the USA was small, whereas it has reduced yields of these crops by 5–8% in Brazil and increased soybean yields in Argentina by about 2%. However, recent events of persistent high temperatures in the USA seem to have had negative effects on yields of both maize and soybean (Schlenker and Roberts, 2009).

2.3 Responses to Climate Change

The biophysical processes of agroecosystems are affected strongly by environmental conditions. The projected increase in greenhouse gases will affect these systems either directly (primarily by increasing photosynthesis at higher CO_2) or indirectly via effects on climate (e.g. temperature and rainfall affecting several aspects of ecosystem functioning) (Table 2.1). The effects of climate change on cropping systems may also be either direct through changes in crop physiology or indirect through effects on soil fertility, crop protection (weeds, pests or diseases) or the ability to perform timely crop management. The exact responses depend on the sensitivity of the particular ecosystem and on the relative changes in the controlling factors.

For intensive cropping systems, concerns are increasing about the effects of such systems on the environment and on other ecosystem services, in addition to food and feed supply. Particular concerns for rainfed systems are linked to the use of fertilizers and the associated nutrient losses that lead to eutrophication of both freshwater and marine ecosystems. Climate change may affect not only the nutrient transfer from intensive agriculture to the aquatic environment but also the responsiveness of these environments to nutrient loading (Jeppesen et al., 2009, 2011). In turn, such changes may affect agroecosystems indirectly through regulation by society aiming to reduce such negative effects from the intensive use of

Table 2.1. Influence of CO_2, temperature, rainfall and wind on various components of the agroecosystem.

Component	Influence of factor		
	CO_2	Temperature	Rain/wind
Plants	Higher CO_2 leads to increased dry matter growth and decrease in water use.	Increase of temperature boosts yield up to a threshold beyond which yield declines.	Decreasing precipitation or increasing wind decreases dry matter growth.
Water	Higher CO_2 conserves soil moisture by reducing transpiration.	Higher temperatures increase evaporation.	Higher rainfall will increase groundwater supply and in some areas increase groundwater levels.
Soil	Higher carbon concentrations of plant residues under higher CO_2 will lead to higher soil carbon contents.	Higher temperatures boost soil organic matter turnover, leading to reduced soil carbon content but temporarily higher nutrient supply for plants.	Drier and more windy environments may lead to enhanced wind erosion, whereas more intense rainfall will enhance water erosion.
Pests/diseases	Higher CO_2 reduces the quality of plant biomass for pests and diseases, leading to fewer pests.	Higher temperatures reduce the generation time of pests and diseases and cause attacks to occur earlier in the year, making pests and diseases more problematic.	Some diseases are spread by wind or rainfall. Therefore, more rainy and windy conditions will favour some diseases.
Weeds	Enhanced CO_2 concentrations will favour crop and weed species differentially, making some weeds more problematic. Higher CO_2 will also reduce the efficacy of some herbicides.	Higher temperatures will lead to invasive weed species in some regions, and will also affect the efficacy of herbicides.	More rainy conditions may make some weed species more difficult to control through herbicides.

fertilizers and pesticides. Further concerns on how agricultural practices may contribute to mitigating greenhouse gas emissions, as well as how this links with the needed adaptation to climate change, may also affect the responsiveness of intensive rainfed cropping systems to climate change (Smith and Olesen, 2010).

2.3.1 Carbon dioxide

Increasing atmospheric CO_2 concentration stimulates yield of crops that have the so-called C_3 photosynthesis pathway, and these include most of the intensively grown crops except maize, which belongs to plants having the C_4 photosynthesis pathway (Fuhrer, 2003). In most C_3 crops, a doubling of atmospheric CO_2 concentration will lead to yield increases of 20–40% (Fig. 2.2). The response is considerably smaller for C_4 plants. Higher CO_2 concentration affects not only photosynthesis but also the water consumption of plants. With higher CO_2 concentration, the density and the openness of stomata will be reduced in both C_3 and C_4 plants. This leads to reduced transpiration and to higher water-use efficiencies, resulting in higher yields under dry or drought

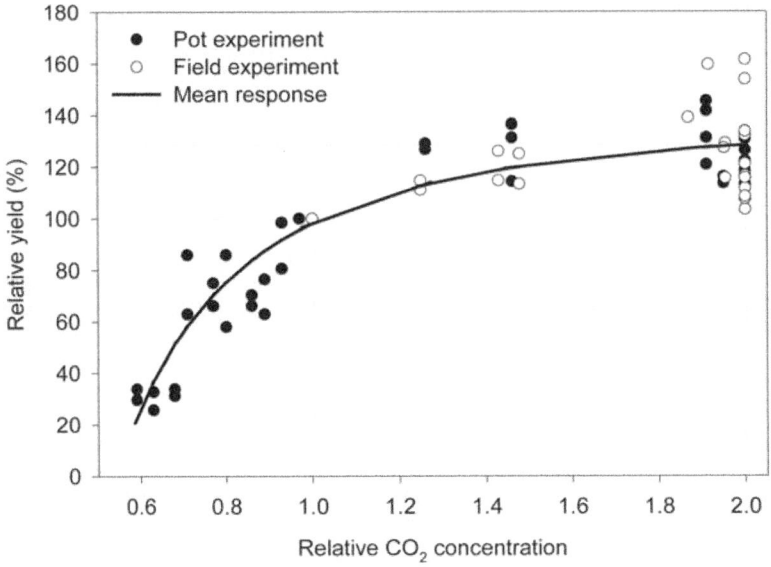

Fig. 2.2. The effects of CO_2 concentration on wheat yield in experiments. Ambient CO_2 is set to 1. Open symbols represent data from field experiments, filled symbols from pot or glasshouse experiments. The solid line shows the mean estimated response. Estimates consistently show that higher CO_2 concentrations in the atmosphere will stimulate the yield of wheat and other crops. In general, similar effects are obtained in both pot and field experiments. (Source: Olesen and Bindi, 2002.)

conditions, and thus this may improve the tolerance of rainfed cropping systems to drier conditions.

Recent estimates of yield benefits from increasing CO_2 are smaller than earlier estimates (Ainsworth and Long, 2005), and the average annual increase over the next decades is marginal compared with what has been achieved through conventional crop management and breeding in the past. Some model studies of CO_2 effects have been based on results from enclosure studies from the 1980s, which exaggerated the effects of increased CO_2 on plant production (Long et al., 2006), although there might also be possibilities for enhancing the responses of plants to higher CO_2 through breeding and changes in crop management (McGrath and Lobell, 2013).

Higher CO_2 concentrations also affect the quality of the plant biomass, because plants accumulate more sugars, thus leading to higher dry matter contents and lower nutrient concentrations of leaves, stems and reproductive organs. This has consequences for the quality of the food and feed, which in some cases are negative. The attraction of plants to pests and diseases will also change, which could make the plants more resistant to attack under higher CO_2. However, weeds will also benefit from increased CO_2, which in some cases will change the need for weed control (Ziska, 2001).

Higher temperatures will also impact on the effects of CO_2 concentration through shifting the optimum temperature for photosynthesis towards higher temperatures at enhanced CO_2. Only the more complex crop models simulate such interactions, and thus they may not always be included in crop yield projections. However, assessments show that such effects are usually small when averaged across growing seasons (Tubiello and Ewert, 2002).

2.3.2 Temperature and rainfall

Temperature affects crops in different ways, partly through affecting the timing of crop phenological phases (i.e. crop development), partly through the efficiency of energy capture, conversion and storage (i.e. crop growth) and partly through crop water supply, since temperature affects evapotranspiration. With warming, the start of active growth is advanced, plants develop faster and the potential growing season is extended. This may have the greatest effect in colder regions (Trnka et al., 2011), and may be beneficial for perennial crops or crops remaining in their vegetative phase, e.g. sugarbeet. However, increased temperature reduces the crop duration of determinate species (plants that flower and mature). For instance, in wheat an increase of 1°C during grain fill reduces the length of this phase by 5%, and yield declines by a similar amount (Olesen et al., 2000). Maize and soybean yields in the USA between 1982 and 1989 decreased by 17% for each 1°C increase in the growing season mean temperature (Lobell and Asner, 2003).

These differential responses of crop yield to increasing temperature in plants with different sensitivities of crop development are illustrated in Fig. 2.3. The largest reduction in grain yield was simulated for winter wheat, where the growth duration was reduced, because any changes in sowing date would have little effect on the duration of the vegetative and reproductive phases in spring and summer. This response concurs well with observed winter wheat yields in Denmark, where the largest reductions were found to be related to high temperatures during the grain-filling phase (Kristensen et al., 2011). However, there may be options for adapting to such changes, since Olesen et al. (2012) found that at higher mean temperatures in Europe, farmers would be growing varieties of cereals with higher thermal requirements of both vegetative and reproductive phases. Figure 2.3 shows a smaller response of spring barley to higher temperatures, because this crop can be sown earlier in spring, thus maintaining a productive growing season. In contrast, simulated yields increase for a grass crop, which represents crops with a non-determinate growth pattern, where the yields depend on the total duration of the

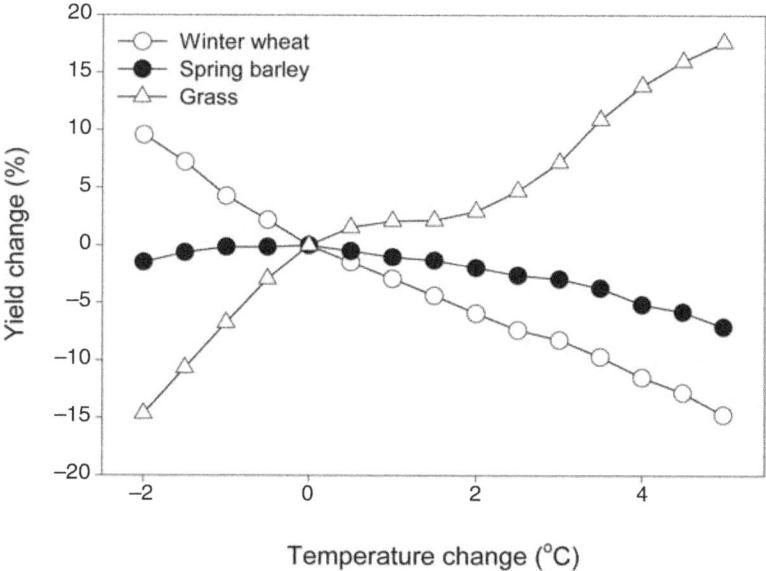

Fig. 2.3. Mean simulated change in yield of winter wheat, spring barley and ryegrass with increasing temperature for a site in Denmark. The simulations were performed with the CLIMCROP model, assuming that water shortages would not affect yields (Olesen et al., 2000; Olesen, 2005).

growing season with suitable temperatures and rainfall.

Crops also show threshold responses to temperature (Porter and Gawith, 1999). Exceeding thresholds may lead to drastic reductions in yield from short episodes of high temperatures during sensitive crop growth phases such as the reproductive period. Temperatures above 35°C during the flowering period can thus affect seed and fruit set severely, and thus reduce yields greatly (Porter and Semenov, 2005). Similar high temperatures during the grain-filling period of small-grain cereals may accelerate leaf senescence, thus also causing large yield losses (Asseng et al., 2011). Currently, yield damages from heat stress episodes are particularly abundant for continental lands at high latitudes (Teixeira et al., 2013), and the areas affected by such events are projected to increase with climate change. Model-based analyses of the effects of climate change on wheat yields in Europe have even shown that the risk of crop losses from drought may decrease but the risk of heat stress may increase, potentially causing substantial yield losses to wheat in the currently highly productive regions in northern Europe (Semenov and Shewry, 2011).

The most important effect of rainfall on crops is through ensuring sufficient water to cover the loss from evapotranspiration during the growing season. Besides rainfall during the growing season, crop water supply depends critically on soil water-holding capacity and plant root development. Agricultural droughts occur when the crop water demand cannot be met by either rainfall or soil water supply. Such drought effects can be enhanced by high-temperature events, which may give rise to large vapour pressure deficits that greatly enhance the evaporative demand beyond what can be sustained through soil water supply. Thus, Lobell et al. (2013) found that in the USA the detrimental effect of temperatures above 30°C on maize yields could be explained by increases in evaporative demand.

Peltonen-Sainio et al. (2010) characterized the coincidence of yield variations with weather variables for major field crops using long-term data sets and tried to find commonalities across European agricultural regions. Long-term national and/or regional yield data sets were used from 14 European countries for spring and winter barley and wheat, winter oilseed rape, potato and sugarbeet. Harmful effects of high precipitation during grain filling in grain and seed crops, and at flowering in oilseed rape, were recorded. In potato, reduced precipitation at tuber formation was associated with yield penalties. Elevated temperature had harmful effects on cereals and rapeseed yields. Similar harmful effects of rainfall and high temperature on grain yield of winter wheat was found by Kristensen et al. (2011) in a study using observed winter wheat yields from Denmark. Rosenzweig et al. (2002) also found that high precipitation events and flooding could lead to crop losses for maize in the USA, and that this might increase greatly with climate change.

2.3.3 Soils

Soils have many functions, of which water and nutrient supply to growing crops are essential for sustaining crop production. However, soils are also important in controlling water and nutrient cycles, for carbon storage, and for regulating greenhouse gas emissions. Soils are also habitats for many organisms that contribute to the functioning of soils and agroecosystems, including effects on crop yield (Pan et al., 2009). Increasing temperatures will speed the decomposition of soil organic matter where soil moisture allows. Hence, direct climate impacts on cropland soils will tend to decrease soil organic stocks, unless this is counterbalanced by larger inputs of organic matter in crop residues (Falloon and Betts, 2010).

Any reduction in soil organic matter stocks can lead to a decrease in fertility and biodiversity, loss of soil structure, reduced soil water infiltration and retention capacity, and increased risk of erosion and compaction. All of this potentially lowers the productivity of crops. Changes in rainfall and wind patterns can lead to an increase in erosion in vulnerable soils, in particular where poor cover from crops and crop residues increase their susceptibility to erosion (Nearing et al., 2004). In addition to depleting soil fertility, erosion may also enhance nutrient runoff to sensitive aquatic ecosystems (Jeppesen et al., 2009).

Faster decomposition of soil organic matter at higher temperatures also increases the mineralization of soil organic nitrogen, which may increase the risk of nitrate leaching during periods of no or little crop cover and precipitation surplus causing runoff or percolation through the soil profile. This may increase the risk of nitrate transport to surface and groundwater systems, thus leading to a decline in environmental quality (Patil et al., 2012).

2.3.4 Crop protection

In current cool regions, higher temperatures favour the proliferation of insect pests, because many insects can then complete a greater number of reproductive cycles (Bale et al., 2002). Warmer winter temperatures may also allow pests to overwinter in areas where they are now limited by cold, thus causing greater and earlier infestation during the following crop season. Climate warming will lead to earlier spring activity of insects and the proliferation of some pest species (Cocu et al., 2005). A similar situation may be seen for plant diseases, thus leading to an increased demand for pesticide control.

Unlike pests and diseases, weeds are influenced directly by changes in atmospheric CO_2 concentration. The differential effects of CO_2 and climate changes on crops and weeds will alter weed–crop competitive interactions, sometimes for the benefit of the crop and sometimes of the weeds. Interaction with other biotic factors and with changing temperature and rainfall may also influence weed seed survival, and thus weed population development (Leishman et al., 2000).

Changes in climatic suitability will lead to the invasion of weeds, pests and diseases adapted to warmer climatic conditions. The speed at which species will invade depends on the rate of climatic change, the dispersal rate of the species and on the measures taken to combat non-indigenous species (Baker et al., 2000). The dispersal rates of pests and diseases are often so high that their geographical extent is determined by the range of climatic suitability. The Colorado beetle (*Leptinotarsa decemlineata*) and the European corn borer (*Ostrinia nubilalis*) are examples of pests and diseases that are expected to show a considerable northward expansion in Europe under climatic warming (Olesen et al., 2011).

2.4 Projected Effects of Climate Change

In temperate regions, a warming of the climate will result in an earlier start of the growing season in spring and a longer duration in autumn. A longer growing season allows the proliferation of species that have optimal conditions for growth and development, and can thus increase their productivity or number of generations (e.g. crop yield, insect population). This will allow for the introduction of new species previously unfavourable due to low temperatures or short growing seasons. This is relevant for the introduction of new crops, for example grain maize or winter wheat in northern Europe (Elsgaard et al., 2012), but will also affect the spreading of weeds, insect pests and diseases (Roos et al., 2010) (see above).

A further lengthening of the growing season, as well as a northward shift of species, is projected as a result of a further increase in temperature across Europe (Olesen et al., 2011). The date of the last frost in spring is projected to advance by about 5–10 days by 2030 and by 10–15 days by 2050 throughout most of Europe (Trnka et al., 2011). The extension of the growing season is expected to be particularly beneficial in northern Europe, where a longer duration for crop growth allows new crops to be cultivated where water availability is generally not restricting growth (Olesen et al., 2011). The warming may, therefore, lead to a further intensification of cropping systems at high latitudes.

The projected impacts of climate change on crop yields depend on the crop type, the emission scenario and the sensitivity of the climate model used to project climate changes (Olesen et al., 2007). Projections for most crops and regions in intensively managed systems show an increase in projected yield during the first decades of the 21st century (Piao et al., 2010; Supit et al., 2012). However, later in the century, projected yield reductions from temperature increases and changes in rainfall would generally exceed the benefits achieved from higher atmospheric CO_2 concentration, thus leading to yield reductions, in particular in cereal crops, whereas continued yield increases are projected in Europe for root and tuber crops such as sugarbeet and potato (Angulo et al., 2013).

There are strong indications emerging that projected climate change would cause increased yield variability of crops under intensive rainfed crop production (Kristensen et al., 2011; Urban et al., 2012). This partly results from changes in inter-annual variability in temperature, but also from non-linearity in the response of crops to temperature and rainfall, thus increasing the risk of low yields. This risk may be particularly large for intensive production systems (Trnka et al., 2012), where the demand for a continued high soil water supply is greater than for low input systems, and where climate-related events that proportionally reduce crop yield have larger absolute effects. Therefore, increases in climatic extremes will have greater effects on intensive rainfed systems than on less intensive and diverse systems (Schaap et al., 2011). Thus, intensive rainfed cropping systems may, in several cases, be vulnerable to climate change, although some of them will also benefit from warming and higher CO_2 concentrations.

There is increasing evidence of the effects of climate change on the occurrence of crop diseases in intensive production systems, as discussed above. Results show projected increases in the occurrence of several diseases with projected warming in the currently cooler parts of intensive cropping regions, such as the UK (Butterworth et al., 2010; Evans et al., 2010), whereas the risk of some diseases may reduce with warming in regions further south, such as for France (Gouache et al., 2013). Such changes in the occurrence of disease may affect not only yield levels but also the quality of the yield; for instance, the occurrence of mycotoxins may increase in northern Europe under projected climate change (Madgwick et al., 2011; Fels-Klerx et al., 2012).

Projections of climate change effects on nitrogen cycling in intensive cropping systems in Europe have generally shown greater risk of nitrate leaching (Olesen et al., 2007; Børgesen and Olesen, 2011), and current measures to reduce it may not be adequately efficient under climate change (Doltra et al., 2013). This, in combination with longer growing seasons, would likely lead to a higher risk of algal blooms and the increased growth of toxic cyanobacteria in lakes (Jeppesen et al., 2011). For greenhouse gas emissions from intensive cropping systems, projections are more uncertain, since the results depend greatly on the balance between the separate effects of temperature, rainfall and CO_2, as well as on their seasonal changes, relative to the effects of changes in crop growth patterns. Therefore, concerns about the effects of climate change on the performance of intensive rainfed crop production systems relate not only to the productivity of these systems and the quality of the products but also to the other ecosystem services to which they contribute.

2.5 Summary and Conclusions

Global food production relies to a large extent on intensive rainfed cropping systems in Europe, East Asia and North and South America, in particular for the production of cereals, oilseed crops and some root and tuber crops. Intensive rainfed crop production systems are characterized by high inputs of fertilizers and pesticides and high resulting crop yields.

Observations over the recent decades show consistent changes in crop phenology and geographical shifts towards higher latitudes of intensive crop cultivation in accordance with observed climate change. The observed effects on crop yield vary from negative to positive effects, with the negative effects dominating at lower latitudes and for cereal and seed crops, whereas the positive effects dominate at higher latitudes and for non-determinate crops. Climate change projections show further shifts of the intensive cropping zones towards higher latitudes. The combined effects of enhanced CO_2 and changes in temperature and precipitation will, in many cases, increase crop productivity at higher latitudes and reduce yields of crops at lower latitudes.

There are consistent results from model-based and empirical studies showing an increased risk of higher inter-annual yield variability with projected climate changes, resulting from changes in inter-annual variability in temperature, but also from non-linearity in the response of crops to temperature and rainfall, thus increasing the risk of low yields. These negative effects on crop yield may be further exacerbated by extreme temperature and rainfall events.

Climate change will also affect crop production indirectly through its impact on soil fertility, weeds, pests and diseases, and nutrient retention in agricultural fields, and in turn, these impacts can lead to nitrate contamination, the release of greenhouse gases and other secondary effects. Thus, climate change will increase the need to reconsider measures for dealing with soil fertility loss, crop protection and nutrient retention in intensive cropping systems. Hence, climate change will have many different impacts on the ecosystem services provided by intensive agricultural systems, and many of these are still poorly understood.

References

Ainsworth, E.A. and Long, S.P. (2005) What have we learned from 15 years of free-air CO_2 enrichment (FACE)? A meta-analytic review of the responses of photosynthesis, canopy properties and plant production to rising CO_2. *New Phytologist* 165, 351–372.

Alexander, L.V., Zhang, X., Peterson, T.C., Caesar, J., Gleason, B., Klein Tank, A.M.G., *et al.* (2006) Global observed changes in daily climate extremes of temperature and precipitation. *Journal of Geophysical Research* 111, D05109.

Angulo, C., Rötter, R., Lock, R., Enders, A., Fronzek, S. and Ewert, F. (2013) Implication of crop model calibration strategies for assessing regional impacts of climate change in Europe. *Agricultural and Forest Meteorology* 170, 32–46.

Asseng, S., Foster, I. and Turner, N.C. (2011) The impact of temperature variability on wheat yields. *Global Change Biology* 17, 997–1012.

Baker, R.H.A., Sansford, C.E., Jarvis, C.H., Cannon, R.J.C., MacLeod, A. and Walters, K.F.A. (2000) The role of climatic mapping in predicting the potential distribution of non-indigenous pests under current and future climates. *Agriculture, Ecosystems and Environment* 82, 57–71.

Bale, J.S., Masters, G.J., Hodkinson, I.D., Awmack, C., Bezemer, T.M., Brown, V.K., *et al.* (2002) Herbivory in global climate change research: direct effects of rising temperature on insect herbivores. *Global Change Biology* 8, 1–16.

Brisson, N., Gate, P., Gouache, D., Charmet, G., Oury, F.X. and Huard, F. (2010) Why are wheat yields stagnating in Europe? A comprehensive data analysis for France. *Field Crops Research* 119, 201–212.

Butterworth, M.H., Semenov, M.A., Barnes, A., Moran, D., West, J.S. and Fitt, B.D.L. (2010) North–south divide: contrasting impacts of climate change on crop yield in Scotland and England. *Journal of the Royal Society Interface* 7, 123–130.

Børgesen, C.D. and Olesen, J.E. (2011) A probabilistic assessment of climate change impacts on yield and nitrogen leaching from winter wheat in Denmark. *Natural Hazards and Earth System Sciences* 11, 2541–2553.

Chloupek, O., Hrstkova, P. and Schweigert, P. (2004) Yield and its stability, crop diversity, adaptability and response to climate change, weather and fertilisation over 75 years in the Czech Republic in comparison to some European countries. *Field Crops Research* 85, 167–190.

Ciais, Ph., Reichstein, M., Viovy, N., Granier, A., Ogée, J., Allard, V., *et al.* (2005) Europe-wide reduction in primary productivity caused by the heat and drought in 2003. *Nature* 437, 529–533.

Cocu, N., Harrington, R., Rounsevell, M.D.A., Worner, S.P. and Hullé, M. (2005) Geographical location, climate and land use influences on the phenology and numbers of the aphid, *Myzus persicae*, in Europe. *Journal of Biogeography* 32, 615–632.

Dai, A. (2013) Increasing drought under global warming in observations and models. *Nature Climate Change* 3, 52–58.

Doltra, J., Lægdsmand, M. and Olesen, J.E. (2013) Impacts of projected climate change on productivity and nitrogen leaching of crop rotations in arable and pig farming systems in Denmark. *Journal of Agricultural Science*, 152, 75–92.

Elsgaard, L., Børgesen, C.D., Olesen, J.E., Siebert, S., Ewert, F., Peltonen-Sainio, P., et al. (2012) Shifts in comparative advantages for maize, oat, and wheat cropping under climate change in Europe. *Food Additives and Contaminants* 29, 1514–1526.

Evans, N., Butterworth, M.H., Baierl, A., Semenov, M.A., West, J.S., Barnes, A., et al. (2010) The impact of climate change on disease constraints on production of oilseed rape. *Food Security* 2, 143–156.

Falloon, P. and Betts, R. (2010) Climate impacts on European agriculture and water management in the context of adaptation and mitigation – the importance of an integrated approach. *Science of the Total Environment* 408, 5667–5687.

Fels-Klerx, H.J. van der, Olesen, J.E., Madsen, M.S., Uiterwijk, M. and Goedhart, P.W. (2012) Climate change increases deoxynivalenol of wheat in north-western Europe. *Food Additives and Contaminants* 29, 1593–1604.

Finger, R. (2010) Evidence of slowing yield growth – the example of Swiss cereal yields. *Food Policy* 35, 175–182.

Fink, A.H., Brücher, T., Krüger, A., Leckebusch, G.C., Pinto, J.G. and Ulbrich, U. (2004) The 2003 European summer heat waves and drought – synoptic diagnosis and impact. *Weather* 59, 209–216.

Fuhrer, J. (2003) Agroecosystem responses to combinations of elevated CO_2, ozone, and global climate change. *Agriculture, Ecosystems and Environment* 97, 1–20.

Gouache, D., Bensadoun, A., Brun, F., Pagé, C., Makowski, D. and Wallach, D. (2013) Modelling climate change impact on *Septoria tritici* blotch (STB) in France: accounting for climate model and disease model uncertainty. *Agricultural and Forest Meteorology* 170, 242–252.

Gregory, P.J. and Marshall, B. (2012) Attribution of climate change: a methodology to estimate the potential contribution to increases in potato yield in Scotland since 1960. *Global Change Biology* 18, 1372–1388.

Hansen, J., Sato, M. and Ruedy, R. (2012) Perception of climate change. *Proceedings of the National Academy of Sciences* 109, E2415–E2423.

Havlik, P., Schneider, U.A., Schmid, E., Böttcher, H., Fritz, S., Skalsky, R., et al. (2011) Global land-use implications of first and second generation biofuel targets. *Energy Policy* 39, 5690–5702.

Haylock, M.R., Hofstra, N., Klein Tank, A.M.G., Klok, E.J., Jones, P.D. and New, M. (2008) A European daily high-resolution gridded data set of surface temperature and precipitation for 1950–2006. *Journal of Geophysical Research* 113(D20119) doi:10.1029/2008JD010201.

Jeppesen, E., Kronvang, B., Meerhoff, M., Søndergaard, M., Hansen, K.M., Andersen, H.E., et al. (2009) Climate change effects on runoff, phosphorus loading and lake ecological state, and potential adaptations. *Journal of Environmental Quality* 38, 1930–1941.

Jeppesen, E., Kronvang, B., Olesen, J.E., Audet, J., Søndergaard, M., Hoffmann, C.C., et al. (2011) Climate change effects on nitrogen loading from catchment: implications for nitrogen retention, ecological state of lakes and adaptation. *Hydrobiologia* 663, 1–21.

Klein Tank, A.M.G. and Können, G.P. (2003) Trends in indices of daily temperature and precipitation extremes in Europe, 1946–99. *Journal of Climate* 16, 3665–3680.

Kristensen, K., Schelde, K. and Olesen, J.E. (2011) Winter wheat yield response to climate variability in Denmark. *Journal of Agricultural Science* 149, 33–47.

Leishman, M.R., Masters, G.J., Clarke, I.P. and Brown, V.K. (2000) Seed bank dynamics: the role of fungal pathogens and climate change. *Functional Ecology* 14, 293–299.

Liu, Y., Wang, E., Yang, X. and Wang, J. (2010) Contributions of climatic and crop varietal changes to crop production in the North China Plain, since 1980s. *Global Change Biology* 16, 2287–2299.

Lobell, D.B. and Asner, G.P. (2003) Climate and management contributions to recent trends in U.S. agricultural yields. *Science* 299, 1032.

Lobell, D.B., Schlenker, W. and Costa-Roberts, J. (2011) Climate trends and global crop production since 1980. *Science* 333, 616–620.

Lobell, D.B., Hammer, G.L., McLean, G., Messina, C., Roberts, M.J. and Schlenker, W. (2013) The critical role of extreme heat for maize production in the United States. *Nature Climate Change* 3, 497–501.

Long, S.P., Ainsworth, E.A., Leakey, A.D.B., Nösberger, J. and Ort, D.R. (2006) Food for thought: lower-than-expected crop yield stimulation with rising CO_2 concentrations. *Science* 312, 1918–1921.

McGrath, J.M. and Lobell, D.B. (2013) Regional disparities in the CO_2 fertilization effect and implications for crop yields. *Environmental Research Letters* 8, 014054.

Madgwick, J.W., West, J.S., White, R.P., Semenov, M.A., Townsend, J.A., Turner, J.A., et al. (2011) Impacts of climate change on wheat anthesis and fusarium ear blight in the UK. *European Journal of Plant Pathology* 130, 117–131.

Monfreda, C., Ramankutty, N. and Foley, J.A. (2008) Farming the planet: 2. Geographical distribution of crop areas, yields, physiological types, and net primary production in the year 2000. *Global Biochemical Cycles* 22, GB1022, doi:10.1029/2007GB002947.

Mueller, N.D., Gerber, J.S., Johnston, M., Ray, D.K., Ramankutty, N. and Foley, J.A. (2012) Closing yield gaps through nutrient and water management. *Nature* 490, 254–257.

Nearing, M.A., Pruski, F.F. and O'Neal, M.R. (2004) Expected climate change impacts on soil erosion rates: a review. *Journal of Soil and Water Conservation* 59, 43–50.

New, M., Todd, M., Hulme, M. and Jones, P. (2001) Precipitation measurements and trends in the twentieth century. *International Journal of Climatology* 21, 1899–1922.

Oenema, O., Witzke, H.P., Klimont, Z., Lesschen, J.P. and Velthof, G.L. (2009) Integrated assessment of promising measures to decrease nitrogen losses from agriculture in EU-27. *Agriculture, Ecosystems and Environment* 133, 280–288.

Olesen, J.E. (2005) Climate change and CO_2 effects on productivity of Danish agricultural systems. *Journal of Crop Improvement* 13, 257–274.

Olesen, J.E. and Bindi, M. (2002) Consequences of climate change for European agricultural productivity, land use and policy. *European Journal of Agronomy* 16, 239–262.

Olesen, J.E., Jensen, T. and Petersen, J. (2000) Sensitivity of field-scale winter wheat production in Denmark to climate variability and climate change. *Climate Research* 15, 221–238.

Olesen, J.E., Carter, T.R., Diaz-Ambrona, C.H., Fronzek, S., Heidmann, T., Hickler, T., et al. (2007) Uncertainties in projected impacts of climate change on European agriculture and ecosystems based on scenarios from regional climate models. *Climatic Change* 81, Suppl. 1, 123–143.

Olesen, J.E., Trnka, M., Kersebaum, K.C., Skjelvåg, A.O., Seguin, B., Peltonen-Saino, P., et al. (2011) Impacts and adaptation of European crop production systems to climate change. *European Journal of Agronomy* 34, 96–112.

Olesen, J.E., Børgesen, C.D., Elsgaard, L., Palosuo, T., Rötter, R., Skjelvåg, A.O., et al. (2012) Changes in flowing and maturity time of cereals in northern Europe under climate change. *Food Additives and Contaminants* 29, 1527–1542.

Pan, G., Pan, W. and Smith, P. (2009) The role of soil organic matter in maintaining the productivity and yield stability of cereals in China. *Agriculture Ecosystems and Environment* 129, 344–348.

Patil, R., Lægdsmand, M., Olesen, J.E. and Porter, J.R. (2012) Sensitivity of crop yield and N losses in winter wheat to changes in mean and variability of temperature and precipitation in Denmark using the FASSET model. *Acta Agriculturae Scandinavica, Section B Plant and Soil* 62, 335–351.

Peltonen-Sainio, P., Jauhianinen, J., Trnka, M., Olesen, J.E., Calanca, P.L., Eckersten, H., et al. (2010) Coincidence of variation in yield and climate in Europe. *Agriculture, Ecosystems and Environment* 139, 483–489.

Piao, S., Ciaia, P., Huang, Y., Shen, Z., Peng, S., Li, J., et al. (2010) The impacts of climate change on water resources and agriculture in China. *Nature* 467, 43–51.

Porter, J.R. and Gawith, M. (1999) Climatic variability and development of wheat: a review. *European Journal of Agronomy* 10, 23–36.

Porter, J.R. and Semenov, M.A. (2005) Crop responses to climatic variation. *Philosphical Transactions of the Royal Society* B 360, 2021–2035.

Ray, D.K., Ramankutty, N., Mueller, N.D., West, P.C. and Foley, J.A. (2012) Recent patterns of crop yield growth and stagnation. *Nature Communications* 3, 1293, doi: 10.1038/ncomms2296.

Reidsma, P., Ewert, F., Boogaard, H. and Diepen, K. van (2009) Regional crop modelling in Europe: the impact of climatic conditions and farm characteristics on maize yields. *Agricultural Systems* 100, 51–60.

Reidsma, P., Ewert, F., Lansink, A.O. and Leemans, R. (2010) Adaptation to climate change and climate variability in European agriculture: the importance of farm level responses. *European Journal of Agronomy* 32, 91–102.

Roos, J., Hopkins, R., Kvarnheden, A. and Dixelius, C. (2010) The impact of global warming on plant diseases and insect vectors in Sweden. *European Journal of Plant Pathology* 129, 9–19.

Rosenzweig, C., Tubiello, F.N., Goldberg, R., Mills, E. and Bloomfield, J. (2002) Increased crop damage in the US from excess precipitation under climate change. *Global Environmental Change* 12, 197–202.

Sakurai, G., Iizumi, T. and Yokozawa, M. (2011) Varying temporal and spatial effects of climate on maize and soybean affect yield prediction. *Climate Research* 49, 143–154.

Schaap, B.F., Blom-Zandstra, M., Hermans, C.M.L., Meerburg, B.G. and Verhagen, J. (2011) Impact of climatic extremes on arable farming in the north of the Netherlands. *Regional Environmental Change* 11, 731–741.

Schlenker, W. and Roberts, M.J. (2009) Nonlinear temperature effects indicate severe damages to US crop yields under climate change. *Proceedings of the National Academy of Sciences* 106, 15594–15598.

Semenov, M.A. and Shewry, P.R. (2011) Modelling predicts that heat stress, not drought, will increase vulnerability of wheat in Europe. *Scientific Reports* 1, 66, doi:10.1038/srep00066.

Smith, P. and Olesen, J.E. (2010) Synergies between mitigation of, and adaptation to, climate change in agriculture. *Journal of Agricultural Science* 148, 543–552.

Supit, I., van Diepen, C.A., de Wit, A.J.W., Kabat, P., Baruth, B. and Ludwig, F. (2010) Recent changes in the climatic yield potential of various crops in Europe. *Agricultural Systems* 103, 683–694.

Supit, I., van Diepen, C.A., de Wit, A.J.W., Wolf, J., Kabat, P., Baruth, B., *et al.* (2012) Assessing climate change effects on European crop yields using the Crop Growth Monitoring System and a weather generator. *Agricultural and Forest Meteorology* 164, 96–111.

Teixeira, E.I., Fischer, G., van Velthuizen, H., Walter, C. and Ewert, F. (2013) Global hot-spots of heat stress on agricultural crops due to climate change. *Agricultural and Forest Meteorology* 170, 206–215.

Trnka, M., Eitzinger, J., Dubrovsky, M., Semeradova, D., Stepanek, P., Hlavinka, P., *et al.* (2010) Is rainfed crop production in central Europe at risk? Using a regional climate model to produce high resolution agroclimatic information for decision makers. *Journal of Agricultural Science* 148, 639–656.

Trnka, M., Olesen, J.E., Kersebaum, K.C., Skjelvåg, A.O., Eitzinger, J., Seguin, B., *et al.* (2011) Agroclimatic conditions in Europe under climate change. *Global Change Biology* 17, 2298–2318.

Trnka, M., Brázdil, R., Olesen, J.E., Eitzinger, J., Zahradníček, P., Kocmánková, E., *et al.* (2012) Could the changes in regional crop yield be a pointer of climatic change? *Agricultural and Forest Meteorology* 166–167, 62–71.

Tubiello, F.N. and Ewert, F. (2002) Simulating the effects of elevated CO_2 on crops: approaches and applications for climate change. *European Journal of Agronomy* 18, 57–74.

Urban, D., Roberts, M.J., Schlenker, W. and Lobell, D.B. (2012) Projected temperature changes indicate significant increase in inter-annual variability of US maize yields. *Climatic Change* 112, 525–533.

Xiao, D., Tao, F., Liu, Y., Shi, W., Wang, M., Liu, F., *et al.* (2013) Observed changes in winter wheat phenology in the North China Plain for 1981–2009. *International Journal of Biometeorology* 57, 275–285.

You, L., Rosegrant, M.W., Wood, S. and Sun, D. (2009) Impact of growing season temperature on wheat productivity in China. *Agricultural and Forest Meteorology* 149, 1009–1014.

Ziska, L.H. (2001) Changes in competitive ability between a C4 crop and a C3 weed with elevated carbon dioxide. *Weed Science* 49, 622–627.

3 Climate Sensitivity of Intensive Rice–Wheat Systems in Tropical Asia: Focus on the Indo-Gangetic Plains

Anil Kumar Singh[1] and Himanshu Pathak[2]

Indian Council of Agricultural Research, IARI Campus, New Delhi, India; [1]*RVS Agricultural University, Gwalior, India;* [2]*Centre for Environment Science and Climate Resilient Agriculture, Indian Agricultural Research Institute, New Delhi, India*

3.1 Introduction

Climatic variability and changes over tropical Asia are of major global concern because of the high population density in the region. Rural populations of this vast continent depend on agriculture for sustenance and are often vulnerable to both the direct impacts of adverse climatic events and the indirect effects of the unpredictability of the climate. Many critical agricultural decisions, ranging from farm to policy level, interact with the climate but must be made several months before the impacts of the climate are realized. Decision makers must, therefore, prepare for the range of possibilities and employ conservative risk management strategies that reduce the negative impacts of climatic extremes. These strategies may be at the expense of reduced average productivity and profitability, inefficient use of resources and accelerated natural resource degradation due, for example, to low investment in soil fertility inputs or conservation measures.

Climate is the primary determinant of agricultural productivity; it influences plant life in many ways and can inhibit, stimulate, alter or modify crop performance. Its components (temperature, solar radiation, rainfall, relative humidity and wind velocity) independently, or in combination, influence crop growth and productivity. It is obvious, therefore, that any significant change in climate will have an impact on agriculture. Climate change is expected to influence crop and livestock production, hydrologic balances, input supplies and other components of agricultural systems. Climate change may also change the types, frequencies and intensities of various crop and livestock pests, the availability and timing of irrigation water supplies and the severity of soil erosion. However, the nature of these biophysical effects and the human responses to them are complex and uncertain.

The intensive rice–wheat system of the Indo-Gangetic Plains (IGP) has evolved rapidly since the 1960s after the introduction of modern high-yielding varieties and access to irrigation, fertilizer and pesticides, and provides staple grain for more than 400 million people in Asia. The system is practised over a variety of soil and climatic conditions, with a wide range of input use and management practices. In addition to soil and climatic conditions, socio-economic considerations and policy issues also govern the magnitude and intensity of the rice–wheat cropping system (Pingali, 1999). The rice–wheat areas are located within subtropical to warm temperate climates, characterized by cool and dry winters and warm and wet summers (Timsina and

Connor, 2001). These occupy about 18 million ha (Mha) of the most productive land in the IGP and China, and are one of the world's largest agricultural production systems (Ladha et al., 2003). In South Asia, the system occupies about 13.5 Mha (10 Mha in India, 2.2 Mha in Pakistan, 0.8 Mha in Bangladesh and 0.5 Mha in Nepal). The rice–wheat cropping system is also practised to a very limited extent in Bhutan and Myanmar. In China, rice–wheat is grown on about 3.4 Mha in the provinces of Jiangsu, Zhejiang, Hubei, Guizhou, Yunnan, Sichuan and Anhui. During the past four decades, the system has contributed substantially towards the food security of South Asia. However, of late, there has been a significant slowdown of the increase in production of this system, and the sustainability of this important cropping system is at stake. Decline in soil fertility, particularly from reduced organic C and N and a deterioration in soil physical characteristics, coupled with delays in the sowing of wheat, decreased water availability and depletion of groundwater, increased soil salinity and water logging and increased pest incidence and the evolution of new, more virulent pests are often suggested as the causes of this reduced productivity. Climate change will have an adverse impact on the productivity of this important cropping system, posing a real threat to food security in Asia.

The likely impact of climate change on rice and wheat production is of paramount importance. As yields in some of the most productive regions of the world are approaching a plateau, or even declining, the likely effect of climate change on crop production adds to this already complex problem. On the one hand, an increase in the CO_2 concentration in the atmosphere would increase crop yield, mainly by stimulating photosynthetic processes and improving water-use efficiency (Matthews et al., 1995), while on the other hand the effect of increased temperature would largely be negative because of increased respiration and shortened vegetative and grain-filling periods. The net effect of an increase in atmospheric CO_2 concentration and temperature is complicated and depends on the relative effects of both variables in a given region. The question, therefore, is whether climate change will reduce yields and pose a risk to cereal production. The sustainability of agricultural production and food security in this region of poor soil fertility depends greatly on the climatic conditions, so any negative change will have serious consequences.

In recent years, climate change has emerged as the most prominent of the global environmental issues. Given the fundamental role of agriculture in human welfare, concern has been expressed regarding the potential effects of climate change on agricultural productivity. This chapter analyses climate change and climatic variability scenarios for South Asia, assesses the vulnerability of the rice–wheat system to climate change and suggests adaptation strategies to climate change for the system, with a focus on the IGP.

3.2 The Indo-Gangetic Plains

The IGP are one of the most populous and productive agricultural ecosystems in the world (Wassmann and Dobermann, 2007). Given the dimensions of the IGP, this region can be further divided into distinct subregions (Fig. 3.1): (i) the Trans-Gangetic Plains or IGP Transects 1 and 2 (areas in Pakistan and parts of Punjab and Haryana in India); (ii) Upper-Gangetic Plains or IGP Transect 3 (most of Uttar Pradesh and parts of Bihar, India, and parts of Nepal); (iii) Middle-Gangetic Plains or IGP Transect 4 (parts of Bihar, India, and parts of Nepal); and (iv) Lower-Gangetic Plains or Transect 5 (parts of Bihar, West Bengal, India, and parts of Bangladesh) (Narang and Virmani, 2001). Solar radiation decreases from IGP Transects 1 to 5 during the rice growing season, while the trend is reversed in the wheat growing season (Table 3.1). The minimum temperature in the rice and wheat seasons increases from IGP Transects 1 to 5. This is also true for maximum temperature in the wheat season, but in the rice season, average maximum temperature remains

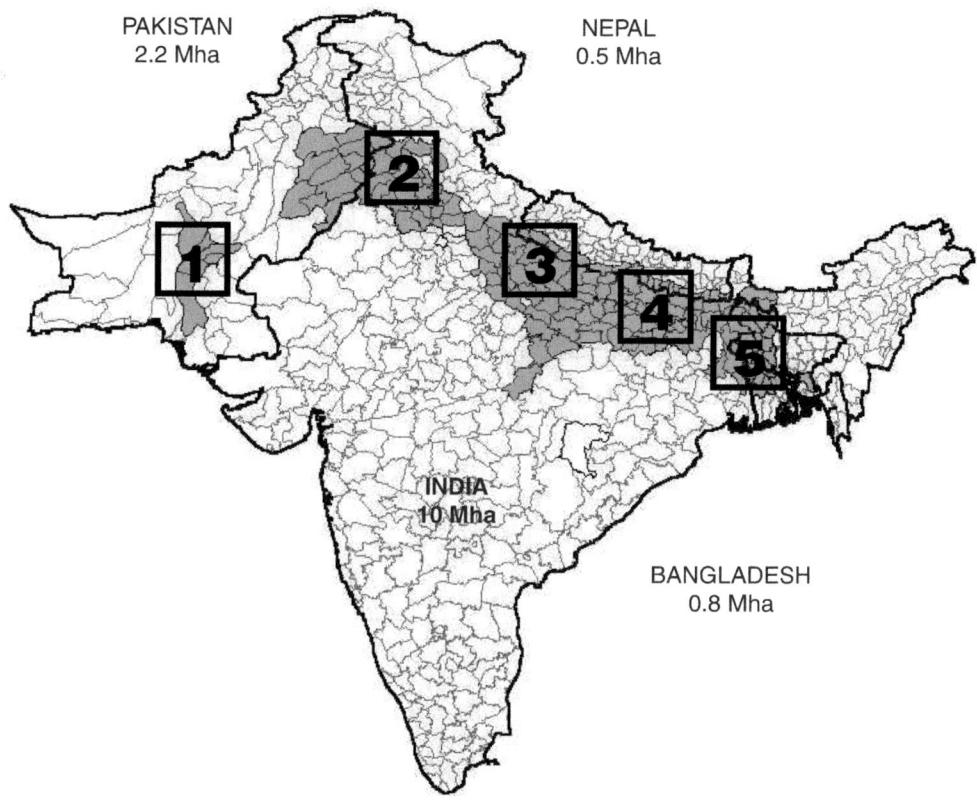

Fig. 3.1. Various transects and areas under rice–wheat cropping systems in the Indo-Gangetic plains. IGP 1, areas in Pakistan; IGP 2, parts of Pakistan and Punjab and Haryana in India; IGP 3, most of Uttar Pradesh and parts of Bihar, India, and parts of Nepal; IGP 4, parts of Bihar, India, and parts of Nepal; and IGP 5, parts of Bihar and West Bengal, India, and parts of Bangladesh. (Source: Wassmann et al., 2009.)

almost similar throughout all the transects of the IGP. Rainfall also follows a distinct pattern of increase from Transects 1 to 5 of the IGP (Table 3.1). Transects 1 and 2 receive only 650 mm of rainfall per annum, while Transect 5 receives more than 2.5 times this rainfall. The climatic parameters, except rainfall, make the upper transects of the IGP more favourable for rice and wheat cultivation. Access to assured irrigation, however, helps to overcome the problem of low rainfall and makes the zone very productive, while less favourable climatic conditions and limited irrigation facilities are the bottlenecks to achieving higher yields in the lower transects of the IGP (Pathak et al., 2003; Wassmann et al., 2009).

3.3 Impact of Climate Change on Yields of Rice and Wheat

The production of rice and wheat, the two most important cereals, is crucial for the food security of Asia. These are the staple food crops of the continent. Production of these cereals in Asia has increased markedly with the introduction of modern crop production technologies based on early maturing, N-responsive, semi-dwarf cultivars since the Green Revolution commencing in the 1960s. To feed the increasing population, the production of cereals has increased, but the challenge is to achieve this while maintaining and enhancing the quality of the land resource on which production

Table 3.1. Soil, climate, irrigation and potential yield of rice and wheat in various transects of the Indo-Gangetic Plains. (Source: Pathak *et al.*, 2003; Wassmann *et al.*, 2009.)

Parameter	IGP transects									
	1		2		3		4		5	
	Rice	Wheat	Rice	Wheat	Rice	Wheat	Rice	Wheat	Rice	Wheat
Rainfall (mm)	550	60	550	60	680	80	950	100	1450	150
Maximum temperature (°C)	34	25	34	24	34	25	33	26	32	28
Minimum temperature (°C)	24	14	22	11	24	11	27	16	25	15
Solar radiation (MJ m^{-2} d^{-1})	20	12	23	14	19	14	20	15	17	16
Potential yield (Mg ha^{-1})	10.7	7.9	10.7	7.9	9.5	7.0	9.2	6.8	7.7	5.2
Irrigated area (%)	99	97	99	98	60	92	40	88	25	73
Soil texture	Loamy sand		Loamy sand		Loam		Silty loam		Sandy loam	
Soil organic C (%)	0.3		0.3		0.3		0.4		0.7	

depends. During the Green Revolution era, production increases resulted from increases in both area and productivity, but little additional land is available at present and increasingly traditional farmlands are lost to urbanization. In addition, since most of the land is already double and even triple cropped, intensification is not a possible option. Therefore, future growth in demand will have to be met mainly through increases in yield per unit land area. Moreover, there has lately been a significant slowdown of the growth rate in area and production, as well as yield (Ladha *et al.*, 2003).

There are three ways in which climate change may affect cereal production. First, increased atmospheric CO_2 concentrations can have a direct effect on the growth rate of crop plants and weeds. Second, CO_2-induced changes of climate may change temperature, rainfall and sunshine, all of which can influence plant growth. Finally, rises in sea level may lead to loss of farmland by inundation and to increasing salinity of groundwater in coastal areas. The first two types of effect have general and global impact, but the effect of the third is more localized.

Several studies have aimed at understanding the nature and magnitude of gains or losses in the yield of particular crops at selected sites in Asia under changed climate. These studies suggest that, in general, areas in mid and high latitudes will experience increases in crop yield, whereas yields in areas in lower latitudes will decrease. Generally, climatic variability and change will endanger sustained agricultural production seriously in tropical Asia in the coming decades. The scheduling of the cropping season, as well as the duration of the growing period of the crop, will also be affected.

3.3.1 Temperature

Temperature is a dominant climatic control on crop growth (see Olesen, Chapter 2, this volume). It determines the potential length of the growing season and generally has a strong effect on the timing of developmental processes and on the rates of plant leaf expansion. The latter, in turn, affects the time at which a crop canopy can begin to intercept solar radiation and, thus, the efficiency with which solar radiation is used to make plant biomass. Plant development does not begin until temperature exceeds a certain threshold; then, the rate of development increases with temperature to an optimum, above which it decreases.

Temperature, particularly high temperature, plays a crucial role in determining the yields of cereals. Crops have varying sensitivity to temperature. There are temperature threshold values beyond which crops become vulnerable to sharp temperature shifts. The effects of differences in mean seasonal temperature on crops are better understood than those of the fluctuating temperatures of many natural environments. For example, the rate of many development processes is a positive linear function of temperature between a base temperature (at and below which the rate of a particular process is zero) and an optimum temperature, and a negative linear function of temperature between this optimum and a ceiling temperature (Table 3.2). Generally, the growth rate of rice increases linearly in the temperature range 22–31°C, while temperature beyond this reduces growth and productivity. During flowering and grain filling, high temperature reduces yield by causing spikelet sterility and shortening of the duration of the grain-filling phase. An increase in leaf surface temperature has significant effects on crop metabolism and yields, and it may make crops more sensitive to moisture stress. Experiments in India showed that higher temperatures and reduced radiation associated with increased cloudiness caused spikelet sterility and reduced yields, to such an extent that any increase in dry-matter production as a result of CO_2 fertilization proved to have no advantage for grain productivity.

One of the most important effects of an increase in temperature, particularly in regions where agricultural production is currently limited by temperature, would be to extend the growing season available for plants. But in tropical Asia, this effect will be largely negative, as it will decrease the length of the growing season and reduce yields. In general, in regions where temperature maxima are near the optimum under current climatic conditions, such as in tropical Asia, increases in temperature will probably lead to decreased yields.

3.3.2 CO_2 enrichment

If increases in atmospheric CO_2 were to occur without the associated changes in climate, then, overall, the consequences for agriculture would probably be beneficial, as increases in CO_2 concentration would increase the rate of plant growth. There are, however, important differences between the photosynthetic mechanisms of different crop plants and, hence, in their response to increasing CO_2. Plant species with the C_3 photosynthetic pathway (the first product in their biochemical sequence of reactions has three carbon atoms) use up some of the solar energy they absorb in a process known as photorespiration; here, a significant fraction of the CO_2 initially fixed into carbohydrates is re-oxidized back to CO_2. C_3 species (e.g. wheat, rice, barley) tend to respond positively to increased CO_2 concentration, because this gas tends to suppress the rates of photorespiration. However, in C_4 plants (those in which the first product in their biochemical sequence of reactions has four carbon atoms), CO_2 is first trapped inside the leaf and then concentrated in the cells which perform the photosynthesis.

Table 3.2. Temperature thresholds and required cumulative daily temperatures above base temperature for rice and wheat. (Modified from Roetter and van de Geijn, 1999.)

Crop	Optimum range (°C)	Lower range (°C)	Upper range (°C)	Emergence to pre-anthesis (day degrees)	Post-anthesis to maturity (day degrees)	Base temperature (°C)	Maximum development (°C)
Rice	25–30	7–12	35–38	700–1300	450–850	8	25–31
Wheat	17–23	0	30–35	750–1300	450–1050	0	20–25

Although more efficient photosynthetically under current levels of CO_2, these plants (e.g. maize, millet, sorghum) are less responsive to increased CO_2 levels than C_3 plants. A doubling of ambient CO_2 concentration causes a decrease in stomatal aperture in both C_3 and C_4 plants, which may reduce transpiration. This will help plants in environments where moisture currently limits growth (e.g. in semi-arid regions), but there remain many uncertainties, such as how much the greater leaf area of plants caused by increased CO_2 will balance the reduced transpiration per unit leaf area. The above discussion suggests that there is no direct risk to crop production due to increased CO_2 in the atmosphere.

3.3.3 Drought

Drought caused by inadequate rainfall is the most important climatic aberration which has influenced agricultural production in the Asian subcontinent since the beginning of settled crop cultivation (Sinha and Swaminathan, 1991). Because of an increase in the area under irrigation and the introduction of early maturing crop varieties characterized by resilience in sowing dates, the impact of drought on India's food production has reduced.

Drought affects the growth of rice plants at seedling, tillering and flowering stages, and causes enormous yield losses. Erratic rainfall during the early part of the monsoon combined with light textured soils exposes upland rice to drought at the vegetative stage. Dry spells during flowering are common in upland rice and inhibit pollen production greatly, leading to sterility. The variability of water supply to crops can cause negative effects on crop production. In rice, water stress at panicle initiation increases the proportion of unfilled grains and decreases mean grain weight (Wopereis et al., 1996). Reduced availability of water at the vegetative stage results in reduced morphological and physiological measurements in rice, including tiller number, leaf area index, apparent canopy photosynthetic rate, leaf nitrogen, shoot and root biomass and root length density (Cruz et al., 1986).

Drought could be a major concern in the future and it is estimated that a 1°C increase in air temperature will lead to 37 mm more potential evapotranspiration south of 40°N. Similarly, frequent droughts not only reduce supplies but also increase the amount of water needed for plant transpiration. When they occur, drier soil conditions suppress both root growth and decomposition of organic matter, and will increase vulnerability to wind erosion, especially if winds intensify.

3.3.4 Sea level rice

With rising sea level, coastal zones and small islands are extremely vulnerable, as flooding and coastal erosion will worsen, damaging key economic sectors and threatening human health. Ocean ecosystems may also be affected by causing serious risk to valuable coastal ecosystems. The most serious physical impacts of sea level rise are: (i) inundation and displacement of wetlands and lowlands; (ii) coastal erosion; and (iii) increased coastal storm flooding and salinization (Barth and Titus, 1984). The impact will vary from place to place, depending on the magnitude of sea level rise, coastal morphology and human modifications.

In South, South-east and East Asia, about 10% of the regional rice production, which is enough to feed 200 million people, is located in these areas, which are susceptible to a 1 m rise in sea level (Hoozemans et al., 1993). Direct loss of land combined with less favourable hydraulic conditions may reduce rice production by 4% if no adaptation measures are taken, endangering the food supply of 75 million people. Saltwater intrusion and soil salinization are other concerns for agricultural productivity. Estimated land loss and population displacement (1990s population) in Asia for various sea level rise scenarios and no adaptation are shown in Table 3.3. It is

Table 3.3. Estimated land loss and population displaced (1990s population) in Asia for various sea level rise scenarios and no adaptation. (Source: Nicholls and Mimura, 1998.)

Country	Sea level rise (cm)	Land loss (km^2)	Population displaced (million)
Bangladesh	100	29,846	14.8
India	100	5,763	7.1
Indonesia	60	34,000	2.0
Malaysia	100	7,000	>0.05
Pakistan	200	1,700	–
Vietnam	100	40,000	17.1

estimated that a 1 m rise could cause loss of about 9.5 Mt of rice in Bangladesh and 16,000 km^2 of rice paddy in Indonesia, while in Vietnam 25,000 km^2 of rice paddy would be subject to annual flooding (Nicholls and Mimura, 1998).

India, with its long coastline of over 7500 km, is susceptible to sea level rise, as many of its human settlements, including infrastructure and agricultural activities, are located along the coastline. Sea level rise will affect many regions of the Andaman and Nicobar islands, the Lakshadweep islands in the south-west, the low-lying deltaic region in West Bengal and the Kutch regions in Gujarat.

3.3.5 Flooding

Mirza et al. (2003) observed that future changes in the peak discharge of the Ganges river are expected to be higher than those for the Brahmaputra river. Peak discharge of the Meghna River may also increase considerably. As a result, significant changes in the spatial extent and depths of inundation in Bangladesh may occur. The mean flooded area may increase in the range of 20–40% for a 6°C rise in temperature. Changes in land inundation categories may introduce substantial changes in rice agriculture and cropping patterns in Bangladesh. In future, 55% of the flooded area may be deeply flooded. Greater changes may occur in the non-flood, moderately and deeply flooded land categories. Cropping intensity and production of high-yielding varieties of rice may be reduced substantially. In terms of population, more people will be vulnerable in future, as an increased number of people will be living in the flood plains of Bangladesh. More houses and infrastructure will be exposed to flooding, and the likelihood of increased damage is high. This underscores the need for strengthening flood management policies and adaptation measures in Bangladesh to reduce increased flood incidence.

3.4 Estimates of Risks of Climate Change on Rice–Wheat in Asia

Estimating the risks of a changing climate on crop production in tropical Asia is difficult, due to the variety of cropping systems and the levels of technology used. However, the use of crop growth models is one way in which these effects can be studied, and probably represents the best method at present for doing so. Although a large number of simplifying assumptions must necessarily be made, these models allow the complex interaction between the main environmental variables influencing crop yields to be understood. There have been a few studies in India to understand the nature and magnitude of yield gains or losses of crops at selected sites under elevated atmospheric CO_2 and the associated climatic change (Lal et al., 1998; Saseendran et al., 1999). The impact of projected global warming on crop yields has been evaluated by indirect methods using simulation models (Aggarwal and Mall, 2002).

3.4.1 Rice

Adverse climatic factors play an important role in determining the yield of rice. Most of the simulation studies showed a decrease in the duration and yield of crops as temperature increased in different parts of India (Aggarwal and Rani, 2009). The effect of climate change on rice production has been studied extensively (Pathak et al., 2003). An increase in the CO_2 concentration in the atmosphere would increase crop yield mainly by stimulating photosynthetic processes and improving water-use efficiency, but the effect of increased temperature would largely be negative because of increased respiration and a shortened vegetative and grain-filling period. The net effect of an increase in CO_2 and temperature is complicated and depends on the relative effects of both variables in a given region. Major rice models indicate a reduction in yield of about 5% per °C rise in the mean temperature above 32°C. This would largely offset any increase in yield as a consequence of increased CO_2. Rice yields decline when temperatures exceed 30°C at flowering, because of spikelet sterility. Rice, being grown in the monsoon season, encounters severe climatic variability, particularly irregular rainfall. If the water supply is not assured, rice is likely to be more affected by water shortage, because of its higher water requirement than other crops such as wheat. In the IGP, water availability for irrigation is decreasing due to: (i) a greater area coming under rice, thus reducing the availability per unit area; (ii) reduced water supply through canal irrigation, which is linked to less capacity in reservoirs because of increasing siltation; and (iii) increasing competition for water for domestic and industrial use (Pathak et al., 2003). Water will become a major constraint to agriculture in the future as the demand from domestic and industrial use increases and new irrigation facilities are not created. Since 1985, Pathak et al. (2003) have estimated the trends of the potential yields of rice across the IGP, and they observed that the rate of change in the potential yield trend of rice ranged from −0.12 to 0.05 Mg ha^{-1} yr^{-1}. Negative yield trends were observed at six of the nine sites, and four of these were statistically significant (P <0.05). The decrease in radiation and increase in minimum temperature were the reasons for the decline in yield. Peng et al. (2004) analysed weather data at the International Rice Research Institute farm from 1979 to 2003, to examine the temperature trends and the relationships between rice yields and temperature. Annual mean maximum and minimum temperatures increased by 0.35°C and 1.13°C, respectively, during the period, and a close linkage between rice grain yield and mean minimum temperature was observed. Grain yield declined by 10% for each 1°C increase in growing season minimum temperature in the dry season, whereas the effect of maximum temperature was insignificant.

3.4.2 Wheat

Predicted effects of climate change on wheat production include reduced grain yield over most of India, with the greatest impacts in the potentially lower areas such as the eastern IGP (Ortiz et al., 2008). Physiological traits that are associated with wheat yield in heat-prone environments are canopy temperature depression, membrane thermostability, leaf chlorophyll content during grain filling, leaf conductance and photosynthesis and senescence.

Hundal and Kaur (1996) showed that in Punjab, India, a temperature increase of 1, 2 or 3°C above present-day conditions would reduce the grain yield of wheat by 8.1, 18.7 and 25.7%, respectively. Aggarwal and Mall (2002) showed that a 2°C increase resulted in a 15–17% decrease in the grain yield of rice and wheat, but with greater increases the decline was very high in wheat. These decreases were compensated for by an increase in CO_2, due to its fertilizing effect on crop growth. However, CO_2 concentration would need to rise to 450 ppm to nullify the negative effect of a 1°C increase in temperature, and to 550 ppm to nullify a 2°C increase in temperature. In the rice–wheat system of eastern India, at least 60% of wheat areas were planted after November

(Chandna et al., 2004), subjecting the crop to suboptimal, often hotter, growing seasons and resulting in lower yield. In many of the dry environments that suffer today from severe heat stress during grain filling, the enzyme starch synthase in wheat appears to be rate limiting at temperatures in excess of 20°C. Furthermore, grain filling of wheat is seriously impaired by heat stress, due to reductions in current leaf and ear photosynthesis at high temperatures. Ortiz-Monasterio et al. (1994) described the dramatic yield-reducing effects of high temperatures around and after heading for wheat crops in South Asia. Samra and Singh (2005) analysed the impact of an abnormal temperature rise in March 2004 on the productivity of wheat. Temperature increases above normal ranged from 1 to 12°C in different parts of northern India, resulting in lost wheat production of 4.6 Mt due to greater incidences of pests and diseases, advanced maturity of wheat by 10–20 days and reduced grain weight. The decrease in yield was pronounced in salt-affected and reclaimed salt-affected land due to an accumulation of salts in the root zone. Pathak and Wassmann (2009) quantified the impacts on wheat yield due to rainfall variability in north-west India and showed that the years with scarce rainfall resulted in only 34% (Ludhiana) and 35% (Delhi) of the baseline yield. In Ludhiana, high rainfall years resulted in 200% yield as compared to the baseline yield, whereas they reached only 105% in Delhi.

It is evident from the above results that an increase in temperature in the future is likely to cause a significant decrease in wheat production in India. A 1°C increase in temperature with no associated CO_2 increase will lead to a decrease of 6 Mt of wheat (7% of the total current wheat production). With a 3°C increase, this loss may touch 19 Mt, and with a 5°C increase this could be almost 27.5 Mt (Aggarwal and Rani, 2009). Simple adaptation strategies such as a change in planting date and variety further reduced the extent of loss caused by high temperatures. Considering the slow process of bridging the yield gap and the costs involved in creating an appropriate environment for bridging yield gaps, it can be concluded that global warming will constrain progress in increasing wheat production in the future, unless some new technologies are evolved. Studies also showed that both potential yields and current yields would be likely to decrease in future, even after considering technological growth in management. Potential yields are likely to decrease relatively more in the future than are current yields, leading to a net reduction in the yield gap.

Adaptation strategies can, however, assist in providing some relief in future, provided these strategies could be operationalized in the field. In the absence of effective adaptation, India at an aggregated level would lose 3.9 Mt of wheat due to climate change by 2020, 11.7 Mt by 2050 and 23.5 Mt of wheat by 2080 (Aggarwal and Rani, 2009). Adaptation strategies, if practised, have the potential to nullify this loss completely in the short term (2020), but not in the long term. Implementing these adaptation strategies even in the short term may, however, be difficult, considering that wheat cannot be planted earlier in most of the IGP because of the late availability of fields after the rice harvest.

3.5 Vulnerability of the Rice–Wheat System due to Climate Change

The vulnerability of the rice–wheat system in different transects of the IGP due to climate change is summarized in Table 3.4. Yield of the IGP rice–wheat system may decrease as per the global warming forecast. This may be aggravated by water scarcity, drought, flood and decline in soil organic C content. Simulation models for rice production indicate a reduction in yield of about 5% per °C rise in the mean temperature above 32°C (Matthews et al., 1995). This would largely offset any increase in yield as a consequence of increased CO_2. Rice is particularly sensitive to high temperature at anthesis, as sterility of some cultivars occurs if temperatures exceed 35°C at anthesis and last for >1 h, and high temperatures cause spikelet sterility in dry and monsoon season crops in parts of Asia (Yoshida, 1981). At anthesis, spikelet fertility is reduced from 90 to 20% by only 2 h exposure to 38°C, and to 0% by <1 h exposure to 41°C. The critical temperature for spikelet fertility (defined as when fertility exceeds 80%) varies between genotypes, but is about 32–36°C. Below 20°C and above about 32°C, spikelet sterility becomes a major factor, even if there is sufficient plant growth. Experiments in India showed that higher temperatures and reduced radiation associated with increased cloudiness caused spikelet sterility and reduced yields to such an extent that any increase in dry matter production as a result of CO_2 fertilization proved to have no advantage to grain productivity (Rao and Sinha, 1994). Hundal and Kaur (1996) reported that if all other climatic variables were to remain constant, a temperature increase of 1, 2 or 3°C would reduce grain yield of rice by 5.4, 7.4 and 25.1%, respectively. Matthews et al. (1997) suggested that rice production in the Asian region might decline by 3.8% under a changed climate. The adverse impacts of likely water shortage would reduce rice yields by 20%. Peng et al. (2004) reported that rice yields declined by 10% for each 1°C increase in growing season minimum temperature in the dry season, whereas the effect of maximum temperature was insignificant.

3.6 Adaptation Strategies in the Rice–Wheat System

According to the recent IPCC (Intergovernmental Panel on Climate Change) assessment, agricultural production in South Asia could fall by 30% by 2050 if no action is taken to combat the effects of increasing temperatures and hydrologic disruption (IPCC, 2007). Adaptive options to deal with the impact of climate change are: (i) developing cultivars tolerant to heat and salinity stress and resistant to flood and drought; (ii) modifying crop management practices, including improving water management; (iii) adopting new farm techniques such as resource conserving technologies (RCTs); (iv) crop diversification; (v) improving pest management; (vi) better weather forecasts and crop insurance; and (vii) harnessing the indigenous and technical knowledge of farmers (Wassmann et al., 2009) (see Webber et al., Chapter 11, this volume). Potential adaptation strategies for the rice–wheat system in different transects of the IGP are listed in Table 3.4.

3.6.1 Crop improvement

It is clear that current technologies alone will not be sufficient to meet the future wheat demand, even if all necessary support is provided for bridging the yield gap. New genotypes with higher yield potential will be required, or change in land use leading to more timely wheat planting will be needed in northern latitudes. In the event of a lack of enough desirable variability, genetic engineering interventions may provide the opportunity for the interspecific and intraspecific introgression of desirable genes. Sources of the novel genes may be from plant, animal or microbial organisms. The coupling of new genomic tools, technologies and resources with genetic approaches is essential to underpin wheat breeding through marker-assisted selection (Habash et al., 2009). The crop improvement programme of India is better equipped now than earlier, and it is essential to exploit this opportunity arising out of challenges posed by climate change and variability. Strategies to develop new genotypes may include the following:

- Improvement of germplasm for heat tolerance is one of the targets of the wheat breeding programme (Ortiz et al., 2008).
- Similarly, it is essential to develop tolerance to multiple abiotic stresses as they occur in nature. The abiotic stress-tolerance mechanisms are quantitative traits in plants. Germplasm with greater oxidative stress tolerance may be exploited, as oxidative stress tolerance is one example where the plant's defence mechanism is targeting several abiotic stresses.

Table 3.4. Vulnerability of the rice–wheat system due to climate change and potential adaptation strategies in different transects of the Indo-Gangetic Plains.[a] (Source: Wassmann et al., 2009.)

Transect	Vulnerability	Adaptation strategies
IGP 1 and 2	High temperature-induced sterility in rice	Heat-tolerant rice cultivar
	Shortage of irrigation water	Water-saving technologies (laser land levelling, direct-seeded rice, no-till rice and wheat)
	Abrupt temperature rise in rabi season	Adjusting sowing date, heat-tolerant cultivar, better weather forecast
	Declining soil organic matter	Residue management
	Rising salinity	Salt-tolerant cultivars
	Increased pests and diseases	Improved pest management
	Late sowing of wheat	No-till wheat
IGP 3	Shortage of irrigation water	Water-saving technologies (laser land levelling, direct-seeded rice, no-till rice and wheat)
	Abrupt temperature rise in rabi season	Adjusting sowing date, heat-tolerant cultivars, better weather forecast
	Rain during maturity of rice	Adjusting planting date, better weather forecast, crop insurance
	Declining soil organic matter	Residue management
	Rising salinity and alkalinity	Salt- and alkali-tolerant cultivars
	Increased pests and diseases	Improved pest management
	Late sowing of wheat	No-till wheat
IGP 4	Widespread flood	Better weather forecast, crop insurance, flood-resistant cultivar
	Frequent drought in some areas	Developing irrigation facilities, drought-resistant cultivar, better weather forecast, crop insurance, indigenous knowledge
	Rain and storm during maturity of rice	Adjusting planting date, better weather forecast, crop insurance
	Shorter wheat season	Crop diversification, no-till wheat
	Rise in temperature during grain filling of wheat	Adjusting sowing date, heat-tolerant cultivar, better weather forecast
	Increased pests and diseases	Improved pest management
	Shortage of irrigation water	Water-saving technologies (laser land levelling, direct-seeded rice, no-till rice and wheat)
IGP 5	Frequent flood	Better weather forecast, crop insurance, flood-resistant cultivar
	Water logging and excess soil moisture in wheat	Crop diversification, no-till wheat
	Occasional drought in some areas	Developing irrigation facilities, drought-resistant cultivar, better weather forecast, indigenous knowledge, crop insurance
	Rain and storm during maturity in rice and wheat	Adjusting planting date, better weather forecast, crop insurance
	Shorter wheat season	Crop diversification, no-till wheat
	Increased pests and diseases	Improved pest management
	Rising salinity	Salt-tolerant cultivar

Note: [a]Refer to Fig. 3.1 for descriptions of different IGP transects.

- Several research efforts are attempting to convert rice from a C_3 to a C_4 crop; such efforts may also be useful for the improvement of radiation-use efficiency of wheat as well.
- Current long-duration varieties may have a reduced phenology when exposed to high temperatures. This may offset the reduction of yield due to shorter crop growth and grain-filling period because of the rise in atmospheric temperature. Exploitation of genetic variability in grain-filling duration may also form one of the strategies to minimize the reduction in grain-filling duration.
- Improvement of water-use and nitrogen-use efficiency has been attempted for a long time. These efforts become more relevant in the climate change scenarios as it is likely that the water resources for agriculture may dwindle in future. Nitrogen-use efficiency may be reduced in climate change scenarios because of high temperatures and heavy precipitation events causing volatilization and leaching losses. To exploit the beneficial effects of elevated CO_2 concentrations, the crop demand for nitrogen will also increase. Thus, it is important to improve the efficiency of roots in absorbing water and mining soil nutrients.
- Exploitation of desirable genes from related wild types.
- Varieties with high revival capacity after stress become important, particularly in view of the increase in climatic variability. Increase in the frequency of events such as heavy precipitation, heatwaves and cold waves pose challenges to standing crops. Exploitation of genetic engineering for 'gene pyramiding' becomes essential to pool all desirable traits in one plant to get the 'ideal plant type' which may also be an 'adverse climate-tolerant genotype'.

3.6.2 Crop management

Farmers can adapt to climate change to some extent by shifting planting dates, choosing varieties with different growth duration, or changing crop rotations. Adjustment of planting dates to minimize the effect of temperature-induced spikelet sterility can be used to reduce yield instability, by avoiding the coincidence of the flowering period with the hottest period. Although shifting of sowing dates is a no-cost decision that can be taken at the farm level, a large shift in sowing dates may interfere with the agrotechnological management of other crops grown during the remaining part of the year (Mall et al., 2006). Seasonal weather forecasts could be one supportive measure to optimize planting dates. However, many of these adaptation mechanisms may result in lower yields. In the IGP, delayed planting is already one of the major causes of reduction in crop yields in rice as well as in wheat. Modification in the choice of crops or cultivars to grow, especially in areas accustomed to high crop diversification, and changes in agronomic management practices including fertilizer use, irrigation and control of pests and diseases are other adjustments for climate change and variability.

Resource-conserving technologies (RCTs) encompass practices that enhance resource- or input-use efficiency and provide immediate, identifiable and demonstrable economic benefits such as reductions in production costs, savings in water, fuel and labour requirements and timely establishment of crops resulting in improved yields. Yields of wheat in heat- and water-stressed environments can be raised significantly by adopting RCTs which minimize unfavourable environmental impacts, especially in small and medium-scale farms. Resource-conserving practices like zero tillage (ZT) can allow farmers to sow wheat sooner after the rice harvest, so the crop heads and fills the grain before the onset of pre-monsoon hot weather. As average temperatures in the region rise, early sowing will become even more important for wheat. Field results show that increasingly RCTs are being adopted by farmers in the rice–wheat belt of the IGP, because of several advantages of labour saving, water saving and early planting of wheat. By the end of 2007, approximately 4.0 Mha were under RCTs and 0.5 million farmers were using these technologies (Erenstein et al., 2007). RCTs in the rice–wheat system also have pronounced effects on the mitigation of greenhouse gas emissions and adaptation to climate change (Aggarwal and Pathak, 2009). Govaerts et al. (2005) observed that under ZT combined with residue retention on the soil surface, C sequestered in the uppermost layer was higher than that for conventional tillage (CT). These approaches to crop management should be coupled with other approaches to crop improvement for wider adaptation to climate change (Kalish et al., 2011; Pathak et al., 2011).

3.6.3 Crop diversification

Diversification (i.e. growing a range of crops suited to different sowing and harvesting times) assists in achieving sustainable productivity by allowing farmers to employ biological cycles to minimize inputs, maximize yields, conserve the resource base and reduce the risk due to both environmental and economic factors. The farmers of the rice–wheat belt have taken the initiative to diversify their agriculture by including short-duration crops such as potato, soybean, urd, mungbean, cowpea, pea, mustard and maize in different combinations. In the eastern Gangetic plains, agriculture is both drought and flood prone. In flood-prone areas, intercropping and the choice of appropriate crop cultivars would be helpful. It is estimated that more than 4 Mha of land used for rice remain as 'rice fallows' in the IGP alone (Chandna et al., 2004). With appropriate crop establishment technologies, it is possible to use this land to raise a second crop of wheat, pulses, maize, or lentils to improve farmers' income and livelihoods in marginal areas. However, there is a need to quantify the impacts of crop diversification on income, employment, soil health, water use and greenhouse gas emissions.

3.6.4 Improved pest management

Changes in temperature and variability in rainfall affect pest and disease incidence and their virulence in major crops. Some of the potential adaptation strategies could be: (i) developing cultivars resistant to pests and diseases; (ii) integrated pest management with more emphasis on biological control and changes in cultural practices; (iii) pest forecasting using recent tools such as simulation modelling; and (iv) alternative production techniques and crops, as well as locations, that are resistant to infestations and other risks.

3.6.5 Better weather forecasts and crop insurance

Forecasting and early warning systems will be very useful in minimizing risks of climatic origin. Information and communication technologies (ICT) could greatly help researchers and administrators to develop contingency plans (see Rickards et al., Chapter 15, this volume). The Indian Meteorological Department has developed improved capabilities for forecasting of the summer monsoon rainfall and its associated variability. Extreme weather events like the Bay of Bengal cyclones are monitored with improved technology. Crop insurance schemes should be put in place to help the farmers in reducing the risk of crop failure due to these events.

3.6.6 Harnessing the indigenous technical knowledge of farmers

Farmers in South Asia are poor and marginalized, and have experienced climatic variability for centuries. There is a wealth of knowledge of a range of measures that can help in developing technologies to overcome climate vulnerabilities. Indigenous knowledge can be harnessed and fine-tuned to suit adaptation to a rapidly changing world (see Ingram, Chapter 16, this volume). The integration of local knowledge into climate change adaptation plans might also result in more culturally appropriate options, and present a more holistic and integrated perspective. Farmers of rice–wheat ecosystems practise an array of adjustments like agronomic practices, family budget adjustments and help from the social set-up and acquisition of loans to cope with adverse events like floods. Locally conducted surveys showed that farmers tried to minimize the losses of recurrent floods by making a change in the sowing and planting time, methods of sowing, change in variety, harvesting time, fertilizer use and seed rate. Using local germplasm highly acclimatized to withstand harsh climates is one innovative adaptation strategy. In Bihar, for example, local farmers grow several rice varieties that are adapted to the depth of flooding. Research on local knowledge for facilitating climate change adaptation requires relatively small resources but might yield a large dividend in furthering sustainable development.

3.7 Summary and Conclusions

Considerable work has gone into questions of just how farming might be affected by climate change in the different regions of the IGP, and by how much the net result may be harmful or beneficial, and to whom. Several uncertainties limit the accuracy of current projections. One relates to the degree of temperature increase and its geographic distribution. Another pertains to the concomitant changes likely to occur in the precipitation patterns that determine the water supply to crops, and to the evaporative demand imposed on the crops. The problem of predicting the future course of agriculture in a changing world is compounded by the fundamental complexity of natural agricultural systems, and of the socio-economic systems governing the world food supply and demand.

Agriculture is sensitive to short-term changes in weather and to diurnal, seasonal, annual and long-term variations in climate. The climate elements, which affect plant growth and development, and hence agriculture as a whole, are carbon dioxide concentration, temperature, radiation, precipitation and humidity. These are likely to be altered with the increased build-up of greenhouse gases in the atmosphere and considerably affect food supply and access through their direct and indirect effects on crops, soils, livestock, fisheries and pests. The increase in atmospheric CO_2 concentration promotes the growth and productivity of C_3 plants. Increase in temperature, on the other hand, can reduce crop duration, enhance crop respiration rates, affect the equilibrium between crops and pests, hasten nutrient mineralization in soils, decrease fertilizer-use efficiencies and increase evapotranspiration. Indirectly, there may be considerable impact on agricultural land use due to snowmelt, availability of irrigation, frequency and intensity of inter- and intraseasonal droughts and floods, soil organic matter transformations, soil erosion, decline in arable areas (due to submergence of coastal lands) and availability of energy. All these changes would have a tremendous impact on agricultural production, and hence food security of any region. The potential impacts of climate change on agriculture include:

- Climate change may reduce average crop yields and may lead to decreased yield stability. However, some plants may increase photosynthesis because of higher temperature and CO_2, but not all plants will benefit.
- The total amount of global rainfall will increase, although there will be regions that may receive less rainfall than before. The El Niño effect, which is manifest in the world's oceans and continents every decade, is suggested to get a big boost by the rise in temperature of the atmosphere.
- Demand for irrigation is likely to increase in all regions. It will lead to higher competition for existing water resources. Increase in temperature may also result in a higher amount of evapotranspiration, which may lead to increased frequency of droughts and demand for irrigation.
- Ranges and populations of agricultural pests currently limited by temperature may change. Higher temperature may increase diseases and heat stress. Some livestock diseases that are limited to tropical countries at present, such as Rift Valley fever and African swine fever, may spread, causing serious economic losses.

- Sea levels will rise by about 18 cm by 2040 and by 48 cm by 2100, attributed mainly to the thermal expansion of water and the melting of glaciers. There are countries such as Bangladesh and the Maldives where much of the population currently lives on the land that would be flooded by a sea level rise of about 50 cm. Sea level rise could also affect fisheries directly and indirectly through the availability of feed.
- Alteration of the energy balance and circulation system in the world's oceans will directly affect the productivity of marine ecosystems.
- Increase in temperature is likely to affect crop calendars in low-latitude regions, particularly where more than one crop is harvested in a year.

In agriculture, climate can reasonably be constructed as a resource, so that climate change will lead to changes in the agricultural productivity of a region. Future challenges include increasing food grain production with little possibility to increase the cultivated area, decline in farm size, and a marginal increase in irrigation and depletion of the resource base. Some steps need to be taken to improve management technologies for 'well-endowed' areas and emphasis needs to be placed on regimes which have climate and soil constraints.

References

Aggarwal, P.K. and Mall, R.K. (2002) Climate change and rice yields in diverse agro-environments in India II. Effect of uncertainties in scenarios and crop models on impact assessment. *Climate Change* 52, 331–343.

Aggarwal, P.K. and Pathak, H. (2009) Can conservation cultivation combat climate change? *Indian Farming* 59, 5–10.

Aggarwal, P.K. and Rani, S.D.N. (2009) Assessment of climate change impacts on wheat production in India. In: Aggarwal, P.K., Singh, A.K., Samra, J.S., Singh, G., Gogoi, A.K., Rao, G.G.S.N, et al. (eds) *Global Climate Change and Indian Agriculture*. ICAR, New Delhi, pp. 1–5.

Barth, M.C. and Titus, J.G. (eds) (1984) *Greenhouse Effect and Sea Level Rise: A Challenge for this Generation*. Van Nostrand Reinhold, New York.

Chandna, P., Hodson, D.P., Singh, U.P., Gosain, A.K., Sahoo, R.N. and Gupta, R.K. (2004) *Increasing the Productivity of Underutilized Lands by Targeting Resource Conserving Technologies – A GIS/Remote Sensing Approach: A Case Study of Ballia District, Uttar Pradesh in the Eastern Gangenic Plains*. CIMMYT, Mexico, DF.

Cruz, R.T., O'Toole, J.C., Dingkuhn, M., Yambao, E.B., Thangaraj, M. and De Datta, S.K. (1986) Shoot and root responses to water deficits in rainfed lowland rice. *Australian Journal of Plant Physiology* 13, 567–575.

Erenstein, O., Farooq, U., Malik, R.K. and Sharif, M. (2007) *Adoption and Impacts of Zero Tillage as a Resource Conserving Technology in the Irrigated Plains of South Asia*. Comprehensive Assessment of Water Management in Agriculture – Research Report 19. IWMI, Colombo, Sri Lanka.

Govaerts, B., Sayre, K.D. and Deckers, J. (2005) Stable high yields with zero tillage and permanent bed planting. *Field Crops Research* 94, 33–42.

Habash, D.Z., Kehel, Z. and Nachit, M. (2009) Genomic approaches for designing durum wheat ready for climate change with a focus on drought. *Journal of Experimental Botany* 60(10), 2805–2815.

Hoozemans, F.M.J., Marchand, M. and Pennekamp, H.A. (1993) *A Global Vulnerability Analysis, Vulnerability Assessments for Population, Coastal Wetlands and Rice Production on Global Scale*, 2nd edn. Delft Hydraulics and Rijkswaterstaat, Delft, the Netherlands.

Hundal, S.S. and Kaur, P. (1996) Climate change and its impact on crop productivity in Punjab, India. In: Abrol, Y.P., Gadgil, S. and Pant, G.B. (eds) *Climate Variability and Agriculture*. Narosa Publishing House, New Delhi, pp. 377–393.

IPCC (2007) *The Physical Science Basis*. Contribution of Working Group I to the Fourth Assessment Report of the Intergovernmental Panel on Climate Change. Solomon, S., Qin, D., Manning, M., Chen, Z., Marquis, M., Averyt, K.B., et al. (eds). Cambridge University Press, Cambridge, UK, and New York.

Kalish. A., Zemek, O., Schellhardt, S. and Pathak, H. (2011) Adaptation in agriculture. In: Porsche, I., Kalish, A. and Fuglein, R. (eds) *Adaptation to Climate Change with a Focus on Rural Areas and India*. GIZ India, New Delhi, pp. 40–83.

Ladha, J.K., Hill, J.E., Duxbury, J.D., Gupta, R.K. and Buresh, R.J. (2003) Improving the productivity and sustainability of rice-wheat systems: issues and impact. In: Ladha, J.K., Hill, J.E., Duxbury, J.D., Gupta, R.K and, Buresh,

R.J. (eds) *American Society of Agronomy Spec. Publ.* 65. ASA, CSSA, SSSA, Madison, Wisconsin, pp. 211.

Lal, M., Singh, K.K., Rathore, L.S., Srinivasan, G. and Saseendran, S.A. (1998) Vulnerability of rice and wheat yields in NW India to future changes in climate. *Agricultural and Forest Meteorology* 89, 101–114.

Mall, R.K., Singh, R., Gupta, A., Srinivasan, G. and Rathore, L.S. (2006) Impact of climate change on Indian agriculture: a review. *Climatic Change* 78, 445–478.

Matthews, R.B., Kropff, M.J. and Bachelet, D. (1995) Introduction. In: Matthews, R.B., Kropff, M.J., Bachelet, D. and van Laar, H.H. (eds) *Modeling the Impact of Climate Change on Rice Production in Asia.* CAB International and International Rice Research Institute, the Philippines, pp. 3–9.

Matthews, R.B., Kropff, M.J., Horie, T. and Bachelet, D. (1997) Simulating the impact of climate change on rice production in Asia and evaluating options for adaptation. *Agricultural Systems* 54, 399–425.

Mirza, M.M.Q., Warrick, R.A. and Ericksen, N.J. (2003) The implications of climate change on floods of the Ganges, Brahmaputra and Meghna rivers in Bangladesh. *Climatic Change* 57, 287–318.

Narang, R.S. and Virmani, S.M. (2001) Rice–wheat cropping systems of the Indo-Gangetic Plains of India. *Rice–Wheat Consortium Paper Series* 11. Rice–Wheat Consortium for the Indo-Gangetic Plain, New Delhi, and International Crops Research Institute for the Semi-Arid Tropics, Patancheru, Andhra Pradesh, India, pp. 36.

Nicholls, R.J. and Mimura, N. (1998) Regional issues raised by sea-level rise and their policy implications. *Climate Research* 11, 5–18.

Ortiz, R., Sayre, K.D., Govaerts, B., Gupta, R.K., Subbarao, G.V., Ban, T., *et al.* (2008) Climate change: can wheat beat the heat? *Agriculture, Ecosystems and Environment* 126, 46–58.

Ortiz-Monasterio, J.I., Dhillon, S.S. and Fischer, R.A. (1994) Date of sowing effects on grain yield and yield components of irrigated spring wheat cultivars and relationships with radiation and temperature in Ludhiana, India. *Field Crops Research* 37, 169–184.

Pathak, H. and Wassmann, R. (2009) Quantitative evaluation of climatic variability and risks for wheat yield in north-west India. *Climatic Change* 93, 157–175.

Pathak, H., Ladha, J.K., Aggarwal, P.K., Peng, S., Das, S., Yadvinder Singh, *et al.* (2003) Climatic potential and on-farm yield trends of rice and wheat in the Indo-Gangetic plains. *Field Crops Research* 80, 223–234.

Pathak, H., Saharawat, Y.S., Gathala, M. and Ladha, J.K. (2011) Impact of resource-conserving technologies on productivity and greenhouse gas emission in rice–wheat system. *Greenhouse Gases: Science and Technology* 1, 261–277.

Peng, S.B., Huang, J.L., Sheehy, J.E., Laza, R.C., Visperas, R.M., Zhong, X.H., *et al.* (2004) Rice yields decline with higher night temperature from global warming. *Proceedings of the National Academy of Sciences* 101, 9971–9975.

Pingali, P.L. (1999) Sustaining rice–wheat production systems: socio- economic and policy issues. In: Pingali, P.L. (ed.) *Rice–Wheat Consortium Paper Series 5.* Rice–Wheat Consortium for the Indo-Gangetic Plains, New Delhi, pp. 99.

Rao, G.D. and Sinha, S.K. (1994) Impact of climatic change on simulated wheat production in India. In: Rosenzweig, C. and Iglesias, I. (eds) *Implications of Climate Change for International Agriculture: Crop Modeling Study.* EPA, Washington, DC, pp. 1–10.

Roetter, R. van de Geijn (1999) Climate change effects on plant growth, crop yield and livestock. *Climate Change* 43, 651–681.

Samra, J.S. and Singh, G. (2005) *Heatwave of March 2004: Impact on Agriculture.* Indian Council of Agricultural Research, New Delhi, pp. 32.

Saseendran, S.A., Singh, K.K., Rathore, L.S., Singh, S.V. and Sinha, S.K. (1999) Effects of climate change on rice production in the tropical humid climate of Kerala, India. *Climate Change* 12, 1–20.

Sinha, S.K. and Swaminathan, M.S. (1991) Deforestation, climate change and sustainable nutrition security: a case study of India. *Climate Change* 19, 201–209.

Timsina, J. and Connor, D.J. (2001) Productivity and management of rice–wheat cropping systems: issues and challenges. *Field Crops Research* 69, 93–132.

Wassmann, R. and Dobermann, A. (2007) Climate change adaptation through rice production in regions with high poverty levels. *SAT eJournal* 4, 1–23.

Wassmann, R., Jagadish, S.V.K., Sumfleth, K., Pathak, H., Howell, G., Ismail, A., *et al.* (2009) Regional vulnerability of climate change impacts on Asian rice production and scope for adaptation. *Advances in Agronomy* 102, 91–133.

Wopereis, M.C., Kropff, M.J., Maligaya, A.R. and Tuong, T.P. (1996) Drought-stress responses of two lowland rice cultivars to soil and water status. *Field Crops Research* 46, 21–39.

Yoshida, S. (1981) *Fundamentals of Rice Crop Science.* International Rice Research Institute, Los Baños, Philippines.

4 Climate Change Challenges for Low-Input Cropping and Grazing Systems – Australia

Steven Crimp,[1] Mark Howden,[1] Chris Stokes,[2] Serena Schroeter[3] and Brian Keating[4]

[1]*CSIRO Climate Adaptation Flagship and Ecosystem Sciences, Canberra, ACT, Australia;* [2]*CSIRO Climate Adaptation Flagship and Ecosystem Sciences, Aitkenvale, Queensland, Australia;* [3]*CSIRO Climate Adaptation Flagship, Highett, Victoria, Australia;* [4]*CSIRO Sustainable Agriculture Flagship, Ecosciences Precinct, Brisbane, Queensland, Australia*

4.1 Introduction

Low-input farming systems have been defined as systems that seek to optimize the management and use of internal production inputs (i.e. on-farm resources) and to minimize the use of production inputs (i.e. off-farm resources), such as purchased fertilizers and pesticides (Parr et al., 1990). In environments with high climatic risk, this is often undertaken when practicable, to lower production costs, reduce overall farm risk and to increase both short- and long-term farm profitability (Parr et al., 1990). In practice, the term 'low input' relates to purchasing fewer off-farm inputs (usually fertilizers and pesticides), while increasing on-farm inputs (i.e. manures, cover crops and especially management). Thus, a more accurate term would be different input or low external input rather than low-input farming (Norman et al., 2000). In the context of this chapter, the term 'low external input farming' has been used to mean extensive rainfed farming activities that endeavour to use minimal external production inputs.

Using this definition, it is estimated that low-input farming comprises around 37% of the world's total land area, providing livelihoods to some 1.3 billion people (FAO, 2012).

Until 2000, agriculture was the mainstay of employment around the world, with growth in agricultural employment accounting for half of all employment growth between 1999 and 2009 in sub-Saharan Africa and approximately 33% of all employment growth in South Asia (FAO, 2012).

Approximately 60% of this land-use is found in developing countries. Low rainfall areas constitute from 75 to 100% of the land area in more than 20 countries in the Near East, Africa and Asia. Farmers in these regions produce more than 50% of the groundnuts, 80% of the pearl millet, 90% of the chickpeas and 95% of the pigeon peas. These dryland areas could continue to produce most of the world's food grains for expanding populations in the years ahead. However, yields are low to extremely low compared with those of higher input humid and sub-humid regions. Many of these areas are typified by a highly fragile natural resource base, with soils that are often coarse textured, low in organic carbon, prone to erosion and inherently low in fertility.

Arable crops directly provide more than 40% of the calories consumed by humans, particularly from cereals such as wheat, maize and rice (FAO, 2012). They also provide feed for livestock, which provides further nutrition (e.g. meat, milk and eggs) and fibre for human consumption and use. While vegetable crops make a smaller contribution to dietary energy supply, they also have an important role in human nutrition. The global demand for grains will increase in the future, as the world population is expected to increase from 7 billion in early 2012 to around 9 billion by 2050 (Lutz and Samir, 2010). The prospects of reducing the proportion of malnourished people in the world, today at ~15%, will further increase grain consumption. As economies continue to develop, ongoing shifts in dietary preferences towards increased consumption of protein, will likely be sustained. A recent study (Kearney, 2010) has estimated that meat consumption in developing countries has increased by 119% over the past four decades and by 15% in developed countries. In addition, increasing utilization of cereals for energy production will further increase the demand for feed grain for livestock production. These trends are already emerging from the global aggregated data available via the FAO.

The increase and changes in global food demand will have significant impacts on low external input farming in Australia. To satisfy these demands, there will be a need to increase the efficiency of current production systems.

The fast pace of climate change projected for the coming decades imposes additional pressure on food supply from low external input farming in Australia, as both crop yield and quality will be affected differently throughout the world's cropping regions (IPCC, 2007). The increase in CO_2 concentration, seasonal temperature, extreme climatic events and shifts in rainfall patterns all point to an unprecedented need for rapid changes to crop management and genotypes used to ensure that local and global demands for food are safely met.

This chapter discusses the mechanisms by which climate change will influence broadacre rainfed farming systems (cropping and grazing) in Australia and critically evaluates possible on-farm adaptation options. These results will be aggregated to national scales to examine the benefits derived from adaptation to climate change.

4.2 Overview of Broadacre Farming Systems in Australia

4.2.1 National importance

In Australia, broadacre rainfed farming systems account for 97 million ha (Mha), or nearly 13% of the Australian continent (BRS, 2001, 2006, 2010). This land use accounts for more than 60% of Australian agriculture in terms of production value (ABARE, 2008).

This type of farming can be characterized predominantly by mixed cropping/livestock systems and includes diverse enterprises where wool production, dual-purpose flocks and prime lamb and beef production are integrated with cereal, pulse and oilseed crops and where forage crops are incorporated into the cropping sequence (Kirkegaard et al., 2011). The intensity of cropping on farms ranges from very low to very high (where it generates more than 80% of farm income), and usually inversely reflects the contribution by livestock (ABARE, 2008).

Broadacre farming contributes approximately AUS$33.1 billion (mean across the 5 years to 2011/12) to Australia's agricultural export earnings (ABARE, 2013), and provides a mean return of around AUS$83,000 per farm (mean across the 10 years to 2011/12) (ABARE, 2013). Broadacre farming is strategically important to the Australian economy because it ensures a reliable domestic and international supply of grain and meat products.

There is considerable diversity in the enterprise mix at the regional and local level (see Plate 4), driven partly by differences in the biophysical resource base and climate (Kirkegaard et al., 2011). In the northern broadacre region defined in Plate 4, beef production makes up approximately 65% of the total volume of production, with sheep and cropping approximately 15 and 20%, respectively (ABS, 2010). In recent decades, there has been a move towards more intensive cropping in the southern and western regions as a result of the collapse in the wool price, and changes in seasonal conditions, with grains making up between 45 and 35% of the total volume of production, respectively (Kirkegaard et al., 2011).

Broadacre farming in Australia is characterized by relatively high productivity when compared with the world average. Wheat yields of approximately 2.2 t grain ha^{-1} are achieved, in spite of harsh climatic and relatively poor soil conditions. In spite of these poor growing conditions, these mean yields are due to farmers' high technological literacy and a culture of embracing innovation that has ensured an increase of 2–3% per year for wheat production since the 1960s (Robertson, 2010), although these rates of increase have reduced over the past decade (ABARES have data on this). These specific characteristics of Australia's broadacre farming systems are strong drivers of regional resilience and the effectiveness of adaptation options to climatic variability and change.

4.2.2 Exposure to climate variability

The economic viability of crops is influenced not only by the average climate of a location but also by climate variability (e.g. Porter and Semenov, 2005). Temperature and rainfall are the main weather variables to drive inter-annual yield variability. Overall, the higher the inter-annual yield variation is, the higher the risks of production. This is illustrated by 30-year mean wheat yields measured across a range of sites in Australia and plotted against the mean climatic variation experienced by these locations. Figure 4.1 shows quite clearly how locations that experience higher levels of climatic variation are also associated with lower mean yields for the same average rainfall. Characteristically, under conditions of higher climatic variability and yield volatility, producers tend to minimize inputs and agronomy. The converse is true where climatic variability, and hence yield volatility, is low (Fig. 4.1).

Changes in rainfall patterns are expected to be a major component of future yield

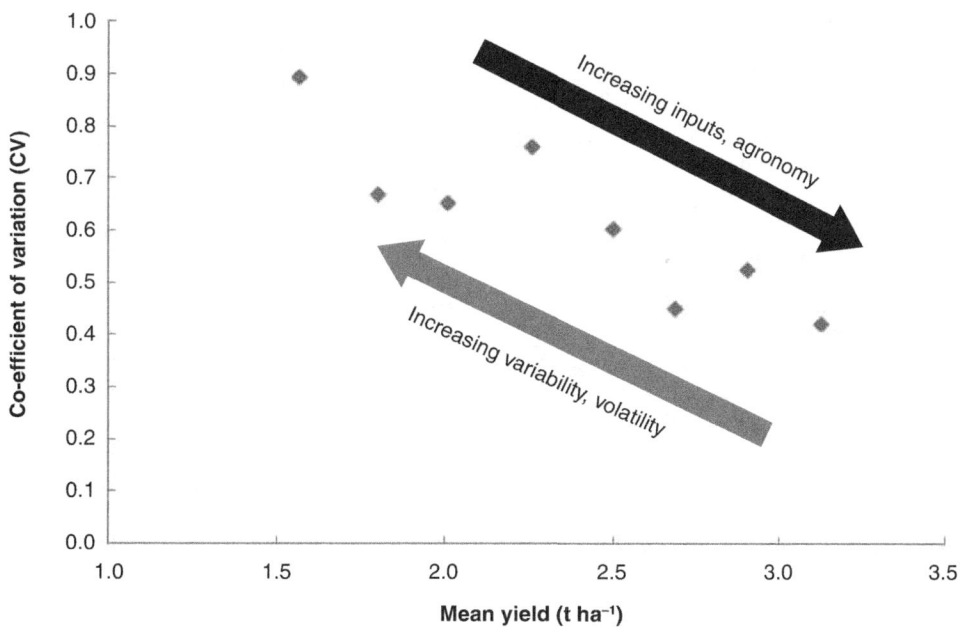

Fig. 4.1. Thirty-year mean yield values for dryland winter wheat plotted against 30-year mean climatic variability. Each point is a location (across the globe) with the same mean annual rainfall. Variation in rainfall amount between years is represented by the coefficient of variation (cv).

variability (Sinclair, 2011) because cropping and grazing respond to both the quantity and the timing of water supply in relation to demand. Climate change is expected to increase this variability further due to more frequent extreme events, thus making farm management a much more challenging prospect in the future.

4.3 Climate Change Impacts on Cropping and Grazing

Changes in climate can impact on the yield and quality of arable crops as well as livestock weight gain and fertility through different mechanisms (see Olesen, Chapter 2, this volume). Increased temperature and its variability, increased atmospheric CO_2 and change in precipitation amounts and distribution are key elements of climate change that will likely affect the performance of broadacre farming systems. Much field, growth chamber, laboratory and modelling research has been performed in Australia and elsewhere to determine the responses of broadacre farming systems to climate change. These processes are generally well understood and can be described according to the effects of temperature, precipitation, extreme events and biotic stress (e.g. Jamieson et al., 2000; Rosenzweig and Wilbanks, 2010; Kimball, 2011). Climate change is likely to add to and exacerbate existing pastoral management challenges such as undesirable grass species, shrub invasions, soil erosion and problems with animal nutrition and health. While technological advances have been developed to address some of these issues, adoption of these technologies is in itself highly variable (Johnston et al., 1996). With climate change, it will become even more important for widespread adoption of these practices in order to maintain the resilience of these farming systems (e.g. Stokes and Howden, 2010). The main principles of cropping and grazing responses to climate change are discussed in the following sections.

4.3.1 Temperature

The impact of temperature on crop yield depends on the changes in two distributional parameters: the mean and the diurnal temperature variance. These can affect the rate of growth and development processes differently, depending on the cardinal temperatures of a given physiological process.

While day-length responses are not affected directly by climate change, the fulfilling of vernalization requirements will be influenced by the likelihood of fewer cold days in the future. Yield decline due to unfulfilment of vernalization requirements was identified as an important climate change impact for cereals grown in high latitudes (Parry et al., 2005), in Mediterranean climates (Guerena et al., 2001) and in cropping areas in China (Xiong et al., 2007). Unfulfilled vernalization requirements impede crop development and can cause low flower bud initiation, ultimately reducing yields (Harrison et al., 2000). The actual impact of climate change on yield through vernalization depends on local background temperatures and the vernalization requirement of specific varieties. For example, yield decline is more likely for highly sensitive cultivars exposed to elevated autumn–winter temperatures.

Small changes in temperature can reduce the production of crops that are already grown close to their temperature optima. The overall effect of temperature on crop growth and yield is therefore the result of a complex and often competing interaction between the effects on photosynthesis, respiration and light interception.

Rising temperatures may benefit pasture production throughout southern Australia in cooler months and more generally across the year in cooler locations, but this is likely to be offset by accelerated phenology and higher spring evaporation rates, thus tending to shorten the length of the overall growing season (Crimp et al., 2010). In regions already predisposed to high temperatures, increased heat stress and increased evaporative demand are likely to result in reductions in pasture productivity (Crimp et al., 2002; McKeon et al., 2009).

The direct effects of increased heat stress on grazing animal physiology will be reduced productivity, decreased reproductive rates and increased concerns about animal welfare in locations where grazing populations are concentrated, such as feedlots. Heat stress on grazing animal physiology is a function of both environmental conditions and forage intake. Livestock need to shed this heat to maintain stable internal temperatures, and increasingly productive livestock means more metabolic heat needs to be transferred to the environment. In humid environments where there is substantial heat gain from high temperatures or solar radiation, or both, this transfer becomes increasingly difficult, either as sensible heat or latent heat through evaporation. This tends to reduce animal intake (e.g. Gaughan and Lees, 2010), thus reducing production. Heat stress in livestock is thus already a key issue in tropical environments (Nwosu and Ogbu, 2011; Sahoo et al., 2013) and there is an expectation that climate change will substantially increase the frequency of heat stress days, reducing livestock productivity, decreasing reproductive rates, increasing susceptibility to pests and diseases and increasing concerns about animal welfare, in particular in locations where grazing populations are concentrated, such as feedlots (Howden and Turnpenny, 1997; Howden et al., 1999; Mader and Davis 2004; Amundson et al., 2006; Gaughan and Lees, 2010). There is evidence of increasing temperature–humidity index (THI) already in some regions (Nidumolu et al., 2013) and that this may, in some cases, already be impacting on animal performance (Petty et al., 1998).

These changes are already altering the ability of livestock systems in the broadacre regions of Australia to carry livestock successfully at a given density, thus posing further challenges for graziers in managing their herd sizes effectively in the future.

4.3.2 Precipitation

The amount and timing of rainfall are important determinants of the inter-annual variability of dryland crops (Lobell and Burke, 2008). Climate change effects on rainfall patterns can have both negative and positive impacts on agricultural production (Sinclair, 2011). Increases in rainfall often increase crop yields in (semi-)arid environments by mitigating the risk of drought stress, while excess rainfall can reduce yield through waterlogging, diseases and nutrient leaching in already wet regions.

Although rainfall patterns have a major impact on inter-annual yield seasonality (Sinclair, 2011), climate model projections of seasonal rainfall change are more uncertain than for other climatic variables, and estimates differ largely among models (Smith et al., 2005). Improvements in climate models have been slow to occur, with seasonal precipitation and the wet-day frequency still depicting large errors, often offsetting observed means and variability beyond 100% (Ramirez-Villegas et al., 2013). Simulated inter-annual climate variability in climate models is of particular concern, given that for a recent study by Ramirez-Villegas et al. (2013) no single climate model matched observations in more than 30% of the areas examined for monthly precipitation and wet-day frequency. Given this level of uncertainty, understanding the likely seasonal variation in future rainfall conditions is unlikely in the near future, and hence remains a significant adaptation challenge moving forward.

Studies of livestock carrying capacity across the rangelands already highlight a non-linear response to rainfall (Wilson, 1982); thus, small spatial changes in rainfall are likely to be associated with large relative changes in livestock carrying capacity. Pasture modelling studies, calculating safe livestock carrying capacity from resource attributes and climate data (Scanlan et al., 1994; Johnston et al., 1996; Day et al., 2000; Alcock et al., 2010), also indicate considerable sensitivity to small variations in rainfall and temperature.

Historical records from past events provide estimates of changes in species composition in response to alterations in rainfall variability and grazing pressures (Park et al., 2003; Hill et al., 2004; McKeon et al., 2004). Bisset (1962) and Orr (1986) found that invasion by undesirable species (e.g. *Aristida latifolia* and *Heteropogon contortus*) occurred during a series of above-average rainfall years, while occurrence of these species declined during drought years.

Future changes in both across- and within-season rainfall variability are likely to lead to changes in species composition. Species differences will be further enhanced by changing species interactions and will result in changes in palatability. Composition changes across much of southern Australia are likely to include both more xeric species as well as increases in legume content; thus, palatability losses may be small (Newton et al., 1994; Alcock et al., 2010). In the northern and western parts of central Australia, less palatable species may dominate pasture communities (Stokes et al., 2010).

Attitudes to the 'desirability' of existing species compositions, and assessments of their suitability to the emerging climate, may need changing, and it may be more productive to recognize, facilitate and direct climate-induced changes in species distributions rather than trying to maintain the status quo. This may include shifts in attitudes, particularly regarding the definition and roles of 'invasive' (needing control) and 'useful' (in terms of production, environmental, biodiversity and aesthetic values) species. Woody weeds, particularly legumes in tropical rangelands, are likely to require more attention with climate change (Webb et al., 2012), as the conditions under which they may invade appear to be enhanced (e.g. Kriticos et al., 2003). Where pasture productivity increases with climate change, there may also be opportunities for more frequent use of fire to control woody weeds (e.g. Howden et al., 2001).

4.3.3 Carbon dioxide

Atmospheric CO_2 diffuses into plants' leaves through stomata and is fixed via photosynthesis to build plant biomass (Qaderi and Reid, 2009). At the same time as CO_2 diffuses into the leaves, water is lost through transpiration. Leaf level water-use efficiency (WUE) is a result of the different rates of these two diffusion processes. Therefore, projected increases in CO_2 are expected to affect both growth and transpiration rates in broadacre crops.

Common to both C_3 and C_4 crops, there is a consistent decrease in water use and an increase in water-use efficiency with increasing CO_2. Several crops have shown an average decrease in stomatal conductance between 20 and 40% at 567 ppm CO_2 from a 365 ppm baseline (Ainsworth and Rogers, 2007). The implication is that at higher CO_2 levels, less water is transpired per biomass produced. This has been proposed as the main reason for observations of positive but variable yield responses from 0 to 12% in C_4 crops at high CO_2 (Kimball, 2011), particularly when growing under water-limited conditions (Ainsworth and Long, 2005).

Although the positive direction of yield response to increased CO_2 is well established for both C_3 and C_4 crops, there is still some uncertainty and scientific discussion is ongoing regarding the magnitude of yield increase that will materialize under farm conditions (Long et al., 2006; Tubiello et al., 2007).

From a grazing perspective, forage production simulated under enhanced CO_2 conditions results in increased aboveground biomass production but declines in forage protein content (Wand et al., 1999), increases in forage non-structural carbohydrates (largely in C_3 species only) (Wand et al., 1999; Lilley et al., 2001) and decreases in digestibility, typically in tropical grasses (Wilson, 1982). Warmer conditions tend to reduce digestibility significantly, while also reducing leaf nitrogen content slightly (Lilley et al., 2001). Decreases in forage nitrogen and increases in digestibility will both tend to decrease intake and animal productivity in the extensive livestock systems typical of inland Australia.

4.3.4 Extreme events

A growing body of evidence suggests that extreme climatic events are likely to increase in both intensity and frequency over coming decades (IPCC, 2007). However, the frequency and intensity of temperature and rainfall events should be interpreted cautiously because climate models still cannot resolve all the key processes that are responsible for the occurrence of these events (Sinclair, 2011; Ramirez-Villegas et al., 2013).

There is large variation in sensitivity to heat stress among crop species and varieties. It has been suggested that the risk of heat stress is greater at high latitudes (~40–60°) and in continental cropping regions (Teixeira et al., 2011).

Paradoxically, climate change might increase the risk of frosts through different mechanisms (Ball et al., 2011). First, while the increase in temperature 'mean' value reduces the risk of frost, an increase in temperature 'variance' has been shown to increase the risk of frost (Rigby and Porporato, 2008). The prospects of faster crop development and earlier sowing in spring could also expose plants to freezing temperatures in a period of greater temperature variability.

4.3.5 Biotic stress

The yield damage caused by biotic factors such as insects, diseases (fungal, bacterial or viral) and weeds on agricultural production is one of the largest sources of crop losses (Oerke, 2006). Globally, it is estimated that more than one-third of the attainable yield of major food crops is lost due to pest damage (Oerke et al., 1999).

Climate change is likely to influence the frequency and intensity of pest damage on arable crops because host and pest development are strongly driven by environmental factors (Coakley et al., 1999; Sutherst et al., 2000; Juroszek and von Tiedemann, 2011). This relationship is evidenced by documented changes in pest ecology (Harrington et al., 2001) and the influence of major climatic patterns on the severity of pest damage in different crops and locations (Chakraborty, 2005).

Historically, the grazing industry has demonstrated some degree of vulnerability to pests, diseases and weed infestation (Sutherst, 1990; Mcleod, 1995; Sutherst et al., 1996). Roundworms, lice and blowflies cost the Australian sheep industry in excess of AUS$552 million annually in control measures and production losses (i.e. approximately 3% of the total production value of Australia's broadacre grazing) (Mcleod, 1995). Bock and de Vos (2001) estimated that the annual on-farm cost of ticks (production losses plus control costs) to the Australian cattle industry was approximately AUS$146 million. Any climate-induced changes in pests and diseases will therefore have significant management and economic implications.

4.4 Climate Adaptation Strategies for Broadacre Agriculture

For the most part, the scientific literature on climate adaptation focuses on exploring the effectiveness of adaptation options specific to location and crop (Easterling et al., 2007). This research often involves the examination of single tactical/incremental adaptations rather than the joint implementation of options (systems/systemic adaptation) or transformational adaptations (Howden et al., 2010), due to the difficulty of evaluating these options in a standard simulation framework (Howden et al., 2013). Many of the tactical-level adaptation options available for broadacre farming systems are largely extensions or intensifications of existing climate risk management or production enhancement activities in response to a potential change in the climate risk profile (Howden et al., 2007).

For grain cropping systems, there are many potential ways to alter management to deal with projected climatic and atmospheric changes. If widely adopted, these adaptation options are shown to offset the negative climate change related impacts associated with modest changes in climate (Fig. 4.2). For example, in a modelling study for Modena in Italy, Tubiello *et al.* (2000) found that simple and feasible adaptations altered negative impacts on sorghum (−48 to −58%) to neutral to marginally positive ones (0 to +12%). In that case, the adaptations trialled were to alter varieties and planting times to avoid drought and heat stress during the hotter and drier summer months predicted under climate change.

However, these options provide limited benefit for more significant changes in climate, and therefore systemic (systems) adaptation options are required that include combinations of a number of the tactical options (e.g. Ghahramani and Moore, 2013), as well as activities that serve to diversify incomes across enterprises or regions (Fig. 4.2). If large climate changes occur in the future, then many of the farming systems already operating close to climatic boundaries may no longer be viable, and transformational change/adaptation will be required. These options include changing livelihoods, changing land use or shifting geographically.

At each successive tier of adaptation, the complexity, cost and risk associated with the implementation of available options is likely to increase. Anticipatory transformational adaptation is considered most complex and

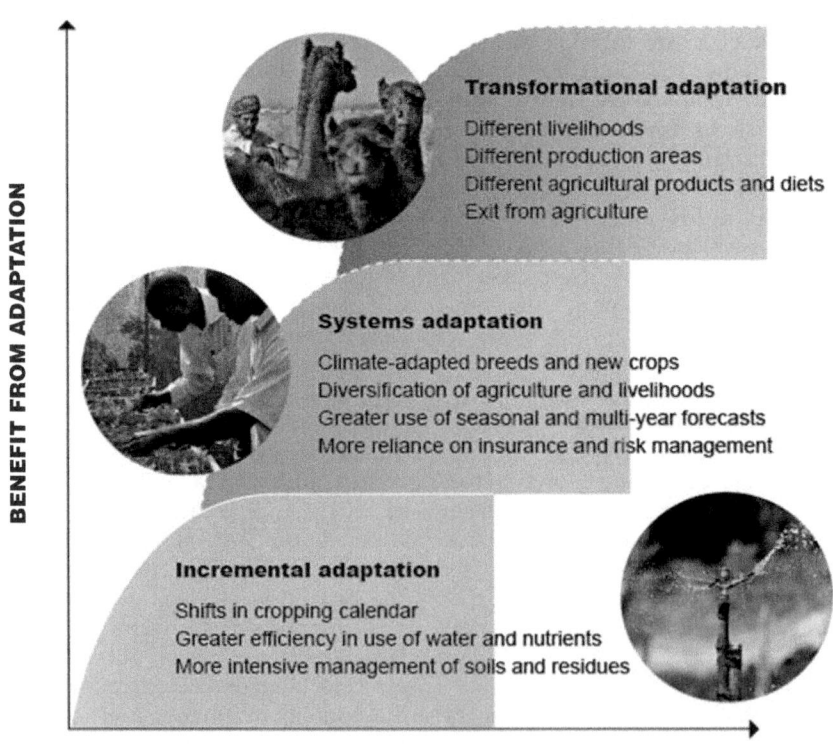

Fig. 4.2. Schematic representation of the benefits of tactical, systemic and transformational adaptation. (Sourced from http://ccafs.cgiar.org/bigfacts/crop-and-farming-adaptation/, accessed 18 November 2013; adapted from Rickards and Howden, 2012.)

difficult to implement because of the uncertainties about climate change risks and adaptation benefits and the high costs of transformational actions and institutional and behavioural actions that tend to maintain existing resource systems and policies (Park et al., 2012; Rickards and Howden, 2012). Implementing transformational adaptation requires effort to initiate and then to sustain the effort over time in the absence of immediate benefit (Kates et al., 2012; Park et al., 2012).

At the incremental or tactical level, the benefit of implementing these sorts of options can be demonstrated and summarized across many adaptation studies globally. There is a tendency for most of the benefits of adapting the existing cropping systems to be gained under moderate warming (<2°C), then to level off with increasing temperature changes. Studies for Australia demonstrate this point clearly (Fig. 4.3; Howden and Crimp, 2005; Howden et al., 2007; Crimp et al., 2012). The adaptations were varietal change and alteration of planting windows – key adaptations previously explored by Howden et al. (1999) with farm-level gross margin analyses. Just

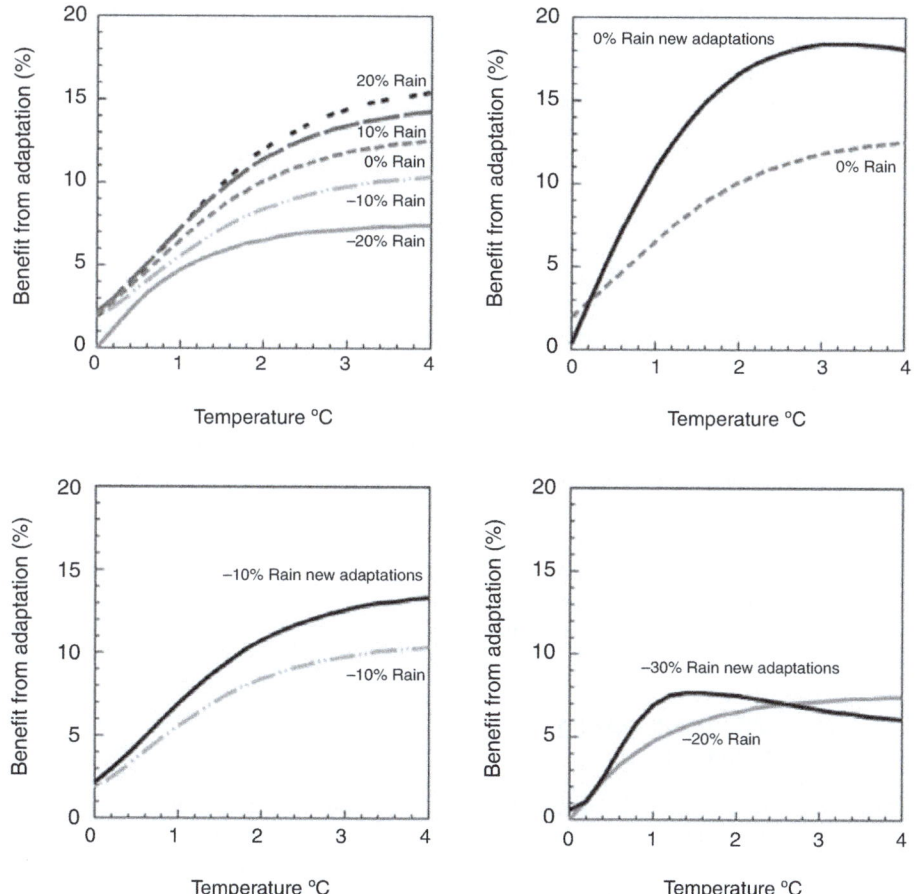

Fig. 4.3. Mean benefit of adapting wheat cropping systems to the impact of temperature and rainfall changes calculated as the difference between per cent yield changes with and without adaptation. The top left graph depicts results from the earlier study by Howden and Crimp (2005). The subsequent graphs show a comparison between the Howden and Crimp (2005) study and the more recent Crimp et al. (2012) study that examined a broader range of adaptation options.

these two adaptations could result in benefits of between 5 and 15%, and save the national wheat industry between AUS$100 million and AUS$500 million each year (in current dollar terms) by maintaining productivity.

These adaptations changed the mean result from being negative (on the balance of probabilities) to positive. Clearly, investment in adaptation is extremely worthwhile for the wheat industry. However, in that study, there remained a large negative 'tail' of results arising from very dry and hot climate change scenarios; hence, adaptation cannot remove all the risk from climate change. Additionally, the yield benefits tend to be greater under scenarios of increased rainfall than those with decreased rainfall, reflecting the fact that there are many ways of using more abundant resources more effectively, whereas there are fewer and less effective options for ameliorating risks when conditions become more limiting. More recent studies (Crimp et al., 2012) examining a broader suite of incremental adaptation options (Fig. 4.3) were compared with the original studies (Howden and Crimp, 2005). The results showed slight improvements in overall benefit from the adaptations considered, but affirm the continued presence of a limit to adaptation benefit once 2°C is exceeded.

Overall, the potential global benefits of management adaptations are substantial, and are similar in temperate and tropical systems (17.9% versus 18.6%) (Howden et al., 2007). The following sections address these management-level adaptations in more detail.

4.4.1 Crop, pasture and grazing management

Temperature increases reduce the duration of the phenological stages of crops, restricting the time they have to accumulate radiation and nutrients. This will generally reduce grain yield, thereby tending to counter the yield increase deriving from the CO_2 fertilization effect. In the absence of adaptive measures, a 1.5–2°C increase in mean temperature would cancel out the grain yield increase in wheat deriving from a CO_2 doubling in the Australian context (Gifford, 1989; Wang et al., 1992; Howden, 2002).

Where there is adequate moisture (wet regions or where climate change increases rainfall), there is likely to be an advantage in breeding and adopting slower-maturing cultivars (greater thermal time requirements) that could capitalize on the earlier date of flowering and a potentially longer, photosynthetically active period before seasonal drought forces maturity.

Where there is likely to be both increases in temperature and reductions in rainfall (e.g. in the strongly Mediterranean climate cropping regions), it may be advantageous to keep varieties with similar or earlier-flowering characteristics than are used currently, as this will allow grain fill to occur in the cooler, wetter parts of the year (Howden et al., 1999; van Ittersum et al., 2003).

Given the importance of water as an increasingly limited resource, improving water-use efficiency is of considerable interest to plant breeders. Attempts are now under way to alter photosynthetic pathways genetically in cereals by introducing aspects of the C_4 pathway into C_3 crops. While the biggest effort currently concentrates on rice (Mitchell and Sheehy, 2006), other crops (e.g. wheat) are also being targeted.

In areas where mixed cropping and grazing land use occurs, efforts to improve pasture productivity, particularly where significant changes in temperature and rainfall are projected, are seen as viable options to improve animal carrying capacity. Past efforts to increase pasture production in more humid rangelands have often relied on removing trees and shrubs to increase the availability of water, nutrients and light for grass growth (Burrows et al., 1988). However, this has been controversial because of the impacts on biodiversity, greenhouse gas emissions and catchment hydrology.

Current management, and particularly rehabilitation, of pastures requires careful grazing management including conservative stocking rates, strategic spelling and responsive adjustments to stocking rates based on seasonal climate forecasts (McKeon and Howden, 1992; Cobon and Clewett, 1999; Johnston et al., 2000) (see Rickards et al., Chapter 15, this volume). These practices will likely become more important with climate change and will be necessary to ensure desirable pasture species establish and are maintained as species ranges shift under climate change. With shifts to rainfall regimes that increase the risk of soil erosion, it will become increasingly important to ensure that ground cover is maintained in mixed cropping and grazing regions. It will also be necessary to redefine safe carrying capacities, pasture utilization levels and grazing management practices, and to review and adjust these continually, in accord with cropping demands and the changing climate (McKeon et al., 2009).

4.4.2 Crop species changes and animal genetics

Higher temperatures may enable the use of summer-growing grain and pulse species such as sorghum in temperate regions where these are not currently used in rotations. The negative impact of a reduction in rainfall is likely to be greater for rotation systems than for single-crop systems, as there is not a fallow in which to store soil moisture. This will impact particularly on crops that show sensitivity to dry conditions, such as canola and certain pulses. However, this will be moderated via the effective monitoring of soil moisture and nutrient levels, effective decision support systems, improved seasonal climate forecasting and continuing improvements in crop management (i.e. zero till, wide rows, variable planting density, canopy management, precision agriculture, etc.) that have relatively recently expanded planting options.

If there is a reduction in rainfall and increased rainfall variability, it will make dryland cropping less attractive, and there is likely to be consideration of a change to a greater proportion of stock in the farm business. Lowered production of annual pastures and crops may be partly offset by a greater planting of perennial species such as lucerne, which would be able to make use of summer rains and generally all available soil moisture.

The use of summer forage crops may be employed increasingly after summer rains – in regions with soils of low water-holding capacity (i.e. much of Western Australia). However, there may be a need for varietal development to increase the reliability of this option. Livestock are also subject to the impacts of drought, requiring active management, such as the early removal of stock from pastures for grain-based finishing. These options may become more important under climate change.

The grazing component of broadacre agriculture will be vulnerable to an increasing incidence of extreme temperatures and declining water availability under climate change. Howden and Turnpenny (1997) and Howden et al. (1999) have shown that the incidence of heat stress has increased significantly since 1957 across large areas of Australia. This suggests that the practice of selecting cattle lines with effective thermoregulatory controls such as feed-conversion efficiency and coat colour (Finch et al., 1984) would need to continue if current levels of productivity are to be maintained. This practice may need to become more common in regions where the frequency of heat stress days increases. Additional adaptation strategies such as modifying the timing of mating could also serve to match the nutritional requirements of cow and calf to periods with favourable seasonal conditions.

For some livestock operations such as stockyards and feedlots, the construction of shading and spraying facilities may represent an economically feasible adaptation measure (see Collier et al., Chapter 7, this volume). It may also be necessary to establish areas of suitable shade trees, and to increase the number of water points. In areas that become more prone to increased flooding, it will be important to provide livestock with access to areas of higher ground.

4.4.3 Planting time and seasonal herd management variation

Higher temperatures are anticipated to reduce frost risks, potentially allowing earlier planting (by a month or more) and consequently increased yields as grain fill is more likely to occur in the cooler months when the likelihood of water stress is lower (see Howden *et al.*, 1999, 2003; van Ittersum *et al.*, 2003). This may require concurrent changes in thermal time requirements of the varieties used, depending on any changes in planting dates.

The most important adaptation by livestock producers to changing climates will be to set stocking rates that are appropriate to the changing levels and patterns of forage production. In the mixed cropping/grazing regions, this may mean using a conservative stocking rate from year to year (Johnston *et al.*, 2000), as well as modifying the timing of mating to match nutritional requirements better with favourable seasonal conditions.

4.4.4 Other crop management options

There are a number of additional crop management practices that could be used in specific circumstances to lower the risks from changed climate conditions (e.g. Easterling *et al.*, 2007). These include:

- adopting zero-tillage practices (especially if there is increased rainfall intensity, as greater infiltration will be needed with fewer but heavier events);
- develop more techniques that minimize disturbance (i.e. seed pushing, all-weather traffic lanes which allow planting while raining);
- using reactive strategies to track climate variation on daily or seasonal time steps (e.g. McKeon *et al.*, 1993);
- extending fallows to capture and store more soil moisture effectively (suitable mostly with heavy soils);
- dry sowing, later plantings or staggering planting times, depending on the rainfall changes.

These options do produce regionally varying benefits and involve trade-offs between production rates, area produced and the ratio of cropping to grazing in any one enterprise (Crimp *et al.*, 2012).

4.4.5 Erosion management

Rainfall intensity is anticipated to increase with climate change, even under scenarios where average rainfall may decrease (e.g. CSIRO, 2007), continuing the current trends to higher-intensity rainfall events in Australia (CSIRO, 2007). This is likely to increase the risks of soil erosion, particularly on soils with high erodability. Key adaptations may be to:

- increase residue retention and maintain crop cover during periods of high risk so as to reduce raindrop damage on the soil surface and allow water to infiltrate;
- maintain erosion control infrastructure (e.g. contour banking, etc.);
- adopt controlled traffic systems up and down slopes.

These actions are already generally implemented in mixed cropping and grazing systems, but their importance is likely to increase over time.

4.4.6 Effective management of pests, diseases and weeds

Pest impacts on crops and livestock are widespread and costly to industry and include many trade access issues related to quality. Many of the pests associated with cropping (such as *Heliothis* moths, armyworms, sucking bugs and diamond-backed moths (on *Brassica*)) respond strongly to climate signals, and their impacts are very dependent on climatic variability (see Collier and Else, Chapter 6, this volume). Similarly, tick populations as well as buffalo fly, sheep blowfly and worms are also sensitive to climate variations.

Adaptations to climate change are likely to happen via an increased understanding of the impacts and potential responses of recent climate variability manifestations (e.g. those of the past 20 years), and may best be delivered via the two key emerging strategies: (i) integrated pest management and (ii) area-wide management (i.e. co-ordinated responses of growers and policy makers across an entire region) (Stokes and Howden, 2010).

Current management practices that respond to, or override, climatic variability include:

- genetic modification of crop plants to create insect- or disease-resistant and herbicide-tolerant varieties (via either conventional plant breeding or genetic modification);
- importation of exotic natural enemies of pests that were previously introduced without them; also, repeated mass (inundative) releases of parasitic wasps to control insect pests;
- isolation and propagation of local natural enemies/diseases (e.g. *Metarhizium* on locusts, termites);
- cultural practices such as crop rotations, mixed crops, use of physical barriers to reduce disease transmission;
- chemical pesticides and increasing biopesticides (e.g. Bt) and biofumigation of soils using *Brassica* sp. as alternate crops;
- monitoring and use of predictive models to improve the timing of interventions to coincide with high-risk periods;
- landscape-scale management involving groups of growers cooperating to reduce communal threats.

Several existing methods will be suitable for combating the spread of rangeland pests and diseases under climate change, including applications of pesticides and chemicals to respond to outbreaks, strategic use of fire to control weeds, biological weed control, vaccinations to enhance resistance to existing pests and diseases, and the selection of tick-resistant cattle (*Bos indicus*) in northern Australia (Stokes and Howden, 2010).

4.5 Broadacre Vulnerability to Climate Change

As confidence in climate change projections increases and trends in weather data become more apparent, the motivation to adapt to climate variability and change will increase. However, the adoption of new property management practices is likely to be highly variable across regions, as a result of the differing capacity of rural households to respond to change (see Ingram, Chapter 16, this volume). This ability to respond to change has often been defined as adaptive capacity, and a number of attempts have been made to quantify it so that the vulnerability of broadacre agriculture to change can be assessed and understood.

Using a top-down approach based on secondary data selected from the Australian Agricultural and Grazing Industries Survey (AAGIS) (ABARE, 2003), a national composite index of the generic adaptive capacity of rural households has been developed (Nelson *et al.*, 2010a,b; Crimp *et al.*, 2012). The sampling frame for the AAGIS survey is all farms in Australia's broadacre cropping, beef and sheep industries with an estimated value of agricultural operations of AUS$40,000 or more (ABARE, 2006), providing a clear definition of 'rural household'. This approach is based on the livelihoods framework developed by Ellis (2000) and conceptualizes adaptive capacity as an emergent property of the diverse forms of human, social, natural, physical and financial capital from which livelihoods are derived. Under this framework, farm households respond to external pressures by transforming, substituting between and utilizing these types of capital. Both the amount of each capital and the balance between them are important in building or maintaining a livelihood, and households with larger asset portfolios (across all five capitals) are considered to have more livelihood options, and thus more capacity to adapt to stress, than those with fewer assets.

This approach provides insights for the development of policies to enhance adaptation at a range of scales, particularly equipping farmers to cope with a range of potential change (e.g. resource availability management, markets, drought, exceptional circumstances) that have been occurring over the past 10–15 years, and also how this approach could be used to think about adapting to future climate changes.

The vulnerability of Australian broadacre farm households to climate change was determined by overlaying a smoothed adaptive capacity measure (Nelson et al., 2010a,b) with an exposure metric (calculated from projections of rainfall and temperature change incorporated into a biophysical production model). A single map representing regions exhibiting consistently high vulnerability, consistently low vulnerability and modest variable vulnerability in response to the multiple climate futures is presented in Fig. 4.4.

Vulnerability is most persistent across South Australia, western New South Wales, western Queensland and parts of the Northern Territory. In the case of South Australia, the high vulnerability is driven strongly due to limitations in physical infrastructure, access to capital and a degree of social isolation. South-west Western Australia, on the other hand, demonstrates consistently low vulnerability, driven strongly by the fact that access to capital and physical infrastructure are not limiting the ability to respond to climate change. This analysis suggests that there are three categories of rural communities most vulnerable to climate variability and change. These are:

1. Agricultural communities currently reliant on extensive grazing industries, particularly sheep in southern and central Australia. These areas have experienced a decline in the wool industry and lack options to diversify into other industries.

2. Farming communities which are affected strongly by declining terms of trade in both livestock and crop production, and rural decline more generally, and social constraints to a natural long-term process of

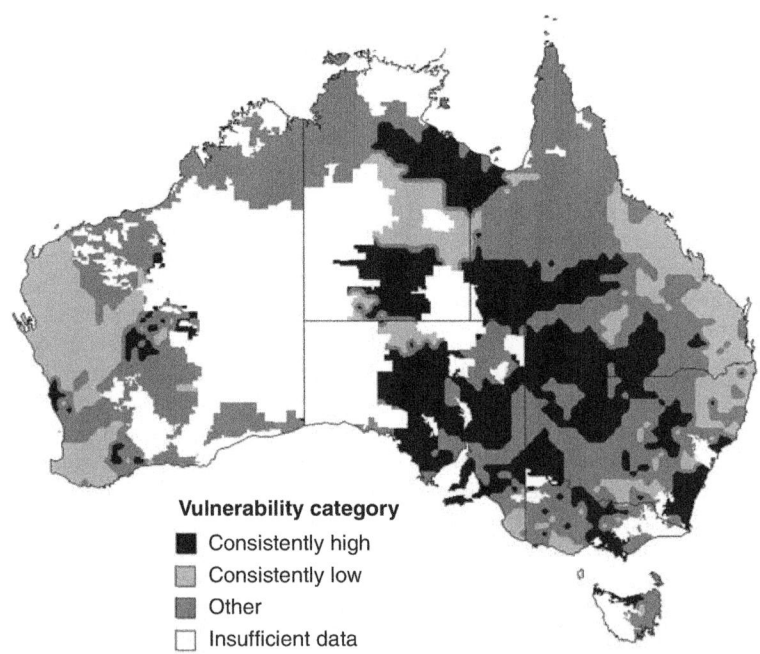

Fig. 4.4. A synthesis map depicting regions of consistently high and low vulnerability as a function of any combination of exposure and adaptive capacity likely out to 2030.

farm consolidation. Agricultural communities vulnerable to climate variability are scattered throughout the wheat–sheep zone, particularly in eastern Australia. These are farming communities affected strongly by declining terms of trade in both livestock and crop production and rural decline more generally, and social constraints to a natural long-term process of farm consolidation.

3. Farming communities on the eastern seaboard in which agriculture is becoming a secondary economic activity due to urbanization.

Rural decline is a recurring theme identified as undermining the ability of broadacre agriculture to adapt to climate change. Ribot et al. (2009) describe the problems in broadacre agriculture regions as products of a chronic lack of resources necessary to hedge against the extreme climatic events of known risk to local farmers. These views are partially reflected in the 'drylands syndrome' discussed in Howden et al. (2002), Reynolds et al. (2007) and Stafford Smith et al. (2009) regarding how communities in dryland areas are marginalized via physical, social and economic constraints.

It will be necessary to develop 'no regrets' options for climate change adaptation (Heltberg et al., 2009), which would support community-based adaptation and help with revitalizing rural communities, and would be good for agriculture and natural resource management generally.

4.6 Concluding Remarks and Key Messages

Climate change is likely to impact negatively the amount and quality of produce, the reliability of production and the natural resource base on which agriculture depends (Crimp et al., 2002; Cobon and Toombs, 2007; McKeon et al., 2009; Stokes et al., 2010). In some areas, new opportunities are likely to emerge, but realizing these potential benefits will require proactive changes to management practices (e.g. McKeon et al., 2004). These challenges require high levels of adaptive responses and will be driven by individuals' propensity and ability to adapt (i.e. adaptive capacity) (see Ingram, Chapter 16, this volume).

Understanding the adaptation process and supporting or enhancing it at various scales will be important for maintaining the effective functioning of social and economic systems in the face of climate-driven changes (Nelson and Lawrence, 2004; Marshall, 2010; Nelson et al., 2010a,b). Important human factors, both in terms of the capacity of managers and the context within which they operate (e.g. policy and institutional support), determine the extent to which adaptation occurs (e.g. Tubiello et al., 2007; Brown et al., 2010).

Climate change will, therefore, impact broadacre agriculture regions against a backdrop of constant economic and social change, and these impacts will occur at multiple scales. Most fundamentally, climate change will affect the relative productivity of alternative land uses, as changes in rainfall and temperature impact crops and livestock differently (Howden et al., 2001).

This review suggests that there is a wide range of adaptive measures to harness the opportunities and reduce the threats of climate change on broadacre farming systems. The effectiveness of adaptive measures is specific to each production system, depending on local weather, crop and animal choices, soil types, available technology and socio-economic conditions.

The viability and vulnerability of alternative agricultural land uses will also depend on the effect of climate change on world commodity prices, as climate changes affect the relative productivity of Australia's trading partners and competitors. All of these changes will take place against a changing institutional context, including changes in greenhouse gas mitigation policy such as carbon trading schemes.

In addition to managing considerable climate and market variability, the continued cost–price squeeze and the introduction of new policies, climate change will represent an additional challenge in terms of both direct impacts and their unpredictable interactions with existing challenges. It is possible that existing response measures will no longer be sufficient or appropriate. Successful adaptation to the combined influences of climate, world markets and policy will require an unusual degree of coordinated adaptation between farmers, communities, industries and governments across those scales.

4.6.1 Key messages

- Effective management of climate variability is already seen as a necessary component of success for low-input crop producers. Adaptation to climate change represents an additional and growing challenge that producers are increasingly responding to and planning for.
- There is a range of tactical adaptation options available to producers to offset the projected impacts of climate change. These include: changed crop management practices, new varieties, changed rotations, diversification of off-farm income, changes in the proportion of cropping and grazing and improved pest and weed control, as well as improved water-use efficiency.
- The main projected challenges for the grazing industry under climate change are declines in pasture productivity, reduced forage quality, livestock heat stress, greater problems with some pests and weeds, more frequent droughts, more intense rainfall events and greater risks of soil degradation.
- At present, many of these tactical adaptation options reflect, and are consistent with, best management practice for climate risk. However, these practices may only be effective for modest changes in both rainfall and temperature.
- The scale of climate change adaptation will need to respond to the extent of projected climate change, with tactical adaptation suitable for modest climate change, systemic adaptation suitable for more extensive climate change and transformative adaptation suitable for significant project change in temperature and rainfall.
- In the Australian context, the most arid and least productive rangelands may be the most severely impacted by climate change, while the more productive eastern and northern rangelands may provide some opportunities for slight increases in production.
- In order to translate climate change information into adaptation action, participatory approaches with groups across the entire agricultural value chain will likely help. Such an approach will carry the analysis iteratively from climate to biophysical impacts on crops and cropping systems, to enterprise level adaptation to regional-level applications, to policy options.

References

ABARE (2003) Australian farm survey report. Australian Bureau of Agriculture and Resource Economics, Canberra.

ABARE (2006) Australian farm survey report. Australian Bureau of Agriculture and Resource Economics, Canberra.

ABARE (2008) Australian farm survey report. Australian Bureau of Agriculture and Resource Economics, Canberra.

ABARE (2013) Australian Bureau of Agriculture and Resource Economics AGSURF database (http://www.abareconomics.com/interactive/agsurf/, accessed 21 March 2013).

ABARES (2011) *Guidelines for Land Use Mapping in Australia: Principles, Procedures and Definitions*, 4th edn. Australian Bureau of Agricultural and Resource Economics and Sciences, Canberra, 133 pp.

ABS (Australian Bureau of Statistics) (2010) *Value of Principal Agricultural Commodities Produced, Australia*, Preliminary, 2009/10, media release 7501.0. ABS, Canberra.

Ainsworth, E.A. and Long, S.P. (2005) What have we learned from 15 years of free-air CO_2 enrichment (FACE)? A meta-analytic review of the responses of photosynthesis, canopy. *New Phytologist* 165, 351–371.

Ainsworth, E.A. and Rogers, A. (2007) The response of photosynthesis and stomatal conductance to rising CO_2: mechanisms and environmental interactions. *Plant, Cell and Environment* 30, 258–270.

Alcock, D., Graham, P., Moore, A.D., Lilley, J.M. and Zurcher, E.J. (2010) GrassGro indicates that erosion risk drives adaptation of southern tablelands grazing farms to projected climate change. In: *Proceedings of the 2010 International Climate Change Adaptation Conference*, 29 June–1 July 2010, Surfers Paradise (www.nccarf.edu.au/conference2010/archives/269, accessed 18 November 2013).

Amundson, J.L., Mader, T.L., Rasby, R.J. and Hu, Q.S. (2006) *Environmental Effects on Pregnancy Rate in Beef Cattle*. Faculty Papers and Publications in Animal Science. Paper 609. University of Nebraska-Lincoln, Lincoln, Nebraska (http://digitalcommons.unl.edu/animalscifacpub/609, accessed 18 November 2013).

Ball, M.C., Harris-Pascal, D., Egerton, J.J.G. and Lenne, T. (2011) The paradoxical increase in freezing injury in a warming climate: frost as a driver of change in cold climate vegetation. In: Storey, K.B. and Tanino, K.K. (eds) *Temperature Adaptation in a Changing Climate: Nature at Risk*. CAB International, Wallingford, UK, pp. 179–185.

Bisset, W.J. (1962) The black spear grass (*Heteropogon contortus*) problem of the sheep country in central western Queensland. *Queensland Journal of Agricultural Science* 19, 189–207.

Bock, R.E. and de Vos, A.J. (2001) Immunity following use of Australian tick fever vaccine: a review of the evidence. *Australian Veterinary Journal* 79, 832–839.

Brown, P.R., Nelson, R., Jacobs, B., Kokic, P., Tracey, J., Ahmed, M., *et al.* (2010) Enabling natural resource managers to self-assess their adaptive capacity. *Agricultural Systems* 103, 562–568.

BRS (Bureau of Rural Sciences) (2001) *Land Use Mapping at Catchment Scales*, 2nd edition. Australian Government Department of Agriculture, Fisheries and Forestry, Canberra.

BRS (2006) Final report: 1992/93, 1993/94, 1996/97, 1998/99, 2000/01 and 2001/02 land use of Australia, Version 3. Australian Government Department of Agriculture, Fisheries and Forestry, Canberra, Australia.

BRS (2010) Guidelines for Land Use Mapping in Australia: Principles, Procedures and Definitions. A technical handbook supporting the Australian Collaborative Land Use and Management Program, 4th edn. Australian Government Department of Agriculture, Fisheries and Forestry, Canberra.

Burrows, W.H., Scanlan, J.C. and Anderson, E.R. (1988) Plant ecological relations in open forests, woodlands and shrublands. In: Burrows, W.H., Scanlan, J.C. and Rutherford, M.T. (eds) *Native Pastures in Queensland: The Resources and Their Management*. Queensland Government Printer, Brisbane, Australia, pp. 72–90.

Chakraborty, S. (2005) Potential impact of climate change on plant–pathogen interactions. *Australasian Plant Pathology* 34, 443–448.

Coakley, S.M., Scherm, H. and Chakraborty, S. (1999) Climate change and plant disease management. *Annual Review of Phytopathology* 37, 399–426.

Cobon, D.H. and Clewett, J.F. (1999) DroughtPlan CD: A compilation of software packages, workshops, case studies and reports to assist management of climate variability in pastoral areas of northern Australia. Queensland Department of Primary Industries, Brisbane, Australia.

Cobon, D.H. and Toombs, N.R. (2007) Practical adaptation to climate change in regional natural resource management. Qld case studies – Fitzroy Basin. Production and natural resource indicators in beef systems under climate change conditions. Final Report to the Australian Greenhouse Office, Queensland Department of Natural Resources and Mines, Brisbane, Australia.

Crimp, S.J., Flood, N.R., Carter, J.O., Conroy, J.P. and McKeon, G.M. (2002) Evaluation of the potential impacts of climate change on native pasture production: implications for livestock carrying capacity. Final Report to the Australian Greenhouse Office. Canberra.

Crimp, S., Kokic, P., McKeon, G., Smith, I., Syktus, J., Timbal, B., *et al.* (2010) A review of appropriate statistical downscaling approaches to apply as part of Phase 2 of the Consistent Climate Projections project. Report prepared by CSIRO on behalf of DAFF's 'Australia's Farming Futures Program' 15 June 2010, Australia.

Crimp, S.J., Laing, A., Alexander, B., Anwar, M., Bridle, K., Brown, P., *et al.* (2012) Developing climate change resilient cropping and mixed cropping/grazing businesses in Australia. A report prepared for DAFF as part of the Australia's Farming Future Program (http://www.daff.gov.au/climatechange/climate/communication/factsheets-case-studies-and-dvds, accessed 18 November 2013).

CSIRO (2007) *Climate Change in Australia*. Technical Report 2007. Commonwealth Scientific and Industrial Research Organisation (CSIRO), Melbourne, Australia.

Day, K.A., Ahrens, D.G., Peacock, A., Rickert, K.G. and McKeon, G.M. (2000) Climate tools for northern grassy landscapes. In: *Proceedings of the Northern Grassy Landscapes Conference*, 29–31 August 2000. CRC for the Sustainable Development of Tropical Savannas, Katherine, Australia, pp. 93–97.

Easterling, W.E., Aggarwal, P.K., Batima, P., Brander, K.M., Erda, L., Howden, S.M., et al. (2007) Food, fibre and forest products. In: Parry, M.L., Canziani, O.F., Palutikof, J.P., van der Linden, P.J. and Hanson, C.E. (eds) *Climate Change 2007: Impacts, Adaptation and Vulnerability.* Contribution of Working Group II to the Fourth Assessment Report of the Intergovernmental Panel on Climate Change. Cambridge University Press, Cambridge, UK, pp. 273–313.

Ellis, F. (2000) *Rural Livelihoods and Diversity in Developing Countries.* Oxford University Press, Oxford, UK.

FAO (2012) *Statisitical Yearbook – World Food and Agriculture.* FAO, Rome.

Finch, V.A., Bennett, I.L. and Holmes, C.R. (1984) Coat colour in cattle: effect on thermal balance, behaviour, and growth, and relationship with coat type. *Journal of Agricultural Science* 102, 141–147.

Gaughan, J.B. and Lees, J.C. (2010) Categorising heat load on dairy cows. In: *Proceedings of the 28th Biennial Conference of the Australian Society of Animal Production.* 28th Biennial Conference of the Australian Society of Animal Production, Armidale, New South Wales, Australia, p. 105.

Ghahramani, A. and Moore, A. (2013) Climate change and broadacre livestock production across southern Australia. 1. Adaptation of climate change on pasture and livestock productivity, and on sustainable levels of profitability. *Global Change Biology* 19, 1440–1455.

Gifford, R.M. (1989) Exploiting the fertilising effect of increasing atmospheric carbon dioxide. In: *Climate and Food Security.* Proceedings of the International Symposium on Climate Variability and Food Security in Developing Countries, 5–9 February 1987. International Rice Research Institute, Manila, pp. 477–487.

Guerena, A., Ruiz-Ramos, M., Diaz-Ambrona, C.H., Conde, J.R. and Minguez, M.I. (2001) Assessment of climate change and agriculture in Spain using climate models. *Agronomy Journal* 93, 237–249.

Harrington, R., Fleming, R.A. and Woiwod, I.P. (2001) Climate change impacts on insect management and conservation in temperate regions: can they be predicted? *Agricultural and Forest Entomology* 3, 233–240.

Harrison, P.A., Porter, J.R. and Downing, T.E. (2000) Scaling up the AFRCWHEAT2 model to assess phenological development for wheat in Europe. *Agricultural and Forest Meteorology* 101, 167–186.

Heltberg, R., Siegel, P.B. and Jorgensen, S.L. (2009) Addressing human vulnerability to climate change: toward a 'no-regrets' approach. *Global Environmental Change* 19, 89–99.

Hill, J.O., Simpson, R.J., Moore, A.D., Graham, P. and Chapman, D.F. (2004) Impact of phosphorus application and sheep grazing on the botanical composition of sown pasture and naturalised, native grass pasture. *Australian Journal of Agricultural Research* 55, 1213–1225.

Howden, M., Crimp, S. and Nelson, R. (2010) Australian agriculture in a climate of change. In: Jubb, I., Holper, P. and Cai, W. (eds) *Managing Climate Change: Papers from the GREENHOUSE 2009 Conference.* CSIRO Publishing, Melbourne, Australia, pp. 101–111.

Howden, M., Nelson, R.A. and Crimp, S. (2013) Food security under a changing climate: frontiers of science or adaptation frontiers? In: Palutikof, J., Boulter, S.L., Ash, A.J., Stafford Smith, M., Parry, M., Waschka M., et al. (eds) *Climate Adaptation Futures.* John Wiley and Sons, Chichester, UK, pp. 56–68.

Howden, S.M. (2002) Potential global change impacts on Australia's wheat cropping systems. In: Doering, O.C., Randolph, J.C., Southworth, J. and Pfeifer, R.A (eds) *Effects of Climate Change and Variability on Agricultural Production Systems.* Springer, the Netherlands, pp. 219–247.

Howden, S.M. and Crimp, S.J. (2005) Assessing dangerous climate change impacts on Australia's wheat industry. In: Zerger, A. and Argent, R.M. (eds) *Proceedings from MODSIM05: International Congress on Modelling and Simulation: Advances and Applications for Management and Decision Making,* Melbourne. Modelling and Simulation Society of Australia and New Zealand, Canberra, pp. 505–511.

Howden, S.M. and Turnpenny, J. (1997) Modelling heat stress and water loss of beef cattle in subtropical Queensland under current climates and climate change. In: McDonald, D.A. and McAleer, M. (eds) *Proceedings Modsim '97 International Congress on Modelling and Simulation,* University of Tasmania, Hobart. Modelling and Simulation Society of Australia, Canberra, pp. 1103–1108.

Howden, S.M., Hall, W.B. and Bruget, D. (1999) Heat stress and beef cattle in Australian rangelands: recent trends and climate change. In: Eldridge, D. and Freudenberger, D. (eds) *People and Rangelands: Building the Future.* Proceedings of the VI International Rangeland Congress Inc, Townsville, Australia, pp. 43–45.

Howden, S.M., McKeon, G.M., Meinke, H., Entel, M. and Flood, N. (2001) Impacts of climate change and climate variability on the competitiveness of wheat and beef cattle production in Emerald, north-east Australia. *Environment International* 27, 155–160.

Howden, S.M., Foran, B.D. and Behnke, R. (2002) Future shocks to people and rangelands. In: Grice, A.C. and Hodgkinson, K.C. (eds) *Global Rangelands: Progress and Prospects*. CAB International, Wallingford, UK, pp. 11–27.

Howden, S.M., Ash, A.J., Barlow, E.W.R., Booth, T., Charles, S., Cechet, R., *et al.* (2003) An overview of the adaptive capacity of the Australian agricultural sector to climate change – options, costs and benefits. Report to the Australian Greenhouse Office, Canberra.

Howden, S.M., Soussana, J.F., Tubiello, F.N., Chhetri, N., Dunlop, M. and Meinke, H.M. (2007) Adapting agriculture to climate change. *Proceedings of the National Academy of Sciences* 104, 19691–19696.

IPCC (2007) *Climate Change 2007: The Physical Science Basis*. Contribution of Working Group 1 to the Fourth Assessment Report of the Intergovernmental Panel on Climate Change. Cambridge University Press, Cambridge, UK.

Jamieson, P.D., Berntsen, J., Ewert, F., Kimball, B.A., Olesen, J.E., Pinter, P.J. Jr, *et al.* (2000) Modelling CO_2 effects on wheat with varying nitrogen supplies. *Agriculture, Ecosystems and Environment* 82, 27–37.

Johnston, P.W., McKeon, G.M. and Day, K.A. (1996) Objective 'safe' grazing capacities for south-west Queensland Australia: development of a model for individual properties. *Rangeland Journal* 18, 244–258.

Johnston, P.W., Buxton, R., Carter, J.O., Cobon, D.H., Day, K.A., Hall, W.B., *et al.* (2000) Managing climate variability in Queenslands grazing lands – new approaches. In: Hammer, G.L., Nicholls, N. and Mitchell, C. (eds) *Applications of Seasonal Climate Forecasting in Agricultural and Natural Ecosystems – The Australian Experience*. Kluwer Academic Press, Amsterdam, pp. 197–226.

Juroszek, P. and von Tiedemann, A. (2011) Potential strategies and future requirements for plant disease management under a changing climate. *Plant Pathology* 60, 100–112.

Kates, R.W., Travis, W.R. and Wilbanks, T.J. (2012) Transformational adaptation when incremental adaptations to climate change are insufficient. *Proceedings of the National Academy of Sciences* 109, 7156–7161.

Kearney, J. (2010) Food consumption trends and drivers. *Philosophical Transactions of the Royal Society of Biological Sciences* 365, 2793–2807.

Kimball, B. (2011) Lessons from FACE: CO_2 effects and interactions with water, nitrogen and temperature. In: Hillel, D. and Rosenzweig, C. (eds) *Handbook of Climate Change and Agroecosystems: Impacts, Adaptation, and Mitigation*. Imperial College Press, London, pp. 87–107.

Kirkegaard, J.A., Peoples, M.B., Angus, J.F. and Unkovitch, M.J. (2011) Diversity and evolution of rainfed farming systems in southern Australia. In: Tow, P., Cooper, I., Partridge, I. and Birch, C. (eds) *Rainfed Farming Systems*. Springer, the Netherlands, pp. 715–754.

Kriticos, D.J., Sutherst, R.W., Brown, J.R., Adkins, S.A. and Maywald, G.F. (2003) Climate change and biotic invasions: a case history of a tropical woody vine. *Biological Invasions* 5, 145–165.

Lilley, J.M., Bolger, T.P., Peoples, M.B. and Gifford, R.M. (2001) Nutritive value and the nitrogen dynamics of *Trifolium subterraneum* and *Phalaris aquatica* under warmer, high CO_2 conditions. *New Phytologist* 150, 385–395.

Lobell, D.B. and Burke, M.B. (2008) Why are agricultural impacts of climate change so uncertain? The importance of temperature relative to precipitation. *Environmental Research Letters* 3, 034007, doi:10.1088/1748-9326/3/3/034007.

Long, S., Ainsworth, E., Leakey, A. and Morgan, P. (2006) Global food insecurity. Treatment of major food crops with elevated carbon dioxide or ozone under large-scale fully open-air conditions suggests recent models may have overestimated future yields. *Philosophical Transactions of the Royal Society of Biological Sciences* 360, 2011–2020.

Lutz, W. and Samir, K.C. (2010) Dimensions of global population projections: what do we know about future population trends and structures? *Philosophical Transactions of the Royal Society of Biological Sciences* 365, 2779–2791.

McKeon, G.M. and Howden, S.M. (1992) Adapting the management of Queensland's grazing systems to climate change. In: Burgin, S. (ed.) *Climate Change: Implications for Natural Resource Conservation*, University of Western Sydney Occasional Papers in Biological Sciences No 1. University of Western Sydney, Sydney, Australia, pp. 123–140.

McKeon, G.M., Howden, S.M., Abel, N.O.J. and King, J.M. (1993) Climate change: adapting tropical and sub-tropical grasslands. In: *Proceedings of 17th International Grassland Congress*, Palmerston North, New Zealand. SIR Publishing, Wellington, New Zealand, pp. 1181–1190.

McKeon, G., Hall, W., Henry, B., Stone, G. and Watson, I. (2004) *Pasture Degradation and*

Recovery in Australia's Rangelands: Learning from History. Queensland Department of Natural Resources, Mines and Energy, Brisbane, Australia.

McKeon, G.M., Stone, G.S., Syktus, J.I., Carter, J.O., Flood, N., Fraser, G.W., et al. (2009) Climate change impacts on rangeland livestock carrying capacity: more questions than answers. Rangeland Journal 31, 1–29.

Mcleod, R.S. (1995) Costs of major parasites to the Australian livestock industries. International Journal for Parasitology 25, 1363–1367.

Mader, T.L. and Davis, M.S. (2004) Effect of management strategies on reducing heat stress of feedlot cattle: feed and water intake. Journal of Animal Science 82, 3077–3087.

Marshall, N.A. (2010) Understanding social resilience to climate variability in primary enterprises and industries. Global Environment Change 20, 36–43.

Mitchell, P.L. and Sheehy, J.E. (2006) Supercharging rice photosynthesis to increase yield. New Phytologist 171, 688–693.

Nelson, R. and Lawrence, L. (2004) Resource use in agriculture. Australian Commodities 11, 27–29.

Nelson, R., Kokic, P., Crimp, S., Meinke, H. and Howden, S.M. (2010a) The vulnerability of Australian rural communities to climate variability and change: Part I – Conceptualising and measuring vulnerability. Environmental Science and Policy 13, 8–17.

Nelson, R., Kokic, P., Crimp, S., Martin, P., Meinke, H., Howden, S.M., et al. (2010b) The vulnerability of Australian rural communities to climate variability and change: Part II – Integrating impacts with adaptive capacity. Environmental Science and Policy 13, 18–27.

Newton, P.C.D., Clark, H., Bell, C.C., Glasgow, E.M. and Campbell, B.D. (1994) Effects of elevated CO_2 and simulated seasonal changes in temperature on the species composition and growth rates of pasture turves. Annals of Botany 73, 53–79.

Nidumolu, U., Crimp, S.J., Gobbett, D., Laing, A., Howden, M. and Little, S. (2013) Spatio-temporal modelling of heat stress and climate change implications for the Murray dairy region, Australia. International Journal of Biometerology, doi:10.1007/s00484-013-0703-6.

Norman, D., Bloomquist, L., Janke, R., Freyenberger, S., Jost, J., Schurle, B., et al. (2000) Sustainable agriculture: reflections of some Kansas practitioners. American Journal of Alternative Agriculture 15, 129–136.

Nwosu, C.C. and Ogbu, C.C. (2011) Climate change and livestock production in Nigeria: issues and concerns. Agro-Science Journal of Tropical Agriculture, Food, Environment and Extension 10, 41–60.

Oerke, E.C. (2006) Crop losses to pests. Journal of Agricultural Science 144, 31–43.

Oerke, E.C., Dehne, H.W., Schonbeck, F. and Weber, A. (1999) Crop Production and Crop Protection – Estimated Losses in Major Food and Cash Crops. Elsevier, Amsterdam.

Orr, D.M. (1986) Factors affecting the vegetation dynamics of Astrebla grasslands. PhD thesis, University of Queensland, Australia.

Park, J.N., Cobon, D.H. and Phelps, D.G. (2003) Modelling pasture growth in the Mitchell grasslands. In: Post, D.A. (ed.) Proceedings of the MODSIM Conference – International Congress on Modelling and Simulation, Townsville, Australia. Modelling and Simulation Society of Australia and New Zealand Inc, Canberra, pp. 519–524.

Park, S.E., Marshall, N.A., Jakku, E., Dowd, A.M., Howden, S.M., Mendham, E., et al. (2012) Informing adaptation responses to climate change through theories of transformation. Global Environmental Change 22, 115–126.

Parr, J.F., Stewart, B.A., Hornick, S.B. and Singh, R.P. (1990) Improving the sustainability of dryland farming systems: a global perspective. Advances in Soil Science 13, 1–8.

Parry, M., Rosenzweig, C. and Livermore, M. (2005) Climate change, global food supply and risk of hunger. Philosophical Transactions of the Royal Society of Biological Sciences 360, 2125–2138.

Petty, S.R., Poppi, D.P. and Triglone, T. (1998) Effect of maize supplementation, seasonal temperature and humidity on the liveweight gain of steers grazing irrigated Leucaena leucocephala/Digitaria erinta pastures in north-west Australia. Journal of Agricultural Science 130, 95–105.

Porter, J. and Semenov, M. (2005) Crop responses to climatic variation. Philosophical Transactions of the Royal Society of Biological Sciences 360, 2021–2035.

Qaderi, M. and Reid, D.M. (2009) Crop responses to elevated carbon dioxide and temperature. In: Sing, S.N. (ed.) Climate Change and Crops, Environmental Science and Engineering. Springer-Verlag, Berlin/Heidelberg, pp. 1–18.

Ramirez-Villegas, J., Challinor, A.J., Thornton, P.K. and Jarvis, A. (2013) Implications of regional improvement in global climate models for agricultural impact research. Environmental Research Letters 8, 024018, doi:10.1088/1748-9326/8/2/024018.

Reynolds, J.F., Stafford Smith, D.M., Lambin, E.F., Turner, B.L., Mortimore, M., Batterbury, S.P.J., et al. (2007) Global desertification: building a science for dryland development. *Science* 316, 847–851.

Ribot, J., Najam, A. and Watson, G. (2009) Climate variation, vulnerability and sustainable development in the semi-arid tropics. In: Lisa, E., Schipper, F. and Burton, I. (eds) *The Earthscan Reader on Adaptation to Climate Change*. Earthscan, London, pp. 117–160.

Rickards, L. and Howden, S.M. (2012) Transformational adaptation: agriculture and climate change. *Crop and Pasture Science* 63, 240–250.

Rigby, J.R. and Porporato, A. (2008) Spring frost risk in a changing climate. *Geophysical Research Letters* 35, doi:10.1029/2008gl033955.

Robertson, M. (2010) Agricultural productivity in Australia and New Zealand: trends, constraints and opportunities. *Proceedings of the New Zealand Grassland Association* 72, LI–LXII.

Rosenzweig, C. and Wilbanks, T.J. (2010) The state of climate change vulnerability, impacts, and adaptation research: strengthening knowledge base and community. *Climatic Change* 100, 103–106.

Sahoo, A., Davendra, K. And Naqvi, S.M.K (eds) (2013) *Climate Resilient Small Ruminant Production*. National Initiative on Climate Resilient Agriculture (NICRA), Central Sheep and Wool Research Institute, Izatnagar, India.

Scanlan, J.C., Hinton, A.W., McKeon, G.M., Day, K.A. and Mott, J.J. (1994) Estimating safe carrying capacities of extensive cattle-grazing properties within tropical, semi-arid woodlands of north-eastern Australia. *Rangeland Journal* 16, 64–76.

Sinclair, T.R. (2011) Precipitation: the thousand-pound gorilla in crop response to climate change. In: Hillel, D. and Rosenzweig, C. (eds) *Handbook of Climate Change and Agroecosystems: Impacts, Adaptation, and Mitigation*. Imperial College Press, London, pp. 179–190.

Smith, S.J., Thomson, A.M., Rosenberg, N.J., Izaurralde, R.C., Brown, R.A. and Wigley, T.M.L. (2005) Climate change impacts for the conterminous USA: an integrated assessment. 1. Scenarios and context. *Climate Change* 69, 7–25.

Stafford Smith, D.M., Abel, N., Walker, B. and Chapin, F.S. III (2009) Drylands: coping with uncertainty, thresholds, and changes in state. In: Chapin, F.S. III, Kofinas, G.P. and Folke, C. (eds) *Principles of Ecosystem Stewardship: Resilience-Based Natural Resource Management in a Changing World*. Springer Science, New York, pp. 171–195.

Stokes, C.J. and Howden, S.M. (2010) *Adapting Agriculture to Climate Change: Preparing Australian Agriculture, Forestry and Fisheries for the Future*. CSIRO Publishing, Collingwood, Australia.

Stokes, C.J., Crimp, S., Gifford, R., Ash, A. and Howden, S.M. (2010) Broadacre grazing. In: Stokes, C.J. and Howden, M. (eds) *Adapting Agriculture to Climate Change: Preparing Australian Agriculture, Forestry and Fisheries for the Future*. CSIRO Publishing, Melbourne, Australia, pp. 153–170.

Sutherst, R.W. (1990) Impact of climate change on pests and diseases in Australia. *Search (Sydney)* 21, 230–232.

Sutherst, R.W., Yonow, T., Chakraborty, S., O'Donnell, C. and White, N. (1996) A generic approach to defining impacts of climate change on pests, weeds and diseases in Australasia. In: Bouma, W., Pearman, G.I. and Manning, M.R. (eds) *Greenhouse: Coping with Climate Change*. CSIRO Publishing, Melbourne, Australia, pp. 190–204.

Sutherst, R.W., Maywald, G.F. and Russell, B.L. (2000) Estimating vulnerability under global change: modular modelling of pests. *Agriculture, Ecosystems and Environment* 82, 303–319.

Teixeira, E.I., Fischer, G., van Velthuizen, H., Walter, C. and Ewert, F. (2011) Global hot-spots of heat stress on agricultural crops due to climate change. *Agricultural and Forest Meteorology* 170, 206–215.

Tubiello, F.N., Donatelli, M., Rosenzweig, C. and Stockle, C.O. (2000) Effects of climate change and elevated CO_2 on cropping systems: model predictions at two Italian locations. *European Journal of Agronomy* 13, 179–189.

Tubiello, F.N., Amthor, J.S., Boote, K.J., Donatelli, M., Easterling, W., Fischer, G., et al. (2007) Crop response to elevated CO_2 and world food supply: a comment on 'Food for Thought...' by Long et al., Science 312, 1918–1921, 2006. *European Journal of Agronomy* 26(3), 215–223, doi:10.1016/j.eja.2006.10.002.

van Ittersum, M.K., Howden, S.M. and Asseng, S. (2003) Sensitivity of productivity and deep drainage of wheat cropping systems in a Mediterranean environment to changes in CO_2, temperature and precipitation. *Agriculture, Ecosystems and Environment* 97, 255–273.

Wand, S.J.E., Midgley, G.F., Jones, M.H. and Curtis, P.S. (1999) Responses of wild C4 and C3 grass (Poaceae) species to elevated atmospheric CO_2 concentration: a meta-analytic test of current theories and perceptions. *Global Change Biology* 5, 723–741.

Wang, Y.P., Handoko, H. and Rimmington, G.M.

(1992) Sensitivity of wheat growth to increased air temperature for different scenarios of ambient CO_2 concentration and rainfall in Victoria, Australia – a simulation study. *Climate Research* 2, 131–149.

Webb, N.P., Stokes, C.J. and Marshall, N.A. (2012) Integrating biophysical and socio-economic evaluations to improve the efficacy of adaptation assessments for agriculture, *Global Environmental Change* (http://dx.doi.org/10.1016/j.gloenvcha.2013.04.007, accessed 18 November 2013).

Wilson, J.R. (1982) Environmental and nutritional factors affecting herbage quality. In: Hacker, J.B. (ed.) *Nutritional Limits to Animal Production from Pastures*. CAB International, Wallingford, UK, pp. 111–131.

Xiong, W., Lin, E., Ju, H. and Xu, Y. (2007) Climate change and critical thresholds in China's food security. *Climate Change* 81, 205–221.

5 Diversity in Organic and Agroecological Farming Systems for Mitigation of Climate Change Impact, with Examples from Latin America

Walter A.H. Rossing,[1] Pablo Modernel[1,2] and Pablo A. Tittonell[1]

[1]*Farming Systems Ecology, Wageningen University, Wageningen, the Netherlands;* [2]*Facultad Agronomía, Universidad de la República, Montevideo, Uruguay*

5.1 Introduction

As the largest global land use, agriculture both contributes to, and is affected by, climate change. According to the Intergovernmental Panel on Climate Change (IPCC, 2007), global warming causing climate change is due to anthropogenic greenhouse gases. Among these, agriculture is the major source of CH_4 and N_2O, and a lesser source of CO_2. The emissions directly emitted by agriculture have been estimated in the fourth IPCC Assessment Report as together constituting 10–12% (5.1–6.1 Gt) of total annual emissions in CO_2 equivalents. A recent overview of indirect emissions resulting from land clearing for agricultural use estimated these to contribute another 4–13% (2.2–6.6 Gt) (Vermeulen *et al.*, 2012). The previous authors estimated the contribution of other components of the food chain to be 4% (2 Gt), with fertilizer production and refrigeration as major items, but noted the major uncertainty surrounding most information (Fig. 5.1).

Farmers are constantly responding to dynamics in the biophysical and socio-economic environments in which not only the climate changes. Farming systems must adapt continuously to respond to changes in demand and resource availabilities that are or are not reflected in prices, to vagaries in weather and to changing social and economic contexts. Such changes operate at timescales often shorter than the decades involved in climate change (Howden *et al.*, 2007; Lin *et al.*, 2008; Darnhofer *et al.*, 2010; Tomich *et al.*, 2011). At the same time, climate change is a key concern when it comes to facing the challenge of feeding 9 billion people by 2050. As agriculture and related land clearance account for 15–25% of all GHG emissions worldwide (estimates for 2008 in Vermeulen *et al.*, 2012), farming systems and methods that contribute to reducing emissions have a key role to play in climate change mitigation. This is why farming systems and methods that rely largely on soil organic matter and biologically mediated processes are often pointed out as promising triple-win agricultural options, as they may simultaneously aid climate change adaptation, mitigation and global food security (e.g. http://climatechange.worldbank.org/sites/default/files/documents/CSA_Policy_Brief_web.pdf). Organic farming, conservation agriculture or traditional indigenous farming are good examples of such systems.

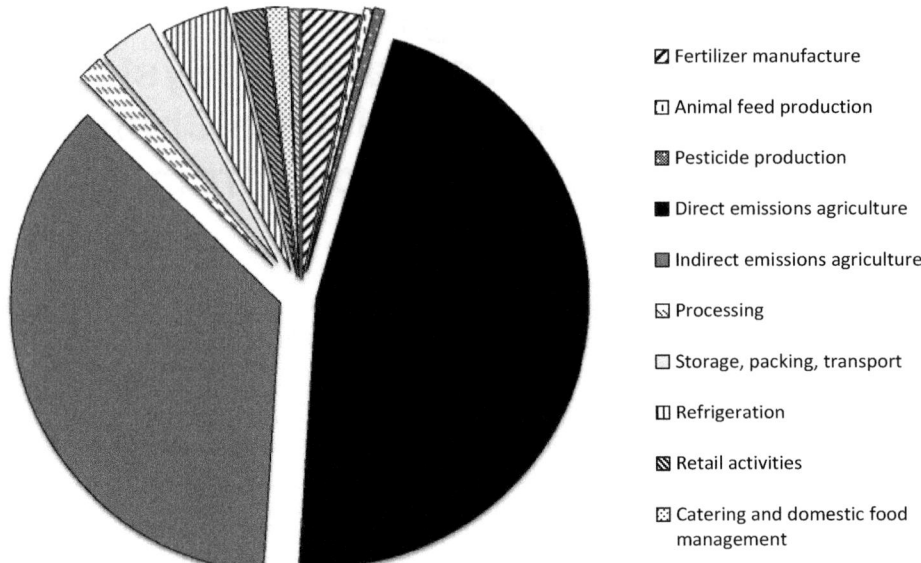

Fig. 5.1. Contribution of different components of the food chain to total annual emissions in CO_2 equivalents. (After Vermeulen *et al.*, 2012.)

This chapter reviews the impact of climate change on, and the adaptation potential of, ecologically intensive systems such as those known in different parts of the world as organic farming, or agroecology. To fit into the rationale of this book, this chapter focuses on Latin America, and on the array of organic and agroecological farming systems (OAFS) that coexist in the region.

Impacts of, and adaptation to, climate change of agricultural production systems find their origins in the way that management by humans at different levels of social organization interacts with agro-ecological processes. At the farm level, differences in management, driven by priorities among values and objectives, give rise to different farm configurations. In addition to the objectives and values of the farmer, management is shaped by external drivers, which are constantly changing. Scholarly work on the evolution of farming systems has emphasized the need to consider uncertainty as a pervasive feature of the farming context and to consider management as a continuous re-balancing of internal and external resources in response to such variability (e.g. Verhoeven *et al.*, 2003; Renting *et al.*, 2009; Darnhofer *et al.*, 2010). An important question for future agricultural systems is, therefore, how to maximize the capacity of farming systems to adapt to variability and change for a given set of internal farm resources.

The rest of the chapter is organized as follows. It starts by defining the farming systems focused on in the chapter. Then climate change predictions for Latin America are reviewed and their impacts assessed. Next, recent evidence from global meta-analyses on the performance of OAFS is summarized. Information from OAFS from Latin America is included in the reviews but is too scant to warrant a specific section. Instead, two specific OAFS from Latin America are focused on: coffee production systems and natural grassland systems. Their current status, challenges in view of climate change predictions and adaptation options proposed in the literature are reviewed. The chapter ends by discussing the key features and impediments of OAFS and how they can help to build farming systems in the face of global and climate change.

5.2 Organic and Agroecological Farming Systems in Latin America

Organic production can be considered as part of two organizational systems (Parrott *et al.*, 2006); on the one hand, a clearly visible, formal certified sector, and on the other hand, a less clearly delineated, informally organized agroecological sector. The former has strong roots in the global north and is described in a set of principles (Luttikholt, 2007; IFOAM, 2013) that have been translated into regulations on the use of crop rotation, green manures, composts and biological pest control, and the absence of synthetic fertilizers, pesticides and GMOs (genetically modified organisms), as well as a range of synthetic food additives. Soil fertility is stimulated by crop rotation, green manures and cover crops, intercropping, mulching and the use of manures. Pest, disease and weed control are based on preventive measures, such as cover cropping, cultivar choice and landscape management, and biological curative measures are used only as a last resort (e.g. Vereijken, 1997; Altieri, 2002; Lotter *et al.*, 2003; Gliessman, 2007). The organic label regulations are enforced by international and national certification bodies (Zorn *et al.*, 2013). In 2011, organic production was found on 37.2 million ha (Mha) (0.86% of total agricultural land) in 86 countries, and the products sold in mainly European and North American markets represented a value of US$62.9 billion. Quadrupling in 12 years, organic products have become an economically interesting niche in the food system (FiBL and IFOAM, 2013). Latin America has 6.9 Mha of certified organic farming, with vast expanses in Argentina (4.2 Mha), Brazil (0.7 Mha) and Uruguay (0.9 Mha) (FiBL and IFOAM, 2013). On average, 85% of what is produced organically is exported to the main organic markets, such as the European Union, the USA and Japan. The main organic export products are coffee, cacao, banana, quinoa, meat and wool. Wild collection of nuts is also of importance for international markets. The degree to which organic farming practices differ from those relying on external inputs may range from simple input substitution to system diversification and implementation of strategies that use multiple cultural and biological means (Shennan, 2008; Oelofse *et al.*, 2011).

A range of farming systems share with certified organic production systems a similar outlook and approach in their practices, but without the formal certification schemes and, therefore, without the international market perspective. Parrott *et al.* (2006) note that here 'definitions of what constitutes organic farming can become quite blurred and the boundaries become ill-defined'. Some system descriptions align with organic farming, while others maintain, or even emphasize, their distance from certified organic systems. Examples of such systems include permaculture, eco-agriculture (Scherr and McNeely, 2008), ecologically intensive farming (as proposed by Doré *et al.*, 2011; but not Cassman, 1999), multifunctional agriculture (e.g. Boody *et al.*, 2009), agroforestry, sustainable agriculture and diversified farming systems (Kremen *et al.*, 2012). These partly conceptual, partly practice-based systems challenge the existing research and policy paradigm, with its focus on maximizing yield at field level by overcoming growth-reducing and -limiting factors through external inputs, and share a more holistic view of agriculture as managed ecosystems (Lin, 2011). And finally, there are production systems that are organic 'by default' in that no artificial fertilizers or pesticides are used for reasons of economic efficiency (e.g. the extensive South American and Australian rangeland systems, or subsistence family farming in vast areas of the humid tropics) or lack of financial resources (cf. Parrott *et al.*, 2006).

In Latin America, 'agroecology' is a notion with connections to 'organic'. Wezel *et al.* (2009) describe the confusion surrounding the term agroecology, and identify connotations as a scientific field, a social movement and a set of practices. As a set of practices, it has been linked strongly to utilization of the indigenous knowledge of traditional farmers in promoting biodiversity, sustaining yield without agrochemicals and conserving ecological integrity

(Altieri, 2004; Altieri et al., 2012). At the same time, the notion was used to lobby for, design and implement policies in, for example, Brazil (Bellon and de Abreu, 2006), geared to supporting small farmers and community development. In recent years, agroecology has been included in Brazil's national research strategy (EMBRAPA, 2006, quoted in Wezel et al., 2009) as a transdisciplinary scientific field in which concepts from systems ecology are linked to agriculture focusing particularly on smallholder farms (Altieri, 2002; Gliessman, 2007; Altieri et al., 2012), using a whole-food system perspective (Francis et al., 2003).

5.3 Projection and Assessment of Climate Change for Latin America

The Fourth Assessment Report of the IPCC (Solomon et al., 2007) and later reports (e.g. Marengo et al., 2008, 2012) predict that all of Central and South America is very likely to warm during the 21st century. In southern South America, the annual mean warming will likely be similar to the global mean warming, but in the rest of the continent it will exceed it. Regional circulation model results for South America, based on both the Special Report on Emissions Scenarios (SRES) A2 and B2 for the end of the 21st century, predict particularly large warming in austral spring for (subtropical) southern Brazil, Paraguay, Bolivia and north-eastern Argentina (Nuñez et al., 2008; Marengo et al., 2012). Annual precipitation is predicted to decrease in most of Central America and in the southern Andes, although with substantial uncertainty due to local circulation patterns in the mountainous areas. Precipitation is predicted to increase in Tierra del Fuego during austral winter and in south-eastern South America (SESA, comprising the La Plata river basin and southern Brazil) during summer (Solomon et al., 2007). Downscaling experiments on climate change scenarios in South America suggest a reduction in Amazonia rainfall and a small increase in SESA rainfall, as well as an increase in dry spells and intense rainfall events in the region of the South American Monsoon System (SAMS), located between the Amazon and the La Plata basin, and in SESA (Neelin et al., 2006; Marengo et al., 2012). This suggests that future rainfall will be concentrated in short periods, followed by longer dry spells. The projected increase in total and heavy rainfall in SESA, particularly after 2050, is consistent with projected changes in atmospheric circulation, and with an increase in dry spells. All these projections suggest potential changes in rainfall extremes in the SESA and SAMS regions (Marengo et al., 2012). Downscaling experiments with the 16 most representative global circulation models (GCMs) for the business as usual (SRES-A2a) emission scenario for Nicaragua showed a decrease in total annual precipitation from 1740 mm to 1610 mm, no change in the number of dry months and a mean annual temperature increase of 2.2°C, along with an increase in temperature range from 10.4 to 10.6°C (Laderach et al., 2009). Schroth et al. (2009) calculated similar values for the Sierra Madre zone in southern Mexico.

Analysis of data over the past 50 years showed that, in many areas, the frequency of heavy precipitation events increased consistent with warming, and widespread changes in extreme temperatures were observed (Marengo et al., 2010). The intensity and frequency of Atlantic tropical cyclones increased significantly (Webster et al., 2005; Elsner et al., 2008). Cold days, cold nights and frost became less frequent, while hot days, hot nights and heatwaves became more frequent (see references in Marengo et al., 2010). These observations and the model predictions suggest that, most immediately, climate change is being experienced as increasing temporal and spatial variability in temperature, precipitation and winds, particularly the incidence and magnitude of extreme events, such as heatwaves, heavy precipitation events and associated floods, tropical cyclone events and extremely high sea levels owing to storm surges (Vermeulen et al., 2012). There is a likely increase of longer dry spells and of the area affected by drought each year. Cold spells and frosts will decrease in frequency and intensity. In the short term, therefore, increasing climate variability has more impact than longer-term change in mean values, and the appropriate focus of adaptation should be climate risk management.

Assessment of the impact of climate change requires approaches that go across levels of biological and social organization, and that take multiple actors and their objectives into account. Distinction of system components and their interrelations within well-defined boundaries at a particular temporal and spatial scale is useful to structure analysis. However, assessment of the impact of drivers that go across scales such as global climate change, and can be influenced at each scale, requires methods for scaling down and up. Scaling down involves the translation of predicted changes in drivers at larger spatio-temporal scales to smaller scales at which the relevant effects and responses are to be evaluated. Downscaling global climate change predictions to local level is wrought with uncertainty, which cannot be ignored in response assessment. Responses at lower levels of organization need to be scaled up to assess global consequences. The challenge here is to take into account non-linear interactions and feedbacks within and between natural and social subsystems, which may result in emergent behaviour. This inherent uncertainty of system dynamics is a central tenet in complexity science (Holling, 2001), and has given rise to increasing interest in the ability of systems to adapt to unexpected change so as to maintain their functional capacity. Dominant current frameworks take fundamental ecological properties of self-sustaining ecosystems such as productivity, stability, reliability, resilience, adaptability, equity and self-reliance (e.g. Holling, 2001; López-Ridaura et al., 2002) as structuring categories and derive context-specific indicators within each category. Other frameworks take a more economic approach to systems analysis, such as the sustainable livelihoods framework, which distinguishes natural, economic, human and social resources or capitals to understand and assess the livelihoods of the earth's poor (Scoones, 1998). Loosely adopted here is the ecosystem services framework (De Groot, 1992; Costanza et al., 1997), which gained global interest through the Millennium Ecosystem Assessment (MEA, 2005). It distinguishes four categories of services to humans provided by ecosystems, including supporting, provisioning, regulating and cultural services that affect, to different degrees, the constituents of well-being security, basic materials for a good life, health, good social relations and freedom of choice and action (MEA, 2005). This framework has been used by both (agro) ecologists and economists (TEEB, 2010), thus providing a useful boundary object for interdisciplinary work.

5.4 Climate Change and Ecosystem Services of OAFS – The State of the Art

The recent literature reveals a range of assessments of organic and agroecological farming systems that are usually compared with conventional systems. Recent reviews with particular interest for global climate change include El-Hage Scialabba and Müller-Lindenlauf (2010), Goh (2011), Gomiero et al. (2011), Lynch et al. (2011), Gattinger et al. (2012), Kremen and Miles (2012) and Tuomisto et al. (2012). Three of these studies (Gattinger et al., 2012; Kremen and Miles, 2012; Tuomisto et al., 2012) used statistical meta-analysis methods for a number of indicators when data availability was sufficient. The other reviews reported on individual studies using vote counts or narratives. The studies published in 2011 and earlier were covered in at least one of the reviews published in 2012. Therefore, only the findings from the 2012 meta-analyses are summarized here, assuming that the results represent the current state of knowledge on the potential of OAFS to mitigate the effects of global climate change (Table 5.1).

Gattinger et al. (2012) analysed the differences in soil organic carbon (SOC) concentrations in 74 comparisons of organic and conventional farming systems, of which 29 comparisons also contained information on SOC concentrations, and 20 of these

Table 5.1. Summary of global evidence from reviews published after 2011 on the mitigation and adaptation potential of organic and agroecological systems (OAFS) compared to locally dominant conventional systems (CS).

	Mitigation or adaptation potential	Nature of evidence	Remarks	Source[a]
Significantly better performance by OAFS	Carbon sequestration to 30 cm depth	One long-term study on carbon sequestration; one long-term study on soil organic matter; one meta-analysis on soil organic matter	Similar effects of OAFS and no-till systems for C sequestration	2
		Meta-analysis of pairwise comparisons in 74 studies	Significant effects of soil organic carbon concentrations, stocks and sequestration rates	1
		Meta-analysis of intensive farming regions; 56 comparisons	Relative median SOM in OAFS 7% higher than conventional. No correlation between OM inputs, ley area or years in OAFS and SOM	3
	Energy-use efficiency	Vote count of 130 studies; analysis per unit area and per unit product; most results exceeded 20% better performance	Mechanical weed control balanced by green manures (systems perspective)	1
		Meta-analysis of intensive farming regions; 34 comparisons; life-cycle analysis per unit product	Particularly due to higher energy requirement of fertilizer production and transport in conventional systems	3
	Soil water-holding capacity	Three long-term studies	Effect size ranging from <25% to >50% more soil water availability	1
	Resilience to drought	Long-term study	Medium to large effects	1
	Resilience to hurricanes and heavy rainfall	Study of impact of hurricanes Mitch, Stan and Ike	Effects not always apparent at farm level	1
Equivocal performance by OAFS and CS	Carbon sequestration to 1 m depth	Two long-term studies	Too few measurements below 30 cm available in the literature	1
	Global warming potential	Vote count of 130 studies; analysis per unit area and per unit product	Uncertainty in measurement of N_2O from soils and manure due to high variability. OAFS had higher GHG emissions per unit product for pork, poultry and dairy due to lower rates of feed conversion and emissions from straw litter	1
		Meta-analysis of intensive farming regions; 23 comparisons; life-cycle analysis per unit product	OAFS had higher GHG emissions per unit product for cereals, pork and dairy, and lower emission for olive, beef and some crops. Few data available per commodity ($N < 7$)	3

Notes: [a]1 = Gattinger *et al.*, 2012; 2 = Kremen and Miles, 2012; 3 = Tuomisto *et al.*, 2012. GHG = greenhouse gases; OM = organic matter; SOM = soil organic matter.

allowed calculation of SOC sequestration rates. Their results show significantly higher concentrations, stocks and rates for organic systems. Similar results for concentrations and stocks appeared from the analysis of a subset with zero net input, but sequestration rates were not significantly different. In a critique of these results, Leifeld *et al.* (2013) argued among others that levels of organic applications differed between the compared systems, and thus a comparison of organic and conventional was biased. This emphasizes the need to be very clear about the system boundaries (Walker *et al.*, 2010). As stated in their reply, Gattinger *et al.* (2012) compared farming systems rather than experimental units to which the *ceteris paribus* hypothesis applied, and one of the inherent characteristics of organic systems was their emphasis on organic matter supplies.

Kremen and Miles (2012) compared the contributions of systems that exhibited diversification across ecological spatial and temporal scales to those of conventional industrialized systems with respect to 12 key ecosystem services. Their analysis gave more weight to meta-analyses or quantitative syntheses using statistical procedures and long-term studies than to studies in which better performance was tallied ('vote counts'). They rated the evidence as strong when medium to large significant trends were revealed, as weak when trends were small and/or significance was low, and equivocal when results were non-significant or studies showed strong positive as well as strong negative trends. Vote-count studies were considered to reveal strong effects when over 75% of studies supported a trend. Their results showed that, compared with conventional farming systems, OAFS supported substantially greater biodiversity, soil quality, carbon sequestration, water-holding capacity in surface soils, energy-use efficiency and resistance and resilience to climate change.

Tuomisto *et al.* (2012) analysed 71 published studies, including 170 cases in which the impacts of organic and conventional farming in intensive farming regions were compared. They selected 11 indicators to include all environmental impact categories and quantified these per unit area and per unit product, using life-cycle analysis when available. They found higher median soil organic matter (no information on concentration or stocks) in organic systems. Although organic systems had higher organic matter inputs and more area under leys than conventional systems, these inputs alone did not explain higher organic matter levels. The authors suggested that less intensive tillage and lower levels of mineral nitrogen additions might have reduced decomposition rates of organic residues. They found significantly lower energy use in organic systems per unit of product, due particularly to the high energy requirement of artificial fertilizer production. Overall results for global warming potential were equivocal, with no significant differences between organic and conventional systems. Per commodity group differences were apparent, but low numbers of case studies precluded statistical comparison. They concluded that, under intensive conditions, organic farming had lower environmental impacts per unit area than conventional systems, but due to lower yields not always per unit of product.

Summarizing, the reviews found that organic and agroecological farming systems provided climate change relevant ecosystem services that were greater than conventional farming systems for carbon sequestration to 30 cm depth, energy-use efficiency, soil water-holding capacity, resilience to drought and resilience to hurricanes and heavy rainfall. No differences between systems were found for carbon sequestration to 1 m depth, due to lack of data, or for global warming potential (see Table 5.1).

5.5 Adaptation Pathways in OAFS: (Re-)Creating Diversity and Managing Risk

5.5.1 Coffee-based systems in Meso-America and Brazil

Coffee production has a history of some 300 years in many areas of Latin America and is deeply entrenched in the culture, ecology and economy (Toledo and Moguel, 2012).

Over 70% of the world's coffee is produced by small-scale family farms, largely grown under rainfed conditions as in most areas irrigation water is not available or accessible. Of the species of coffee, the commonly grown Robusta coffee (*Coffea canephora*), which is considered of lower quality than Arabica coffee (*C. arabica*), can tolerate more elevated temperatures and is therefore found at lower altitudes. Traditionally, coffee has been grown as one of the components of diverse agroforestry systems (Fig. 5.2). In pursuit of greater coffee yields, and stimulated by government intervention programmes (e.g. Cardoso et al., 2001; Schroth et al., 2009), these systems have been modified throughout Latin America to reduce competition for light and nutrients. Broadly speaking, and allowing for substantial local variation, modifications involved reducing tree density and associated shade cover and adapting tree species composition to facilitate shade management by pruning. The intensification of coffee production has been associated with the substantial use of fertilizers and herbicides. For instance, Costa Rican intensive coffee farmers are recommended to apply 50–55% more N than taken off by production, resulting in fertilizer costs accounting for up to 55% of total variable costs and major environmental impacts as compared to shade systems (Castro-Tanzi et al., 2012). Various authors have questioned the consistency of higher yields in sun systems as compared to moderate shade (e.g. Muschler, 2001; Lin et al., 2008), and pointed out the higher quality associated with more shade, which decreases temperatures, and hence berry-filling rates, and

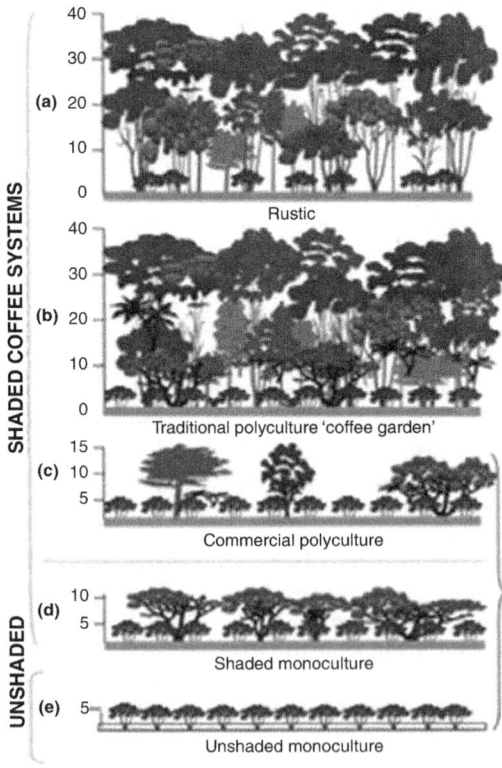

Fig. 5.2. The five main coffee-producing landscapes distinguished by vegetation structure, species variety and composition, as well as by the impact of human manipulation. (Reproduced from Toledo and Moguel, 2012, with permission from Taylor and Francis.)

increases organoleptic properties (Vaast et al., 2006; Lin et al., 2008). Comparing temperatures in sun coffee systems in Minas Gerais, Brazil, with shaded coffee systems and reference forest, De Sousa et al. (2012) found that mean maximum daily temperatures in sun coffee reached values over 30°C that were detrimental to growth, about 6°C greater than in shaded coffee.

The general features of climate change for tropical Latin America (i.e. decreased precipitation and increased temperatures, along with greater variability) have been shown to have potentially grave consequences for coffee production. Schroth et al. (2009) compared crop suitability in the Mexican Sierra Madre region based on 19 bioclimatic variables using historical climate data and climate predicted by 15 global circulation models under the A2a emissions scenario for 2040–2069 at 1 × 1 km grid cells. Their results show that increasing temperatures will cause current coffee production at altitudes below 1700 m to be forced to move to higher altitudes in response to drought stress caused by a decrease in dry season rainfall combined with higher evaporative demand. On the other hand, at higher altitudes, conditions for coffee production were found to improve due to a combination of decreasing rainfall in the cold, wet season and an associated increase in sunny days, as well as increasing temperature. For the Sierra Madre, the coffee production limit now located at 600 m altitude would shift to 850–900 m. The shift to higher altitudes will decrease the total area suitable for coffee production, although quantitative estimates of the loss of productive area are lacking. Laderach et al. (2009) conducted a similar study for Nicaragua and found losses in suitability of 20–60% at a 1 km² scale for many departments, and only small areas with an increase in suitability of 20–30%.

Where climate continues to evolve negatively, coping strategies may ultimately give way to either abandonment of coffee or investment in adapted varieties and improved infrastructure. As pointed out by Schroth et al. (2009), such adaptation may lead to the destruction of existing agroforestry systems when farmers intensify production by reducing shade and increase inputs to recover their investments by higher yields, or by clearing land for other uses, for instance for cattle pastures or sugarcane, as has been observed in lower altitude areas in Mexico. In their study for 2050, Laderach et al. (2009) assessed options for 30 crops other than coffee. Their results suggested that suitability for other crops improved across large areas, even though that for coffee declined.

Where adaptation to climate change is possible, an initial option will be to decrease the input of labour and capital, as has been found to happen in response to low world market prices (Eakin et al., 2006). Also, the inclusion of other crops, in particular for household food security, has been observed. A general adaptation strategy emerging from the literature is the diversification of habitat structure. In a participatory study of the impact of hurricane Mitch in Nicaragua and Honduras, Holt-Giménez (2002) found that economic losses on farms with agroforestry techniques generally were less than on conventional coffee farms. In a study of the impact of hurricane Stan in Chiapas, Philpott et al. (2008) showed that structurally more complex farms had significantly smaller fractions of their area affected by landslides. As extreme weather events including hurricanes are expected to become more prevalent in the future climate, these results provide important guidelines for adaptation measures.

Agronomic adaptation and traditional means of risk insurance by the farm family, including habitat and crop diversification, maintaining financial reserves, reliance on off-farm income and involving children to work on the farm, can be accompanied by supporting policies including minimum price guarantees, payments for ecosystem services and speciality market niches such as fair trade with price premiums (Schroth et al., 2009). Micro-insurance, either as an explicit insurance product or as government intervention when prices fall below a minimum level, has recently been proposed as a means to provide additional risk management options for smallholders (Diaz Nieto et al., 2012).

Schroth et al. (2009) conclude that Meso-American agriculture will suffer severe impacts as a result of climate change, but they also show that opportunities are likely to appear if farmers have the access and information to change varieties and, if necessary, their crops. They point out that when crops are grown for cash, this adaptation may be much easier than for traditional crops of large cultural importance, such as occurs with most staple crops. In a study of traditional maize seed systems, Bellon et al. (2011) found that the exchange of seedlots between farmers in eastern Mexico typically took place in a 10 km radius around each farm and across a range of altitudes. They concluded that these exchange systems would allow effective local adaptation of maize varieties to climate change. Coffee has an intermediate position between traditional staple crops and modern cash crops in that this cash crop has been cultivated for such a long time that it has become part of the local culture. Moreover, many of the Latin American coffee growing regions fall within areas identified as biodiversity hotspots (Myers et al., 2000), and the industry has an impact on biodiversity which is disproportional to its area (Toledo and Moguel, 2012). Climate change adaptations, therefore, may benefit from the ecological diversity of, and local knowledge on, polyculture coffee systems and from the market niche of agroecological coffee among affluent consumers. Ecological diversity appears as a key element to manage short-term variability, as well as to mitigate temperature and humidity changes in the longer term. In addition, the ecological values may be developed into marketable products alongside coffee or may add value to coffee in certification schemes. Although they found highly successful examples, Toledo and Moguel (2012) also said that there was still much to learn about effective governance in certification schemes before coffee farmers were effectively rewarded for the multiple values their diverse coffee systems provided. Such governance will need to transcend the boundaries of the individual farm and move to the landscape scale, which may provide benefits in terms of social cohesion, biodiversity conservation, market position and sourcing ability from the perspective of retailers (Kissinger et al., 2013).

5.5.2 Natural grassland-based livestock farming

The Pampas and Campos, situated between 24° and 35° southern latitude in Argentina, Brazil, parts of Paraguay and Uruguay are a 500,000 km^2 biodiversity hotspot, estimated to harbour 3000–4000 vascular plant species including some shrubs and trees (Overbeck et al., 2007), of which 450 grass species and 150 legumes are considered forage for domestic animals (Boldrini, 2002). Different soil types and regional topography create highly diverse habitats, which foster the co-occurrence of C_3 and C_4 grasses and leguminous herb species. The region comprises 500,000 farms with 65 million cattle, mostly Zebu-based followed by European breeds and sheep that were mostly introduced in the 17th century. In addition to providing high-quality beef, wool and ecological services of local and global relevance, these natural grassland-based livestock systems sustain rural livelihoods and prevent mass migration to urban centres. The relatively low productivity level of the natural grassland-based livestock systems makes them vulnerable to the expanding agricultural frontiers in the world, fuelled by the increasing global demands for food, feed and energy crops. It has even been argued that these soils constitute a 'reservoir' of land and nutrients for the expansion of agriculture (especially soybean) and forestry to respond to such demands. To maintain these systems and the benefits associated with them, new forms of intensification are needed that allow increasing productivity while delivering valuable ecological services in the face of global and climate change.

Overstocking, forage off-take beyond the annual regrowth capacity of the vegetation, is considered the major reason for low farm productivity (Nabinger et al., 2000). High stocking rates are stimulated by the farmer's traditional sense of wealth, exacerbated in Brazil by legislation that links ownership rights to minimum stocking rates, irrespective of carrying capacity (Carvalho and Batello, 2009). Over the past 50 years, the dominant strategy to increase production has consisted of replacing the natural highly diverse vegetation with non-endemic species, sown as single species annual swards or as mixtures of two or three perennial species. Nabinger et al. (2009) reported rates of decrease of natural grasslands of 440,000 ha per annum. This process led to a homogenization of landscapes and increased use of external inputs (fertilizers, pesticides and fossil fuels), and therefore production costs. There is widespread concern that the replacement of the species-rich native vegetation by a few exotic species makes systems more vulnerable to droughts (Tilman et al., 2006), the frequency of which is predicted to increase in the next decades (Bidegain et al., 2012). In addition, replacing native grasslands with crops or exotic grass species has resulted in the rapid loss of soil carbon (Sala and Paruelo, 1997). This process has occurred under both tillage and no-tillage management, where reductions in root biomass, active carbon stocks and in soil biota have been reported (DuPont et al., 2010). For other areas in the region, such loss of soil carbon has been shown to decrease soil water-holding capacity (Alliaume et al., 2013; Dogliotti et al., 2013). Heat stress for animals has become an issue in areas where landscape clearing has taken place, and is likely to affect productivity rates and animal welfare negatively across wider areas under climate change (see Collier et al., Chapter 7, this volume). Thus, current productivity and the expected impact of climate change raise question about the dominant pathways for increasing animal production, and for managing the resource base to sequester carbon, to preserve biodiversity and to maintain water-storage capacity (Roesch et al., 2009; Carvalho et al., 2011).

Productivity of the natural pastures in the region follows an annual pattern, with higher production levels in spring and summer, declining in autumn to become lowest in winter. This variation is related to temperature more than to precipitation, and is associated with a change in the share of C_3 and C_4 species in total biomass; the C_4 species dominating in summer and the C_3 in the cooler winter. Overgazing is especially destructive in winter, when production is low, and leads to a decline in C_3 species relative to C_4, which sets off a vicious cycle of decreasing forage productivity. The ecophysiological consequences of climate change for the Campos and Pampas vegetation have received limited attention. Research in Australia (reviewed by McKeon et al., 2008) indicates that rising temperatures, along with increases in summer rainfall, as predicted for 2050 in the Campos and Pampas region (AIACC, 2006), are likely to shift the balance between C_3 and C_4 in favour of C_4. The most certain global change, the increase in atmospheric CO_2 levels, will stimulate growth of both C_3 and C_4 species through the direct effects on photosynthesis, exacerbated by the indirect effects of water saving due to larger stomatal resistance. The result may be changes in vegetation composition and vegetation quality, the latter due to the dilution of protein by more vigorous growth and increases in non-structural carbohydrates. A range of other potential effects, such as soil acidification by enhanced productivity of leguminous species, deeper soil moisture profiles, changes in rooting pattern, and shifts between woody and herbaceous species, render the overall outcome highly uncertain and location specific, as suggested by simulation studies for Queensland, Australia, where sustainable forage harvesting rates changed between +3 to +45% in response to different combinations of 10% changes in temperature, CO_2 and rainfall (Hall et al., 1998).

Long-term experiments in Brazil reviewed by Carvalho et al. (2011) demonstrate how lifting the accompanying limiting factors can increase animal productivity from the current 60 kg meat equivalent per hectare (Table 5.2). They describe 'moderate grazing' as the basis for increasing the productivity of rangelands, with off-take of around 30% of biomass of the preferred species equivalent to a forage offer of 12–14% of animal live weight (Table 5.3). Linking grazing pressure to vegetation production leads to seasonal modulation of stocking rates, with spring as an especially critical period. The result is a more than three times higher production of meat per hectare than under traditional systems. A next intensification step involves greater control over excessive vegetation heterogeneity by maintaining tussock cover to below 30%, which would allow 230 kg meat equivalent per hectare. Up to this point, no additional external inputs are necessary. A third step would involve the addition of the most limiting nutrients, often phosphorus and potassium, and increasing pH by liming. At field level, these measures allow the system to produce 400 kg meat equivalent per hectare. When N is also added, 700 kg meat equivalent per hectare can be produced.

Carvalho et al. (2011) suggest that such measures, which would affect vegetation composition severely, could be taken on specific paddocks to provide a buffer for overcoming periods of forage shortage on the natural pastures in cases where farmers cannot conserve forage. Further interventions by including productive species up to the level of complete re-sowing would allow further production increases, at the sacrifice, however, of much of the ecosystem services the rangelands provide under moderate grazing: increased soil organic matter levels, water infiltration rates and macronutrient (Mg, Ca) concentrations.

The intensification trajectory presented by Carvalho et al. (2011) offers an exciting new outlook on the use of agroecological concepts to support agriculture and conservation, and demonstrates where major trade-offs among values need to be struck

Table 5.2. Potential intensification pathway of rangeland cattle farming in the Campos and Pampas (after Carvalho et al., 2011). Current production levels are around 60 kg meat equivalent per hectare.

Limiting factor	Management action	Estimated system productivity (kg meat equivalent per hectare)
Light interception by vegetation	Seasonal modulation: adjusting stocking rate to pasture productivity, particularly in spring	170
Light utilization by vegetation	Controlling dominance of grass tussocks	230
Nutrient supply (P, K) and pH	Application of fertilizer and lime	400
Nutrient supply (N)	Application of fertilizer	700
Species productivity	Inter-seeding of legumes and grasses with higher production potential	1000
Species productivity	Sown pastures replace rangelands	1500

Table 5.3. Average impact of different levels of forage offer on productivity and carbon sequestration in the Campos biome. Results based on a long-term experiment in Río Grande do Sul (Brazil).

	Forage offer (per cent of live weight)			
Ecosystem service	4	8	12	16
Grassland production (kg DM ha^{-1} yr^{-1})[a]	2705	3488	3723	3393
Live weight gain (kg ha^{-1})[b]	80	120	140	135
Carbon sequestration (kg C ha^{-1} yr^{-1})[c]	113	131	146	156
Diversity index[d]	2.6	3.3	3.5	3.4
Energy efficiency conversion (MJ kg LW^{-1}) gained per hectare[e]	0.009	0.016	0.017	0.013

Notes: [a]Maraschin et al., 1997; [b]Maraschin, 2001; [c]Conceição et al., 2005; [d]Carvalho et al., 2003; Nabinger et al., 2011; [e]Nabinger et al., 2000.

or overcome. What they developed at paddock level now needs to be evaluated at the level of individual farms, each with their own resource endowments and specific constraints. A number of farm-oriented projects in Brazil and Uruguay have been funded, and the challenge is to explore the feasibility and effectiveness of harnessing ecosystem services alongside livelihood requirements. Adapting to climate variability is a strong motivation in these projects, with better soil management both to reduce erosion and to harvest water as key ingredients. In Uruguay, these innovative processes can build on earlier experiences with the co-innovation of more agro-ecologically intensive systems by collaboration between farmers, researchers and other important actors in rural areas (Dogliotti et al., 2013). Whether there is net benefit in a landscape approach to management and marketing, as has been described for the coffee system example, remains a question to be answered.

5.6 Outlook

This chapter began by stating that agriculture was an important driver of predicted global climate change. Although in the short term agricultural activities result in increased climate variability, there is no doubt that agriculture can (and must) be part of the solution to the problem of global warming. Organic and agroecological farming systems (OAFS) offer a promising avenue. Let us examine the key elements that allow OAFS to adapt to and to contribute to mitigate climate change, along with a number of potential hindrances.

OAFS are different from other ways of farming in their purposeful incorporation of diversity across multiple spatial and temporal scales. By using crop rotations, green manures and intercropping, local varieties and breeds, recycling manure and other inputs to organic matter and creating and maintaining an ecological infrastructure surrounding fields and at landscape scale, a variety of ecological niches are used as part of agricultural production. This leads to short-term or tactical flexibility, as well as longer-term or strategic flexibility (Darnhofer et al., 2010). In the examples of coffee and natural grasslands, well-managed multi-species assemblages can provide resilience to external fluctuations in weather, markets, or input supplies, as evidenced by the enhanced recovery after hurricane impact in diverse compared to simplified coffee-based systems. From a longer-term perspective, farms in Meso-America that have diversified coffee with other crops will have more room to manoeuvre when adaptation options to increasing temperatures are insufficient to continue coffee production on at least the lower altitude parts of a farm. This requires farmers to be flexible not only in their choice of crops and cropping practices but also in their relation with markets. In the natural grassland systems, spatial variation in grazing pressure, including leaving a part of the farm ungrazed as a set-aside for winter, and prudent management of vegetation structure to allow the regeneration of productive species are key constituents for an ecological intensification. Most of the meat originating from these smallholder systems is currently sold on the 'bulk' market. Social organization to enable regional branding may contribute to better prices and enhanced economic resilience to climate variability. In both cases, socio-economic and ecological diversity rather than the amount of external inputs constitutes the key to the mitigation of climate change impact in these systems. Note that this required a definition of system performance broader than 'yield level', in line with Walker et al. (2010), who postulated the principle that pursuing narrowly defined efficiency reduces resilience.

How much diversity is needed for adaptability to uncertain climate change? The persistence of the traditional food systems based on maize in Meso-America (e.g. the Milpa system) and in the tropical Andes show that highly diversified systems are capable of adaptation to changes in drivers over centuries (e.g. Pretty et al., 2006). These systems not only supported societies with food but also became part of the sociocultural systems, thus providing a full range of ecosystems services. Low resilience to changing socio-economic context, i.e. strong demographic growth and the advent of an urban-oriented market economy, has in many places prompted system simplification and impoverishment. In the case of Mexico, and since the signature of the NAFTA free trade agreement, maize produced under the climate-resilient Milpa system is being replaced on the national market by maize imports from the USA, making food security dependent on large-scale industrialized agriculture.

However, reducing diversity is not necessarily synonymous with reducing resilience. For instance, reducing shade levels in cacao agroforestry from 80 to 40% led to small decreases in biodiversity while doubling income (Steffan-Dewenter et al., 2007). Further reducing diversity by deforestation – the conventional alternative – may lead to plantation cropping (e.g. of oil palm, cacao, or coffee) that supplies neither local food nor stable income due to global price fluctuations and high risks of soil quality loss or lack of resilience to climate variability (Jackson et al., 2012).

Among the hypotheses recently reviewed to understand landscape effects on biodiversity patterns (Tscharntke et al., 2012), two may be applied to the question of how much diversity is needed for resilience to climate variability and change. The 'dominance of beta diversity hypothesis' postulates that the negative local effects of habitat fragmentation (i.e. decreasing alpha diversity) on overall biodiversity are overridden by the (landscape-moderated) dissimilarity of local communities (i.e. increasing beta diversity). The 'landscape-moderated insurance hypothesis' suggests that complexity in landscape structure and composition provides spatial and temporal insurance, i.e. the higher resilience and stability of ecological processes in changing environments. In brief, the insurance hypothesis states that more landscape diversity is better, while the dominance of the beta diversity hypothesis allows for local simplification to be compensated by diversification in mosaics at the intermediate scale, suggesting that as long as sufficient and sufficiently large centres of diversity remain, functional diversity can be maintained. The implication is that reducing diversity at the local scale should be part of changes managed carefully at landscape level.

In both case studies presented in this chapter, the trend is towards the simplification of systems that traditionally relied on biodiversity for their functioning, and increasing reliance on fossil fuel-based inputs. This resulted in evidence of reduced resilience to climate variability. At the same time, remaining levels of biodiversity and local knowledge on managing diversity are still substantial, providing entry points for the recovery of resilience. The 'intermediate landscape complexity hypothesis' of Tscharntke et al. (2012) suggests that it is in such impoverished landscapes that the greatest gain in biodiversity-related ecosystem services can be expected, as compared to complex landscapes or cleared landscapes.

Organic farming and agroecology, seen as a set of practices and a movement in Latin America, have been considered jointly by taking a perspective focused on practices and aspirations, rather than on regulations. The regulatory framework of organic agriculture in Europe and North America may be restrictive for smallholders in other parts of the world in terms of the costs of certification, and has little sensitivity to local conditions. The development of local participatory certification schemes which allow for the adaptation of generic regulations and local self-control may provide a way out of overly rigid schemes. Policy support for diversified farming systems is indispensable, and the Brazilian government had already decided in 2006 to appoint, alongside the minister for large-scale agriculture, a minister specifically for family farmers, which constitute 85% of farmers. This will stimulate rethinking the land property rights based on minimum stocking density, as the high levels imposed are deleterious to the Campos biome (Carvalho and Batello, 2009). In Uruguay, a Round Table for Rangelands was initiated recently to bring together major stakeholders and develop and spread sustainable grazing systems.

Assessment of the performance of the various land-use systems in the scientific literature, particularly those involving organic agriculture, is often associated with polarization and value judgements (Vanloqueren and Baret, 2009). We subscribe to an increasing number of observations (e.g. Shennan, 2008; Tomich et al., 2011; Kremen and Miles, 2012) that the amount of research funding available for investigating and further developing ecologically diverse strategies has been, and is, substantially less than that for strategies aimed at mainstream industrial production systems. As a result, research data that allow assessment of the adaptation potential of diversified systems are coming available only slowly. In view of the negative environmental and social externalities associated with current production systems, as well as the potential shown by those diversified systems that have received research attention, this should be a concern to research policy makers.

References

AIACC (2006) Final report project LA 27. Climate change and variability in the mixed crop/livestock production systems of the Argentinean, Brazilian and Uruguayan Pampas. Washington, DC, 54 pp (http://www.aiaccproject.org/Final%20Reports/Final%20Reports/FinalRept_AIACC_LA27.pdf, accessed 20 November 2013).

Alliaume, F., Rossing, W.A.H., García, M., Giller, K.E. and Dogliotti, S. (2013) Changes in soil quality and plant available water capacity following systems re-design on commercial vegetable farms. *European Journal of Agronomy* 46, 10–19.

Altieri, M.A. (2002) Agroecology: the science of natural resource management for poor farmers in marginal environments. *Agriculture Ecosystems and Environment* 93, 1–24.

Altieri, M.A. (2004) Agroecology versus eco-agriculture: balancing food production and biodiversity conservation in the midst of social inequity (http://www.wildfarmalliance.org/resources/ECOAG.pdf, accessed 24 April 2013).

Altieri, M.A., Funes-Monzote, F.R. and Petersen, P. (2012) Agroecologically efficient agricultural sysems for smallholder farmers: contributions to food sovereignty. *Agronomy for Sustainable Development* 32, 1–13.

Bellon, M.R., Hodson, D. and Hellin, J. (2011) Assessing the vulnerability of traditional maize seed systems in Mexico to climate change. *Proceedings of the National Academy of Sciences* 108, 13432–13437.

Bellon, S. and de Abreu, L. (2006) Rural social development: small-scale horticulture in São Paulo, Brazil. In: Holt, G.C. and Reed, M. (eds) *Sociological Perspectives of Organic Agriculture: From Pioneer to Policy.* CAB International, Wallingford, UK, pp. 243–259.

Bidegain, M., Crisci, C., del Puerto, L., Inda, H., Mazzeo, N., Taks, J., et al. (2012) Clima de cambios: Nuevos desafíos de adaptación en Uruguay. Project FAO-MGAP. TCP URU/3302 (http://www.fao.org/climatechange/80141/es/, accessed 24 April 2013).

Boldrini, I.I. (2002) Campos sulinos: caracterização e biodiversidade. In: Araújo, E.L., Noura, A.D.N., Sampaio, E.V.S.B., Gestinari, L.M.S. and Carneiro, J.M.T. (eds) *Biodiversidade, Conservação e Uso Sustentável da Flora do Brasil.* Sociedade Botânica do Brasil, Recife, Brazil, pp. 95–97.

Boody, G., Vondracek, B., Andow, D.A., Krinke, M., Westra, J., Zimmerman, J., et al. (2009) Multifunctional agriculture in the United States. *BioScience* 55, 27–38.

Cardoso, I.M., Guijt, I., Franco, F.S., Carvalho, A.F. and Ferreira Neto, P.S. (2001) Continual learning for agroforestry system design: university, NGO and farmer partnership in Minas Gerais, Brazil. *Agricultural Systems* 69, 235–257.

Carvalho, P.C.F. and Batello, C. (2009) Access to land, livestock production and ecosystem conservation in the Brazilian Campos biome: the natural grasslands dilemma. *Livestock Science* 120, 158–162.

Carvalho, P.C.F., Soares, A.B., Garcia, É.N., Boldrini, I.I., Pontes, L.S., Velleda, G.L., et al. (2003) Herbage allowance and species diversity in native pastures. In: *Proceedings of the VIIth International Rangelands Congress.* Durban, South Africa, pp. 858–859.

Carvalho, P.C.F., Nabinger, C., Lemaire, G. and Genro, T.C. (2011) Challenges and opportunities for livestock production in natural pastures: the case of Brazilian Pampa Biome. In: Feldman, S., Oliva, G.E. and Sacido, M.B. (eds) *Proceedings of the IX International Rangeland Congress: Diverse Rangelands for a Sustainable Society,* Rosario, Argentina. Fundación Argentina, Buenos Aires, pp. 9–15.

Cassman, K.G. (1999) Ecological intensification of cereal production systems: yield potential, soil quality, and precision agriculture. *Proceedings of the National Academy of Sciences* 96, 5952–5959.

Castro-Tanzi, S., Dietsch, T., Urena, N., Vindas, L. and Chandler, M. (2012) Analysis of management and site factors to improve the sustainability of smallholder coffee production in Tarrazu, Costa Rica. *Agriculture, Ecosystems and Environment* 155, 172–181.

Conceição, P.C., Amado, T.J.C., Mielniczuk, J. and Spagnollo, E. (2005) Qualidade do solo em sistemas de manejo avaliada pela dinâmica da matéria orgânica e atributos relacionados. *Revista Brasileira de Ciência do Solo* 29, 777–788.

Costanza, R., d'Arge, R., de Groot, R.S., Farber, S., Grasso, M., Hannon, B., et al. (1997) The value of the world's ecosystem services and natural capital. *Nature* 387, 253–260.

Darnhofer, I., Bellon, S., Dedieu, B. and Milestad, R. (2010) Adaptiveness to enhance the sustainability of farming systems. A review. *Agronomy for Sustainable Development* 30, 545–555.

De Groot, R.S. (1992) *Functions of Nature: Evaluation of Nature in Environmental Planning, Management and Decision Making.* Wolters-Noordhoff, Groningen, the Netherlands.

De Sousa, H.N., de Goede, R.G.M., Brussaard, L., Cardoso, I.M., Duarte, E.M.G., Fernandes, R.B.A., et al. (2012) Protective shade, tree diversity and soil properties in coffee agroforestry systems in the Atlantic Rainforest biome. *Agriculture, Ecosystems and Environment* 146, 179–196.

Díaz Nieto, J., Fisher, M., Cook, S., Läderach, P. and Lundy, M. (2012) Weather Indices for designing micro-insurance products for smallholder farmers in the tropics. *PLoS ONE* 7, e38281, doi:10.1371/journal.pone.0038281.

Dogliotti, S., García, M.C., Peluffo, S., Dieste, J.P., Pedemonte, A.J., Bacigalupe, G.F., et al. (2013) Co-innovation of family farm systems: a systems approach to sustainable agriculture. *Agricultural Systems* (http://www.sciencedirect.com/science/article/pii/S0308521X13000280, accessed 25 November 2013).

Doré, T., Makowski, D., Malézieux, E., Munier-Jolain, N., Tchamitchian, M. and Tittonell, P. (2011) Facing up to the paradigm of ecological intensification in agronomy: revisiting methods, concepts and knowledge. *European Journal of Agronomy* 34, 197–210.

DuPont, S.T., Culman, S.W., Ferris, H., Buckley, D.H. and Glover, J.D. (2010) No-tillage conversion of harvested perennial grassland to annual cropland reduces root biomass, decreases active carbon stocks, and impacts soil biota. *Agriculture, Ecosystems and Environment* 137, 25–32.

Eakin, H., Tucker, C. and Castellanos, E. (2006) Responding to the coffee crisis: a pilot study of farmers' adaptations in Mexico, Guatemala and Honduras. *Geography Journal* 172, 156–171.

El-Hage Scialabba, N. and Müller-Lindenlauf, M. (2010) Organic agriculture and climate change. *Renewable Agriculture and Food Systems* 25, 158–169.

Elsner, J.B., Kossin, J.P. and Jaggar, T.H. (2008) The increasing intensity of the strongest tropical cyclones. *Nature* 455, 92–95.

EMBRAPA (Empresa Brasileira de Pesquisa Agropecuária) (2006) *Marco referencial em agroecologia.* Ministério da Agricultura, Pecuária e Abastecimento, Brasília, 70 pp.

FiBL and IFOAM (2013) The world of organic agriculture 2013. Frick and Bonn (http://www.organic-world.net/fileadmin/documents/yearbook/2013/web-fibl-ifoam-2013-25-34.pdf, accessed 29 April 2013).

Francis, C., Lieblein, G., Gliessman, S., Breland, T.A., Creamer, N., Harwood, R., et al. (2003) Agroecology: the ecology of food systems. *Journal of Sustainable Agriculture* 22, 99–118.

Gattinger. A., Muller, A., Haeni, M., Skinner, C., Fliessbach, A., Buchmann, N., et al. (2012) Enhanced top soil carbon stocks under organic farming. *Proceedings of the National Academy of Sciences* 109, 18226–18231.

Gliessman, S.R. (2007) *Agroecology: Ecological Processes in Sustainable Agriculture*, 2nd edn. Lewis Publisher, Boca Raton and New York.

Goh, K.M. (2011) Greater mitigation of climate change by organic than conventional agriculture: a review. *Biological Agriculture and Horticulture: An International Journal for Sustainable Production Systems* 27, 205–229.

Gomiero, T., Pimentel, D. and Paoletti, M.G. (2011) Environmental impact of different agricultural management practices: conventional vs. organic agriculture. *Critical Reviews in Plant Sciences* 30, 95–124.

Hall, W.B., McKeon, G.M., Carter, J.O., Day, K.A., Howden, S.M., Scanlan, J.C., et al. (1998) Climate change and Queensland's grazing lands: II. An assessment of the impact on animal production from native pastures. *Rangeland Journal* 20, 177–205.

Holling, C.S. (2001) Understanding the complexity of economic, ecological, and social systems. *Ecosystems* 4, 390–405.

Holt-Giménez, E. (2002) Measuring farmers'

agroecological resistance after Hurricane Mitch in Nicaragua: a case study in participatory, sustainable land management impact monitoring. *Agriculture, Ecosystems and Environment* 93, 87–105.

Howden, M.S., Soussana, J.-F., Tubiello, F.N., Chhetri, N., Dunlop, M. and Meinke, H. (2007) Adapting agriculture to climate change. *Proceedings of the National Academy of Sciences* 104, 19691–19696.

IFOAM (2013) http://www.ifoam.org/en/organic-landmarks/principles-organic-agriculture (accessed 25 November 2013).

IPCC (Intergovernmental Panel on Climate Change) (2007) Synthesis report. In: Metz, O.R.D., Bosch, P.R., Dave, R. and Meyer, L.A. (eds) *Fourth Assessment Report: Climate Change 2007*. Cambridge University Press, Cambridge, UK.

Jackson, L.E., Pulleman, M.M., Brussaard, L., Bawa, K.S., Brown, G.G., Cardoso, I.M., et al. (2012) Agrobiodiversity and intensification of agriculture: insights from eight landscapes in several biomes. *Global Environmental Change* 22, 623–639.

Kissinger, G., Brasser, A. and Gross, L. (2013) *Scoping Study. Reducing Risk: Landscape Approaches to Sustainable Sourcing*. Landscapes for People, Food and Nature Initiative, Washington, DC.

Kremen, C. and Miles, A. (2012) Ecosystem services in biologically diversified versus conventional farming systems: benefits, externalities, and trade-offs. *Ecology and Society* 17, 40.

Kremen, C., Iles, A. and Bacon, C. (2012) Diversified farming systems: an agroecological, systems-based alternative to modern industrial agriculture. *Ecology and Society* 17, 44.

Laderach, P., Jarvis, A., Ramirez, J., Eitzinger, A. and Ovalle, O. (2009) The implications of climate change on Mesoamerican agriculture and small-farmers' coffee livelihoods. Tropentag 2009, University of Hamburg, 6–8 October 2009 (http://www.tropentag.de/2009/proceedings/, accessed 20 November 2013).

Leifeld, J., Angers, D.A., Chenuc, C., Fuhrer, J., Kätterer, T. and Powlson, D.S. (2013) Organic farming gives no climate change benefit through soil carbon sequestration. *Proceedings of the National Academy of Sciences* 110, E984.

Lin, B.B. (2011) Resilience in agriculture through crop diversification: adaptive management for environmental change. *Bioscience* 61, 183–193.

Lin, B.B., Perfecto, I. and Vandermeer, J. (2008) Synergies between agricultural intensification and climate change could create surprising vulnerabilities for crops. *BioScience* 58, 847–854.

López-Ridaura, S., Masera, O. and Astier, M. (2002) Evaluating sustainability of complex socio-environmental systems, the MESMIS framework. *Ecological Indicators* 2, 135–148.

Lotter, D.W., Seidel, R. and Liebhart, W. (2003) The performance of organic and conventional cropping systems in an extreme climate year. *American Journal of Alternative Agriculture* 18, 146–154.

Luttikholt, L.W.M. (2007) Principles of organic agriculture as formulated by the International Federation of Organic Agriculture Movements. *NJAS – Wageningen Journal of Life Sciences* 54, 347–360.

Lynch, D.H., MacRae, R. and Martin, R.C. (2011) The carbon and global warming potential impacts of organic farming: does it have a significant role in an energy constrained world? *Sustainability* 3, 322–362.

McKeon, G.M., Stone, G.S., Syktus, J.I., Carter, J.O., Flood, N.R., Ahrens, D.G., et al. (2008) Climate change impacts on northern Australian rangeland livestock carrying capacity: a review of issues. *Rangeland Journal* 31, 1–29.

Maraschin, G.E. (2001) Production potential of South American grasslands. In: Gomide, J.A., Mattos, W.R.S. and Silva, S.C. (eds) *Proceedings of the XIX International Grassland Congress*, Piracicaba, Brazil. São Pedro, São Paulo, Brazil, pp. 5–18.

Maraschin, G.E., Moojen, E.L., Escosteguy, C.M.D., Correa, F.L., Apezteguia, E.S., Boldrini, I.I., et al. (1997) Native pasture, forage on offer and animal response. In: Buchanan-Smith, J.G., Bailey, L.D. and McCaughey, P. (eds) *Proceedings of the XVIII International Grassland Congress*, Winnipeg, Canada. Association Management Centre, Saskatoon, Canada, pp. 27–28.

Marengo, J.A., Rusticucci, M., Penalba, O. and Renom, M. (2010) An intercomparison of observed and simulated extreme rainfall and temperature events during the last half of the twentieth century: part 2: historical trends. *Climatic Change* 98, 509–529.

Marengo, J.A., Nobre, C., Tomasella, J., Oyama, M., Sampaio, G., Camargo, H., et al. (2008) The drought of Amazonia in 2005. *Journal of Climate* 21, 495–516.

Marengo, J.A., Liebmann, B., Grimm, A.M., Misra, V., Silva Dias, P.L., Cavalcanti, I.F.A., et al. (2012) Recent developments on the South American monsoon system. *International Journal of Climatology* 32, 1–21.

MEA (Millennium Ecosystem Assessment) (2005) *Ecosystems and Human Well-being: Synthesis*. Island Press, Washington, DC.

Muschler, R. (2001) Shade improves coffee quality in a sub-optimal coffee-zone of Costa Rica. *Agroforestry Systems* 51, 131–139.

Myers, N., Mittermeier, R.A., Mittermeier, C.G., da Fonseca, G.A.B. and Kent, J. (2000) Biodiversity hotspots for conservation priorities. *Nature* 403, 853–858.

Nabinger, C., de Moraes, A. and Maraschin, G.E. (2000) Campos in Southern Brazil. In: Lemaire, G., Hodgson, J., de Moraes, A., Nabinger, C. and Carvalho, P.C. de F. (eds) *Grassland Ecophysiology and Grazing Ecology*. CAB International, Wallingford, UK, pp. 355–376.

Nabinger, C., Ferreira, E.T., Freitas, A.K., Carvalho, P.C.F. and Sant'Anna, D.M. (2009) Produção animal com base no campo nativo: aplicações de resultados de pesquisa. In: Pillar, V.P., Müller, S.C., Castilhos, Z.M.S. and Jacques, A.V.A. (eds) *Campos Sulinos: Conservação e Uso Sustentável da Biodiversidade*. Ministério do Meio Ambiente, Brazil, pp. 214–228.

Nabinger, C., Carvalho, P.C.D.F., Pinto, E.C., Mezzalira, J.C., Brambilla, D.M. and Boggiano, P. (2011) Servicios ecosistémicos de las praderas naturales: ¿es posible mejorarlos con más productividad? Ecosystems services from natural grasslands: is it possible to enhance them with more productivity? *Archivos Latinoamericanos de Producción Animal* 19, 27–34.

Neelin, J.D., Münnich, M., Su, H., Meyerson, J.E. and Holloway, C.E. (2006) Tropical drying trends in global warming models and observations. *Proceedings of the National Academy of Sciences* 103, 6110–6115.

Nuñez, M.N., Solman, S.A. and Cabré, M.F. (2008) Regional climate change experiments over southern South America. II: Climate change scenarios in the late twenty-first century. *Climate Dynamics* 32(7–8), 1081–1095.

Oelofse, M., Høgh-Jensen, H., Abreu, L.S., Almeida, G.F., El-Araby, A., Yu Hui, Q., et al. (2011) Organic farm conventionalisation and farmer practices in China, Brazil and Egypt. *Agronomy for Sustainable Development* 31, 689–698.

Overbeck, G.E., Müller, S.C., Fidelis, A., Pfadenhauer, J., Pillar, V.D., Blanco, C.C., et al. (2007) Brazil's neglected biome: the South Brazilian Campos. *Perspectives in Plant Ecology, Evolution and Systematics* 9, 101–116.

Parrott, N., Olesen, J.E. and Høgh-Jensen, H. (2006) Certified and non-certified organic farming in the developing world. In: Halberg, N., Alrøe, H.F., Knudsen, M.T. and Kristensen, E.S. (eds) *Global Development of Organic Agriculture: Challenges and Promises*. CAB International, Wallingford, UK, pp. 153–179.

Philpott, S.M., Lin, B.B., Jha, S. and Brines, S.J. (2008) A multi-scale assessment of hurricane impacts on agricultural landscapes based on land use and topographic features. *Agriculture, Ecosystems and Environment* 128, 12–20.

Pretty, J.N., Noble, A.D., Bossio, D., Dixon, J., Hine, R.E., Penning de Vries, F.W.T., et al. (2006) Resource-conserving agriculture increases yields in developing countries. *Environmental Science and Technology* 40, 1114–1119.

Renting, H., Rossing, W.A.H., Groot, J.C.J., Van der Ploeg, J.D., Laurent, C., Perraud, D., et al. (2009) Exploring multifunctional agriculture. A review of conceptual approaches and prospects for an integrative transitional framework. *Journal of Environmental Management* 90 Supplement 2, S112–S123.

Roesch, L.F.W., Vieira, F.C.B., Pereira, V.A., Schünemann, A.L., Teixeira, I.F., Senna, A.J.T., et al. (2009) The Brazilian Pampa: a fragile biome. *Diversity* 1, 182–198.

Sala, O.E. and Paruelo, J.M. (1997) Ecosystem services in grasslands. In: Daily, G.C. (ed.) *Nature's Services: Societal Dependence on Natural Ecosystems*. Island Press, Washington, DC, pp. 237–251.

Scherr, S.J. and McNeely, J.A. (2008) Biodiversity conservation and agricultural sustainability: towards a new paradigm of 'ecoagriculture' landscapes. *Philosophical Transactions of the Royal Society B-Biological Sciences* 363, 477–494.

Schroth, G., Laderach, P., Dempewolf, J., Philpott, S., Haggar, J., Eakin, H., et al. (2009) Towards a climate change adaptation strategy for coffee communities and ecosystems in the Sierra Madre de Chiapas, Mexico. *Mitigation and Adaptation Strategies for Global Change* 14, 605–625.

Scoones, I. (1998) Sustainable rural livelihoods: a framework for analysis. IDS Working Paper 72 (http://graduateinstitute.ch/files/live/sites/iheid/files/sites/developpement/shared/developpement/mdev/soutienauxcours0809/Gironde%20Pauvrete/Sustainable%20Rural%20Livelihhods%20-%20Scoones.pdf, accessed 25 November 2013).

Shennan, C. (2008) Biotic interactions, ecological knowledge and agriculture. *Philosophical Transactions of the Royal Society B-Biological Sciences* 363, 717–739.

Solomon, S., Qin, D., Manning, M., Chen, Z., Marquis, M., Averyt, K.B., et al. (eds) (2007) *Climate Change 2007: The Physical Science Basis*. Contribution of Working Group I to the Fourth Assessment Report of the Intergovernmental Panel on Climate Change.

Cambridge University Press, Cambridge, UK, and New York.

Steffan-Dewenter, I., Kessler, M., Barkmann, J., Bos, M.M., Buchori, D., Erasmi, S., et al. (2007) Trade-offs between income, biodiversity, and ecosystem functioning during tropical rainforest conversion and agroforestry intensification. *Proceedings of the National Academy of Sciences* 104, 4973–4978.

TEEB (2010) *The Economics of Ecosystems and Biodiversity Ecological and Economic Foundations*. Kumar, P. (ed.) Earthscan, London and Washington, DC.

Tilman, D., Reich, P.B. and Knops, J.M.H. (2006) Biodiversity and ecosystem stability in a decade-long grassland experiment. *Nature* 441, 629–632.

Toledo, V.M. and Moguel, P. (2012) Coffee and sustainability: the multiple values of traditional shaded coffee. *Journal of Sustainable Agriculture* 36, 353–377.

Tomich, T.P., Brodt, S., Ferris, H., Galt, R., Horwath, W.R., Kebreab, E., et al. (2011) Agroecology: a review from a global-change perspective. *Annual Review of Environment and Resources* 36, 193–222.

Tscharntke, T., Tylianakis, J.M., Rand, T.A., Didham, R.K., Fahrig, L., Batáry, P., et al. (2012) Landscape moderation of biodiversity patterns and processes – eight hypotheses. *Biological Reviews* 87, 661–685.

Tuomisto, H.L., Hodge, I.D., Riordan, P. and Macdonald, D.W. (2012) Does organic farming reduce environmental impacts? A meta-analysis of European research. *Journal of Environmental Management* 112, 309–320.

Vaast, P., Bertrand, B., Perriot, J.J., Guyot, B. and Genard, M. (2006) Fruit thinning and shade improve bean characteristics and beverage quality of coffee (*Coffea arabica* L.) under optimal conditions. *Journal of the Science of Food and Agriculture* 86, 1–197.

Vanloqueren, G. and Baret, P.V. (2009) How agricultural research systems shape a technological regime that develops genetic engineering but locks out agroecological innovations. *Research Policy* 38, 971–983.

Vereijken, P. (1997) A methodical way of prototyping integrated and ecological arable farming systems (I/EAFS) in interaction with pilot farms. *European Journal of Agronomy* 7, 235–250.

Verhoeven, F.P.M., Reijs, J.W. and Van Der Ploeg, J.D. (2003) Re-balancing soil–plant–animal interactions: towards reduction of nitrogen losses. *NJAS – Wageningen Journal of Life Sciences* 51, 147–164.

Vermeulen, S.J., Campbell, B.M. and Ingram, J.S.I. (2012) Climate change and food systems. *Annual Review of Environment and Resources* 37, 195–222.

Walker, B., Sayer, J., Andrew, N.L. and Campbell, B. (2010) Should enhanced resilience be an objective of natural resource management research for developing countries? *Crop Science* 50, 10–19.

Webster, P.J., Holland, G.J., Curry, J.A. and Chang, H.R. (2005) Changes in tropical cyclone number, duration, and intensity in a warming environment. *Science* 309, 1844–1846.

Wezel, A., Bellon, S., Doré, T., Francis, C., Vallod, D. and David, C. (2009) Agroecology as a science, a movement and a practice. A review. *Agronomy for Sustainable Development* 29, 503–515.

Zorn, A., Lippert, C. and Dabbert, S. (2013) An analysis of the risks of non-compliance with the European organic standard: a categorical analysis of farm data from a German control body. *Food Control* 30, 692–699.

6 UK Fruit and Vegetable Production – Impacts of Climate Change and Opportunities for Adaptation

Rosemary Collier[1] and Mark A. Else[2]

[1]Warwick Crop Centre, School of Life Sciences, University of Warwick, Wellesbourne, Warwick, UK; [2]East Malling Research, East Malling, Kent, UK

6.1 Introduction

Outdoor horticultural crops grown in the UK are particularly sensitive to changes in climate due to the impact of increasing temperatures, changing rainfall patterns and increased frequency of extreme events (Knox et al., 2010a). It is clear that climate change will offer both opportunities and threats to UK horticultural production (Knox et al., 2010b). The complex interactions between the variables make accurate predictions of the effects of climate change on agricultural and horticultural production notoriously difficult, and recent predictions in the UK Climate Change Risk Assessment (CCRA) published in January 2012 (CCRA, 2012) have stimulated much debate (e.g. Knox and Wade, 2012; Semenov et al., 2012). This chapter considers the impact of climate change on the production of tree fruit, soft fruit and field vegetables, with a focus on apple, strawberry and the main types of vegetable grown in the UK. Most of these crops are grown 'conventionally' and relatively small areas of land are farmed organically. The multiple retailers are the main market, and they require a consistent supply of high-quality produce. Growers generally supply produce according to a predetermined programme and so they have to plan their cropping schedules carefully to ensure that they are able to fulfil their commitments.

6.2 Fruit Production

The main tree fruit crops grown in the UK are apple, pear and plum, with the area planted to sweet cherry increasing rapidly. These crops are grown mainly by specialist growers located towards the south of the country, on an area of about 18,000 ha. Traditional orchards are being replaced with high-intensity growing systems (see Plate 5), for which irrigation is essential to aid establishment and to deliver the yields and quality needed for a profitable business. The main soft fruit crops are strawberry, raspberry and currant. Again, these are grown mainly by specialists, on an area of about 10,000 ha in the South-east, the east and the west Midlands although a considerable proportion of the crop is grown in Scotland. Most soft fruit produced commercially in the UK is grown under cover. The values of the marketed crop (home production marketed) in 2011 were estimated to be £156 million and £441 million for the tree and soft fruit industries, respectively (Defra, 2012). There is little published information on the predicted effects of climate change on the economics of UK fruit production (Else and Atkinson, 2010; Atkinson et al., 2013).

6.3 Field Vegetable Production

A diverse range of vegetable crops is grown outdoors in the UK on an area of about

120,000 ha. The largest areas are devoted to the production of *Brassica* and *Allium* crops, carrot and lettuce (Defra, 2012), and the main crops are grown over a wide geographical spread, from south-west England to central Scotland. The total value of home production marketed is about £900 million. These crops are grown mainly by specialists, many of whom grow crops in, or source crops from, different parts of the country, and this provides continuity of supply in many instances. Inevitably, due to seasonal variations in temperature and day length, a greater variety of vegetable crops are grown and harvested in the summer than in the winter, but a number of crops are overwintered successfully outdoors, including several types of *Brassica* crop, leek and carrot. Most crops are grown without any form of protection, although polythene or fleece covers are used to protect certain young crops in the spring. Some crops are grown without irrigation, while others are heavily reliant on irrigation to maximize yield and quality. Potato crops are grown on approximately 150,000 ha and have a value of about £500 million (Anon., 2012). Potato production is undertaken in a number of regions and is heavily dependent on irrigation.

6.4 Crop Responses to Elevated CO_2 and Elevated Ozone

The main predicted effects of elevated CO_2 that might occur up to 2050 are to increase photosynthesis and improve the efficiency of water use, with the potential to increase crop yield (Easterling *et al.*, 2007; Defra 2009a), provided that other essential resources such as soil water and nutrient availability are not limiting (see Olesen, Chapter 2, this volume). The ability of crops to benefit also depends on crop genotype (cultivar) and management. Plants grown at elevated CO_2 can have significant compositional differences (higher carbohydrates and lower nitrogen), with consequences for product quality, and fertilizer management programmes, for example, will need to be adjusted to avoid nutrient imbalances. Any increase in ozone levels in the UK is predicted to be small, so the impact will be limited. Specific effects on field vegetable crops are summarized in Table 6.1.

6.5 Effects of Changes in Temperature on Crops

6.5.1 Impacts of temperature on cropping in everbearing strawberry

High temperatures (30/20°C day/night) during the growing season can reduce the cropping potential (thermodormancy) in everbearing strawberry varieties, due to the effects on pollen germination and pollen tube growth (Karapatzak *et al.*, 2012), leading to flower abortion post anthesis (Wagstaffe and Battey, 2006). In the latter study, flower initiation was monitored by crown dissection and was not affected by high temperatures. Interestingly, cool night temperatures (13°C) reduced the severity of thermodormancy, and changes in commercial tunnel venting practices to help maintain lower night temperatures may help to ameliorate the impact of heat-induced reductions in cropping potential in the short term.

6.5.2 Impacts of climate change on dormancy and bud break in perennial fruit

In temperate climates with cold winters, perennial plants avoid low temperature damage during winter and early spring by entering a period of dormancy in which meristematic growth is suspended and physiological processes are halted or slowed. After a specific duration of cold conditions (chilling), endodormancy is broken and the plant is ready to resume growth in spring. To avoid frost damage to developing flowers and leaves, the resumption of growth must be coordinated to coincide with the onset of warmer temperatures, and many perennial fruit crops have a 'heat sum requirement' which has to be met for buds to break. The longer the chill accumulation, the lower the heat sum required. Perennial crop plants

Table 6.1. Responses of crops to CO_2 increase (yield and quality) and elevated ozone. (Sources: see Defra, 2009a.)

Crop	Effect of CO_2 on yield	Effect of CO_2 on quality	Ozone
Potato	Increased dry weights and tuber numbers. Reports of increased tuber numbers and higher proportion of commercially desirable tubers.	Positive impact includes increased starch content and reduced glycoalkaloids. Negative impacts are higher reducing sugars (browning potential) and decreased nitrogen (protein) content in tubers.	Causes foliar injury but comparatively less reduction in yield. Quality effects are small and variable.
Salads/leafy vegetables	Growth is enhanced. For lettuce, final head weight is increased.	Percentage N is decreased in lettuce leaves at elevated CO_2, iron levels increase.	Lettuce is described as an ozone-sensitive crop. Ozone causes loss of yield and visible quality.
Brassica	Limited information but CO_2 increases dry weight of cauliflowers and is likely to do the same for other vegetable brassicas.	Dry weight:fresh weight ratio of curd increased. N content likely to be reduced but no data available.	Broccoli is described as moderately sensitive. Slight reduction in yield but resistant to foliar injury.
Carrot	CO_2 has a positive effect on carrot growth and yield. Increased fixed carbon is allocated preferentially to roots. Dry weight increase is proportionately greater than fresh weight.	Little information available.	Causes chlorosis of the leaves and some loss of yield.
Onion	Fresh weight of onion bulb is increased, and yield increases can be significant.	Bulb/neck ratio is not affected. No information on compositional changes.	Causes flecking and tip senescence in green leaves but minor impact on bulbs. Salad onions may therefore be more susceptible than bulb onions.

must fulfil their chilling and heat sum requirements in order to break dormancy and so they have evolved mechanisms to enable them to sense temperature and to integrate periods of cold and warm over winter (Horvath et al., 2003). Combinations of differences in temperature during the dormant phase and the duration of the chilling period generally satisfy plants' chilling requirements in temperate climates (Jacobs et al., 2002). The effects of insufficient winter chilling on the productivity of perennial fruit crops have been reviewed recently by Campoy et al. (2011), Luedeling (2012) and Atkinson et al. (2013).

In the absence of effective chilling, floral bud development is hampered and anthesis may be protracted and unsynchronized with the life cycles of pollinators. Often, floral structures are of 'poor quality' and fail to attract pollinators or produce viable ovules and/or pollen. This is true for many economically important temperate fruit crops, including apple, pear, plum, sweet cherry, strawberry, raspberry and other non-traditional UK crops such as apricot and peach. It is well known that within each species, different cultivars have different chilling requirements (Guerriero et al., 2010), but the mechanistic basis of these differing requirements remains poorly understood.

The amount of winter chill occurring in the UK has declined and is predicted to continue to do so, based on future climate change scenarios described in the UK Climate Impacts Programme. There is already incontrovertible evidence that climate change is inducing longer growing seasons; for example, bud break of blackcurrant in northern Europe has already advanced by 6 days, while autumn leaf fall has been delayed by 5 days (Sunley et al., 2006). Effective chill accumulation appears to occur only after significant leaf fall and so longer growing seasons could exacerbate the effect of warmer winters on bud break and flower quality due to insufficient chilling (Chandler, 1960). Indeed, the amount of winter chill in the 1990s declined by 12% in Kent, while the number of spring frosts was reduced by 50% (Atkinson et al., 2004). The timing of chill accumulation is also important; low temperatures in early winter appear less effective than low temperatures later in winter, in terms of satisfying chill requirements. Partially chilled apple require a larger heat sum prior to flowering compared with fully chilled trees. The interaction between heat sum requirements and chilling received is also important (Atkinson et al., 2013). Warm autumn temperatures are also likely to delay flowering and reduce flower number in subsequent cropping years in some tree species; pear, for example (Atkinson and Taylor, 1994; Atkinson and Lucas, 1996).

6.5.3 Opportunities to ameliorate the risks associated with reduced winter chill in perennial fruit

There are some opportunities to ameliorate the potentially negative impacts of insufficient chilling on perennial fruit production. Strategies include exploiting the genotypic variability within perennial crops to develop low chill cultivars and changing crop management practices and growing systems to circumvent the effects of periods of warmer weather interrupting effective chilling temperatures. Inclusion of low chilling requirements as an explicit target in breeding programmes would be the most productive approach. However, such efforts are currently hampered by limited understanding of the physiological, molecular and genetic bases of winter chill requirement and dormancy-related environmental factors which affect perennial crop growth, productivity and produce quality (Atkinson et al., 2013). Orchard management practices such as irrigation and shading may influence orchard microclimates favourably, and other cultural practices, such as artificial defoliation, also have the potential for reducing chilling requirements (Luedeling, 2012).

Although lack of winter chilling will have serious repercussions for UK perennial fruit crops, this may be partially compensated for by gains caused by less frequent frosts. Early flowering fruit, such as plum and pear, produced in the UK are well known for their vulnerability to low temperature injury (<3°C) during the early stages of flowering and fruit development. However, such events have declined in recent decades and are predicted to decline further, given the effects of climate change on late winter temperatures (Atkinson et al., 2004; Sunley et al., 2006).

6.5.4 Growth and quality of vegetable crops

The rate of growth and the quality of field vegetable crops is highly dependent on temperature. In addition, spring and autumn temperatures (together with day length) determine the length of the growing season, particularly for frost-sensitive crops such as lettuce. In general, warmer spring and summer weather is likely to extend the growing season and shorten the period from sowing to harvest (Wurr et al., 1996). This will have an impact on crop programmes developed to provide continuity of supply, but since the change in the mean temperature is likely to be gradual, such programmes can be adapted. Programmes can be based on the accumulation of thermal units (day degrees) from sowing or planting to harvest, or on 'effective day degrees' which take account of both temperature and solar radiation (Scaife et al., 1987; Wurr et al., 1988).

The effects of temperature can be more complex in crops such as cauliflower, where distinct phases of growth are affected by temperature in different ways. Cauliflower plants have three phases of growth: a juvenile phase, where plants produce leaves, a curd induction phase and a curd growth phase. Curd induction requires relatively low temperatures and eventually results in curd initiation; high temperatures are likely to delay initiation. Mathematical models have been produced to describe the effect of temperature on juvenility, induction and curd growth. In a study by Wurr et al. (2004), models for the three stages were linked to simulate what might happen to winter cauliflower in different parts of the UK using a range of climate change scenarios. The overall effect was to advance maturity in all situations, except for Roscoff types grown in Cornwall. More recently, the models described by Wurr et al. (2004) were used to investigate the effect of climate change using the Roscoff X Walcheren cross cv Renoir (Collier et al., 2008). Future scenarios were produced by the Long Ashton Research Station-Weather Generator (LARS-WG1) linked with the UK Climate Impacts Programme (UKCIP02) projections of future climate, as described in Semenov (2007). The model was run from seven sequential starting dates defining the beginning of the induction phase, from early July to September. As predictions were made further into the future, there was a trend for the differences in maturity time to diminish, so that by the 2050s, the continuity of production was predicted to be lost.

High temperatures may affect the quality of vegetable crops. For example, broccoli growers are aware that if the temperature rises above 23–25°C for a period of 3–5 days during head development, there can be a loss of quality postharvest. To predict the possible impact of climate change, the incidence of runs of 3 days or more when the maximum temperature would be 23°C or greater over the entire harvest season (from 1 May to 31 October) was calculated based on UKCP09 predictions for the 2030s and 2050s (Defra, 2010). This was done for the three main growing areas in the UK. The increase was most marked in Lincolnshire (Kirton), less in Cornwall (Camborne) and least at Leuchars, the Scottish location (Fig. 6.1). This suggests that, should this factor become critical, moving production areas may be a possible adaptation.

Table 6.2 summarizes some of the key effects of warmer winters and summers on field vegetable crops. In most of the examples given, it should be possible to overcome problems with continuity of supply by changing to varieties with altered temperature requirements; for example, varieties that are more robust with regards to higher growing temperatures. However, further research will be required to identify or breed such adapted varieties. A good example of successful breeding for, in this case, relatively extreme climatic adaptation is the carrot cultivar, Brasilia, which is

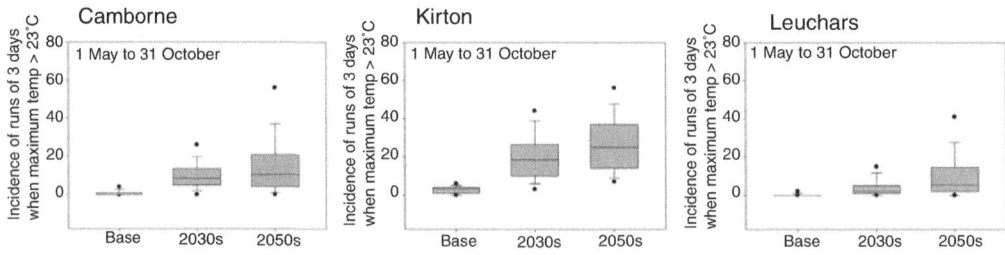

Fig. 6.1. Predicted incidence of 3 days at or above 23°C between May and October for Camborne in Cornwall, Kirton in Lincolnshire and Leuchars in Fife, for base, 2030s and 2050s time slices. (After Defra, 2010.)

Table 6.2. Responses of crops to warmer winters and warmer summers with increased heatwaves. (Sources: see Defra, 2009a.)

Crop	Winter temperature	Summer temperature
Potato	No impact on growth and yield but higher winter temperatures may cause loss of quality in ambient stores.	Warmer summers in high latitudes are associated with higher yields.
Salads/leafy vegetables	Winter salads sourced from overseas.	Crops develop more quickly at higher temperatures and mature earlier. Yield increases for early season plantings but decreases for late summer plantings. Premature bolting can impair quality.
Brassica	Warm winters can delay curd initiation and affect scheduling.	High temperature has a negative impact on yield and quality, e.g. blindness and buttoning. Seed production impaired at high temperatures.
Carrot	Unlikely to have an impact.	Increased temperature has a positive impact on growth and yield, which is enhanced at elevated CO_2.
Onion	Onions are in storage over winter.	Increased temperature accelerates development and reduces yield, offset by elevated CO_2.

adapted to tropical and subtropical conditions atypical for most areas of carrot production (Nascimento et al., 2003).

6.6 Effects of Changes in Precipitation

The effects of drought or limited soil water availability on crop productivity are relatively well known, but the effects of flooding episodes on yields and quality are, perhaps, less widely appreciated. Transient soil flooding and the associated reduction in oxygen availability (hypoxia) in the rooting zone can have serious implications for plant vigour and resilience to subsequent abiotic and biotic stresses. In England and Wales, around 50,000 ha of high-quality horticultural and arable land are at risk of flooding frequently (i.e. at least once every 3 years), and this is projected to increase to around 200,000 ha by the 2080s (or 1% of total agricultural land) (CCRA, 2012). Around 200,000 ha of agricultural land is currently at risk from less frequent flooding (i.e. 1 in 10 year events) and this is projected to increase to about 400,000 ha by the 2050s and over 500,000 ha by the 2080s. Flooding damage to crops may also occur following excessive irrigation, especially on heavier soils, and high water tables following persistent rain can also limit root growth and function.

6.6.1 Impact of soil flooding on fruit crop production

Transient flooding of tree and soft fruit crops during the growing season and longer-term flooding during the dormant season will become more frequent due to the predicted impact of climate change on rainfall patterns. Paradoxically, the most immediate crisis following soil flooding is a decreased ability to take up water; this is evident within a few hours of the start of flooding (Kramer, 1969; Schildwacht, 1989; Else et al., 2001). This lowered root hydraulic conductance may be due to depletion of oxygen (Mees and Weatherley, 1957) and/or accumulation of carbon dioxide (Kramer, 1940; Kitaya et al., 1984; Smit and Stachowiak, 1988) that can occur in the rhizosphere within 1 h of inundation (Else et al., 1995) and is brought about by proton-induced conformational changes to water channel proteins and reduced expression of plasma membrane intrinsic protein (PIP) aquaporin genes (Tournaire-Roux et al., 2003; Rodríguez-Gamir et al., 2011). Shoot dehydration and death quickly follow unless water loss from leaves is minimized. In some species, any reduction in shoot water status caused by the lowered root hydraulic conductance is ameliorated by stomatal closure, which commences within 4 h (Else et al., 1995, 2001) and intensifies over the following 48 h flooding (Jackson et al., 1978; Bradford and Hsiao, 1982). The xylem-borne signal that triggers the onset of stomatal closure in flooded herbaceous and woody plants is not yet known (Else et al., 1996, 2001; Blanke and Cooke, 2004; Rodríguez-Gamir et al., 2011), but a novel anti-transpirant produced in hypoxic tomato roots may help to limit shoot water loss and dehydration and improve tolerance to short-term flooding episodes (Else at al., 2006).

Prolonged flooding during the growing season results in many changes in soil biochemistry (Ponnamperuma, 1972) that can be damaging to plants. These changes result from microbial respiration that utilizes inorganic ions as alternative electron acceptors to oxygen in order to sustain energy generation. Consequently, the rhizosphere of flooded soils becomes chemically reduced and the redox potential is lowered (Gambrell et al., 1991). Potentially phytotoxic organic compounds, such as ethanol and acetaldehyde, are released by stressed plant roots and soil microflora (Drew, 1990), and accumulations of methane, hydrogen sulfide and ethylene may further exacerbate injury to the plant (Setter and Belford, 1990). Under anoxic conditions, plant metabolism switches from respiration to fermentation to generate energy (Drew, 1997; Gibbs and Greenway, 2003). The ability to operate alcoholic fermentation under hypoxic or anoxic conditions is regarded as an essential requirement of the flooding tolerance of plants because energy metabolism is maintained, and NAD+ is regenerated, in spite of the lack of mitochondrial respiration (Kreuzwieser et al., 2004). Nevertheless, an energy deficit often develops (Gibbs and Greenway, 2003) which, combined with a suboptimal supply of nutrients from flooded soils, can inhibit shoot growth. The effect of flooding on growing branches is an important response that could affect fruit yields and quality adversely in subsequent years. Prolonged soil flooding can also increase the susceptibility of roots to soil-borne pathogens (Stolzy and Sojka, 1984). For example, the incidence of Phytophthora root rot increased significantly in Prunus species grown for fruit production when plants were flooded simultaneously or immediately after infection (Wilcox and Mircetich, 1985).

6.6.2 Differing sensitivities of fruit crops to soil flooding

The impact of short-term flooding on fruit crop productivity varies according to species, genotype, duration of flooding, temperature and the stage of crop development (Bailey-Serres and Voesenek, 2008). Strawberry plants appear to be affected more by drought than by flooding (Blanke and Cooke, 2004) and there are very few reports on the responses of raspberry to flooding stress. The flooding sensitivity of fruit trees is mediated mostly by the characteristics of the rootstock; for example, 'Red Delicious' scions on MM.111 and seedling rootstocks were more sensitive to June flooding than those on M.27 and MM.106 rootstocks; trees on M.26 rootstocks were least affected (Rom and Brown, 1979). The same authors reported that flooding during the dormant season had very little effect on plant growth, irrespective of rootstock; however, prolonged winter flooding of Prunus avium resulted in severe damage and death to many trees (see Plates 6 and 7).

6.6.3 Opportunities to ameliorate the risks associated with soil flooding

Although differing sensitivities to soil flooding in fruit tree species have been described (e.g. Rowe and Beardsell, 1973; Anderson et al., 1984), the morphological, physiological and biochemical adaptive responses that confer this improved flooding tolerance are not widely reported. However, tolerance to flooding is often associated with the formation of new adventitious roots on the submerged part of stems that are usually thicker, with more intercellular spaces (lysigenous aerenchyma) than those growing in well-aerated soils (Kozlowski, 1997; Gibberd et al., 2001; Malik et al., 2003), which could be expected to guarantee continued root activity (i.e. mitosis and nutrient uptake) in anaerobic soils (Mancuso and Boselli, 2002). The level of carbohydrate reserves, or the capacity to maintain their transport, may be crucial factors in the tolerance of long-term flooding (Parent et al., 2008). Swelling of submerged portions of the lower shoot and the development of swollen stems and hypertrophied lenticels may also enhance flooding tolerance by promoting tissue gas exchange, and stem hypertrophy occurs in many herbaceous dicots and woody crop species such as apple (Hook and Brown, 1973).

Breeding to develop new hybrids that carry flooding tolerance is under way (Dichio et al., 2004; Xiloyannis et al., 2007; Amador et al., 2009; Pistelli et al., 2012), and linking tolerance with identifiable beneficial phenotypic traits, such as the formation of adventitious roots and aerenchyma, may help guide evaluation and selection processes associated with breeding. Rootstock progenies that segregate for beneficial traits associated with flooding tolerance now need to be phenotyped and the associated quantitative trait loci (QTL) identified and mapped.

In addition to more judicious irrigation scheduling and improved drainage of agricultural land to avoid over-wet soils, there are also agronomic manipulations that conceivably could improve the tolerance of fruit crops to short-term soil flooding. Remedial effects can be observed after applying nitrogen fertilizers as a foliar spray or to the flooded root system; in the latter case, the applied nitrate may enter anaerobically damaged roots by passive means and be translocated to the shoot. The nitrate may also act simply to replace that leached in the drainage water or transformed by anaerobic denitrifying bacteria (Trought and Drew, 1981). Furthermore, arbuscular mycorrhizal infection has been shown to confer limited tolerance to flooding in peach seedlings. This could be due to improved plant nutrition, the suppression of ethanol accumulation in roots and/or the extension of the duration of root activity in flooded soil (Rutto et al., 2002).

6.6.4 Irrigation requirements for tree fruit and soft fruit production

As most soft fruit produced commercially in the UK is grown under cover, irrigation is essential to ensure that quality at market date matches the specifications demanded by retailers and consumers (Defra, 2007). The rapid expansion of soft fruit cropped areas has led to a sustained increase in the demand for irrigation. Soft fruit growers abstracted 7 million m^3 of water in 2005 (Weatherhead, 2006), which is relatively low compared to other horticultural sectors such as field vegetables and hardy nursery stock. However, the underlying growth in the volume of water applied (Knox et al., 2009) and the increase in cropped area since 2005 means that annual water abstraction by the soft fruit sector is probably nearer to 10 million m^3 (Else and Atkinson, 2010).

Irrigation is essential for the successful establishment and continued productivity of high-intensity tree fruit growing systems, and so although water abstraction by the tree fruit industry was estimated to be 0.7 million m^3 in 2005 (Weatherhead, 2006), the switch towards more intensive growing systems over the past 10 years means that the sector is increasingly reliant on irrigation to deliver the yields and quality demanded by retailers and consumers.

6.6.5 Future water availability for irrigation of UK fruit crops

A combination of greater crop water use and lower summer rainfall will undoubtedly increase the demand for irrigation in regions of the UK that are already suffering from water stress (Environment Agency, 2009). The major tree fruit and soft fruit growing areas in England are in the south-east and the east and in the west Midlands, where public, industrial and agricultural demands on water supplies are already high. Less water is available per capita in the south-east of England than in parts of the Mediterranean (Environment Agency, 2009), and the predicted rise in summer temperatures and extended growing seasons will exacerbate the situation. It is estimated that 76% and 84% of tree fruit and soft fruit growers, respectively, currently farm in areas already classified as being under water stress (Knox et al., 2009). Abstraction rates in the major fruit growing areas are currently unsustainable and are predicted to rise by a further 30% by 2050 (King et al., 2006). Higher temperatures and longer growing seasons will increase crop evapotranspiration rates, which, when combined with the lower summer rainfall, will place increasing pressure on finite freshwater supplies, especially in the south-east and east of England. Estimated increases in agricultural water demand in the 2050s in England and Wales range from 25% to 180% of current demand (Weatherhead et al., 2008, cited by Knox et al., 2012). A useful indicator of aridity is the potential soil moisture deficit (PSMD), which represents the balance between rainfall and potential crop water use over the year. Forecasts of changes in PSMD in the major tree fruit and soft fruit growing regions in England suggest there will be an increasing dependence on irrigation (Else and Atkinson, 2010). Other estimates indicate that, in the south-east, the average annual maximum PSMD may increase, with deficits that currently occur every 5 years occurring every 2 years by 2080, and deficits that currently occur every 15 years occurring every 5 years (Wade and Counsell, 2013).

Matching plant demand for water with supply can be difficult in changeable summer weather and at different stages of crop development. If growers are to maintain or increase yields against a backdrop of increasing summer temperatures, dwindling water supplies and governmental demands for greater environmental protection, new production methods based on outputs from scientific research that improve water- and fertilizer-use efficiency are needed.

6.6.6 Improved irrigation scheduling

The efficiency with which growers use their irrigation water will vary widely within each sector and one contributory factor is the ability to schedule irrigation effectively. Significant water savings can be achieved without reducing Class 1 yields when some form of scientific irrigation scheduling tool is used (Else and Atkinson, 2010). Identifying and promoting current industry 'best practice' via benchmarking is a vital first step towards improving on-farm water-use efficiencies (Knox et al., 2013).

6.6.7 Opportunities to ameliorate the risks associated with limited soil water availability

In the short term, there are opportunities to improve irrigation system design and increase the consistency of delivery, to increase the water-holding capacity of soils by incorporating organic matter such as green wastes (Else et al., 2011) and to identify and promote 'irrigation best practice' across the fruit industry (ADAS, 2012). In the medium term, deficit irrigation offers another attractive management option in areas where water resources are already limited and likely to worsen. The techniques involve applying less water than the plant needs at each irrigation event. Regulated deficit irrigation (RDI) and partial rootzone drying (PRD) are the two most widely used forms of deficit irrigation and offer great potential to optimize water-use efficiency while maintaining or improving marketable yields and quality, provided that they are applied and maintained judiciously (Dodds et al., 2007). In the long term, pre-breeding to improve resource-use efficiency is a key component of the UK Government's strategy to deliver sustainable intensification of UK agriculture and horticulture. The selection of new soft fruit cultivars with an improved tolerance to moderate soil drying would provide a longer-term opportunity to improve the economic and environmental sustainability of UK soft fruit production. Several strawberry breeding programmes continue to release new cultivars with desirable attributes. However, performance under low-input regimes is generally not assessed during the selection process, and so cultivars that crop consistently under water- and fertilizer-saving regimes are not always identified and selected.

6.6.8 Impact of wet soils on vegetable production

As with fruit, flooding for more than a short period can have adverse impacts on the yield and quality of vegetable crops. In addition, wet conditions can prevent machinery from getting on to the land to prepare the ground for sowing or planting, or to harvest crops. For example, during 2000, heavy rainfall late in the season caused widespread problems for potato lifting. By Christmas 2000, 20% of the crop (25,000 ha) was still in the ground. In contrast, in a year such as 2003, where lifting went to plan, the whole crop was lifted for storage by early November (Collier *et al.*, 2008).

6.6.9 Impact of limited soil water availability on vegetable production

Very dry ground can cause problems with growth and prevent sowing or transplanting. During the summer of 2006, there were periods when it was impossible to transplant horticultural crops in the UK because the soil was bone-dry and the weather was very hot. Even where irrigation was available, the demand for water from the transplants made it impossible to proceed (Collier *et al.*, 2008).

For transplanted crops such as lettuce, irrigation is important for crop establishment and most vegetable crops require irrigation to maximize yield and quality. Of the total amount of water used for crop irrigation in England in 2005, 56% was used for the irrigation of potato crops and 27% for the irrigation of vegetables (Weatherhead, 2006). For potato, half of the irrigation applied is used to control common scab (*Streptomyces* spp.), while the other half is required for optimizing tuber yield (Defra, 2007). The production of *Brassica* crops has been a major exception to date, in that many crops do not receive irrigation either after transplanting or during growth. This is particularly the case for crops grown in the silt soils of south Lincolnshire.

The same concerns about the impact of climate change on the availability of water for irrigation apply to vegetables as to fruit. Vegetables (including potatoes) are grown in some regions of the UK where future water supplies are likely to be constrained. One approach may be to alter the distribution of such crops, with crops that require high amounts of irrigation being grown only in areas where water is available readily. However, this may not be easy because production occurs in certain areas for other reasons, such as soil type and favourable temperatures. In addition, a considerable infrastructure has been developed in these areas to support the production, processing and distribution of these crops. Water is also used to wash and process vegetables, although the total amount is relatively small in comparison to total use (Defra, 2007). There is considerable interest for vegetable production, as with other sectors, in reducing water use (Defra, 2007, 2009b).

6.7 Pests and Pathogens

The yield and quality of fruit and vegetable crops grown in the UK can be affected adversely by a wide range of pests and pathogens. The main vertebrate pests are birds, rabbits, hares and deer. However, the impact of climate change on these organisms will not be considered in this chapter. The key groups of pest insects that infest fruit and vegetable crops in the UK are the larvae of flies, moths and butterflies, adult and larval beetles, aphids, bugs and thrips. Crops can also be infested by mites, molluscs (slugs and snails) and nematodes.

6.7.1 Effects of changes in temperature on invertebrate pests

Most invertebrate pests have temperature-dependent rates of development and particular threshold temperatures defining the temperature range within which development can occur. Some pest species may be able to complete more generations with higher temperatures (if these are below lethal high temperatures). For other species, the number of generations may be limited by factors such as an obligate diapause (a period of suspended development during unfavourable environmental conditions), although activity may still occur earlier in a warm year. An example of this would be beetle pests of *Brassica* crops such as flea beetles and pollen beetles (*Meligethes* spp.), which overwinter in the adult stage and complete one generation a year. There is a number of models that can be used to predict generation times, which range from relatively simple day-degree models based on linear regression (e.g. Butts and McEwen, 1981; Pitcairn *et al.*, 1992) to more complex process-based simulation models (e.g. Phelps *et al.*, 1993), and these may be used to predict the effects of climate change.

Insect survival is affected by biotic and abiotic factors, but extremes of rainfall or temperature often have a major impact. Most species overwinter in the UK and their sensitivity to winter conditions, principally low temperature but also rainfall, depends on the robustness of the overwintering stage. *Delia radicum* (cabbage root fly) pupae overwinter in diapause and can survive extremely low temperatures (the species is also endemic in North America), while aphids that overwinter in the active stages are more vulnerable.

Most fruit and vegetable crops are colonized by one or more species of aphid. There is abundant information about the effects of temperature on aphids and also about the effects of global change (Hullé *et al.*, 2010). For aphids such as peach-potato aphid (*Myzus persicae*) that can overwinter as active adults and nymphs (anholocyclic life cycle), warmer conditions in winter lead to earlier and larger spring migrations to crop hosts (Harrington *et al.*, 2007). For species that overwinter as eggs in diapause, often on a woody host (holocyclic life cycle), warmer conditions may also lead to earlier infestation, as a result of more rapid post-diapause development. It is possible to forecast the spring migration of pest aphids, although the approach varies depending on the overwintering strategy adopted, and these forecasts can be used to predict the impact of climate change on phenology and, in some cases, abundance. In general, increases in mean temperature are likely to lead to earlier infestations, more generations and a potential increase in population size. In terms of the possible impact of climate change, dates by which the first alate *M. persicae* would be captured in the Rothamsted suction trap located at Kirton in Lincolnshire were predicted to be 9 days earlier by the 2020s, and 20 days earlier by the 2050s, than the date of 26 May predicted from the baseline data (Collier *et al.*, 2008).

Forecasting models developed for non-aphid pests of fruit and vegetable crops can also be used to assess the impact of climate change. A program for simulating the patterns of egg laying by populations of *D. radicum* was used to model the effects of climate change on future infestations by *D. radicum* (Collier *et al.*, 1990). An increase of 3°C in mean daily temperature would cause the cabbage root fly to become active about a month earlier compared with the baseline. Under such conditions, the emergence of flies from the overwintering population would be less synchronized, as the completion of diapause and post-diapause development would occur at the same time in different individuals within the population. However, there would continue to be only three generations of fly each year, even in the south of England. Temperature increases of 5°C or 10°C would lead to four generations each year, and aestivation (summer dormancy in the pupal stage) would seriously disrupt the pattern of egg laying.

The diamond-back moth (*Plutella xylostella*) is a pest of *Brassica* crops and is of global significance. At present, this species does not survive UK winters in large numbers and new infestations are generally the result of migration from continental Europe and possibly further south (Chapman et al., 2002). This species, and other migrant Lepidoptera, generally originate from due south of the UK and will have flown over many kilometres of open sea. Wind direction, and high-altitude wind in particular, is important (Chapman et al., 2002). Overall, there has been increased migration of moths and butterflies to the UK during the last two decades, and recent research has linked this to a pattern of rising temperatures in south-west Europe (Sparks et al., 2007). The study suggests that for every 1°C rise in temperature in south-west Europe, 14 new species of moth or butterfly can be expected to arrive on the south coast of the UK, although, obviously, most of these will never become pests.

The possible effects of climate change on *P. xylostella* infestations in the UK were investigated using data from published studies (Collier et al., 2008). Predictions were made using three different times for the start of egg laying. First, the assumption was made that the majority of moths still migrated into the UK from abroad and there were two scenarios: eggs laid on 1 June (close to the current situation) or eggs laid 1 month earlier (due to climate change). The third scenario assumed that, in future, the moths would overwinter successfully in the UK and start laying eggs from 1 February onwards if temperatures were high enough. As egg laying occurred earlier, and predictions were made further into the future, the number of generations completed within the calendar year increased. *P. xylostella* is a difficult pest to control, because of its relatively short generation time and its propensity to develop resistance to insecticides; if it is able to start development earlier and complete additional generations in a year, it may become more of a threat to UK crops in the future.

Extreme temperatures and the increased frequency of heatwaves will undoubtedly be deleterious to some current UK pests and, if sufficiently high, cause mortality. Temporary exposure of populations to extreme temperature may induce dormancy and delay the subsequent generation (Harrison and Barlow, 1972; Finch and Collier, 1985). While temperature optima are often studied, temperatures above 30°C are not tested routinely for temperate organisms. Consequently, there are limited data published about the likely effects of periods of unusually warm weather on UK pests.

6.7.2 Effects of changes in precipitation on invertebrate pests

The effects of water (drought or periods of heavy rainfall) on pest populations will depend entirely on the species and the timing of the event. Water-intolerant pests, which may prosper in periods of drought, include cutworms (*Agrotis segetum*). Wet conditions, as a result of heavy rainfall, cause mortality of the early larval instars (Esbjerg, 1988). A descriptive population model for *A. segetum* (Bowden et al., 1983) was run using 150 years of synthetic weather data for two locations (Collier et al., 2008). The model estimates larval survival to third instar based on temperature–rate relationships, and larval mortality is attributed to daily rainfall. In both locations, the predicted numbers of surviving batches increase into the future (2020s and 2050s).

Slugs are important pests of a number of horticultural crops including potato, and their development and activity benefits from high levels of moisture. A simple model was developed to forecast the number of days when slugs might be active using a threshold of ≥ 2 mm of water remaining after potential evapotranspiration was subtracted from the daily rainfall, and a minimum temperature ≥ 5°C (Defra, 2010). The predictions were made using UKCP09 projections. The output indicated that the higher spring and autumn rainfall in Cornwall provided more days that were favourable for slug activity than in the other locations studied (Lincolnshire and Angus). There is no indication that this number of days will vary appreciably between the 2030s and the 2050s, with the possible exception of a slight increase for Angus in the 2050s. Thus, it is predicted that no additional adaptation responses to the changing climate will be required beyond the control measures necessary under current conditions.

6.7.3 Effects of changes in CO_2 levels on invertebrate pests

Although temperature and rainfall are key factors determining insect phenology and abundance, insects are also affected by the environment through their host plants. Increases in CO_2 concentration stimulate plant growth but can decrease the nutritional quality of plants for insects (Lincoln et al., 1993).

6.7.4 Effects on plant pathogens

The key pathogens of fruit and vegetable crops are fungi, bacteria and viruses, and as would be expected, the severity of plant disease epidemics is affected greatly by the weather, especially temperature and rainfall (Evans et al., 2008).

6.7.5 Effects of changes in temperature on plant pathogens

Warmer winters will increase the winter survival of plant pathogens, and generally higher temperatures throughout the year will accelerate pathogen life cycles, thereby increasing, for example, sporulation and infection efficiency of fungal foliar pathogens (Coakley et al., 1999; Harvell et al., 2002). Similarly, *Fusarium oxysporum* f. sp. *cepae*, which is a soilborne fungus that infects the roots and basal plate of onion bulbs, causing bulb rot, and can also cause damping off and delayed emergence of onion seedlings, is likely to become a greater problem in the future, according to current climate change models (Cramer, 2000). This is because fusarium basal rot is favoured by high temperatures, with an optimum temperature of 28–32°C.

6.7.6 Effects of changes in precipitation on plant pathogens

The availability of free water is extremely important in the life cycle of many microorganisms. For fungi, both infection and sporulation often require close to 100% relative humidity. Such moist conditions occur most commonly overnight. Therefore, a favourable temperature in this particular time period is important (Harvell et al., 2002). There is often an interaction between the conditions of temperature and wetness that allow the development of disease. Hong et al. (1996) showed that when oilseed rape pods or leaves were inoculated with spore suspensions of *Alternaria brassicae* (a pathogen which also infects horticultural *Brassica* crops), more favourable wetness conditions compensated to a certain extent for less favourable temperature conditions, and vice versa. So, for example, the development of lesions took the same amount of time (4 days) following exposure to 12 h wetness at 15°C as for 24 h wetness at 10°C.

Several approaches have been used to model the effects of climate change on plant disease (Garrett et al., 2006). However, there has been limited research in this respect on the pathogens that infect horticultural crops in the UK. Forecasting models have been developed for a number of important fungal pathogens of fruit and vegetable crops: for example, *Botrytis cinerea* on strawberry (Xu et al., 2000); *Venturia inaequalis* and *Podosphaera leucotricha* on apple (Berrie and Xu, 2003; Xu and Robinson, 2005); *Peronospora destructor* ((De Visser, 1998; Gilles et al., 2004); *Sclerotinia sclerotiorum* (Clarkson et al., 2007); *B. squamosa* (Clarkson et al., 2000); and several fungal pathogens of *Brassica* (Evans et al., 2008; Minchinton et al., 2013). Some of these are based on the type of data described for *A. brassicae* above. One reason for the limited use of models to predict the impact of climate change may be that the effects of free water are often incorporated into forecast models using measurements of leaf wetness rather than rainfall (e.g. Minchinton et al., 2013). Unfortunately, there is no simple mechanism to calculate leaf wetness from the rainfall data predicted by a number of climate change models. Consequently, it is more difficult to integrate the climate models with pathogen models to predict the effects of climate change on diseases. One exception is a model for phoma (*Leptosphaeria maculans*) (Evans et al., 2008), which is a fungal disease of oilseed rape and other *Brassica* crops, although currently relatively unimportant in horticultural brassicas. The models to describe the development of phoma stem canker epidemics were based on field data rather than laboratory studies and used inputs of temperature and rainfall. The start of phoma leaf spotting in autumn predicted for 2020 and 2050, respectively, was only 5–10 or 10–15 days earlier than in the 1960–1990 (baseline) period. As the start of leaf spotting is dependent on both temperature and rainfall, the effects of increasing summer temperature were counteracted by the effects of decreasing summer rainfall. However, there was a large effect of predicted climate change on the start of phoma stem canker in spring, with predicted dates often 80 days earlier than during 1960–1990.

6.7.7 Effects of climate change on dispersal of plant pathogens

Pathogen dispersal may also be affected by climate change. For example, rain splash is used by many pathogens as a passive means of physical dispersal of spores or other propagules (e.g. Pielaat et al., 2002), and such dispersal may be affected by a change in the frequency, distribution and intensity of rainfall events.

For those pathogens which are dispersed by invertebrate vectors, then the effect of weather conditions on the phenology and abundance of the vector is important for virus transmission (already discussed for aphids which are common vectors of plant viruses). However, temperature can have additional effects on the interaction. For example, temperature can affect the retention time of non-persistently transmitted viruses on aphid mouthparts, with implications for the spread of viral epidemics (Jurik et al., 1987). In terms of overwintering, the success of viruses will depend on the amount and condition of overwintering hosts and the survival of vectors.

6.7.8 Effects of changes in CO_2 levels on plant pathogens

Increased levels of CO_2 can have an impact on a plant host and its pathogens in multiple ways (Coakley et al., 1999). Some of the observed effects on disease may counteract others. Researchers have shown that the higher growth rates of leaves and stems of plants grown under high CO_2 concentrations may result in denser canopies with higher humidity that favour pathogens. Lower plant decomposition rates observed at higher CO_2 concentrations could increase the amounts of crop residue on which disease organisms can overwinter, resulting in higher inoculum levels at the beginning of the growing season. Pathogen growth itself can be affected by higher CO_2 concentrations, resulting in greater fungal spore production. However, increased CO_2 can also result in physiological changes to the host plant that can increase host resistance to pathogens.

6.7.9 New pests and diseases

Invasive species will respond to climate change, and their responses will have ecological and economic implications (e.g. Hellman et al., 2008). The advent of milder winters and warmer summers, more typical currently of other parts of Europe, has implications for the survival and reproduction of new pests and diseases. Non-indigenous pests and diseases may initially become established in protected crops under glass. Despite national and international restrictions and procedures, sometimes unavoidable circumstances occur that allow new pests and diseases to establish and spread. A current, and very new, threat to soft fruit crops in Europe is *Drosophila suzukii* (Calabria *et al.*, 2012), and this has been found recently in the UK (J. Cross, personal communication). Future novel crops may be sources of pests and diseases new to the UK, some of which may have host ranges which include current UK crops.

6.7.10 Effects of climate change on the ability to control pests and diseases

The most economically important pests and diseases are often controlled through combinations of monitoring and forecasting, applications of pesticides, host-plant resistance, use of scheduling times and good husbandry. In considering the effects of climate change, it should not be forgotten that this might also affect the efficacy of current control measures.

Control with pesticides can be affected by weather; for example, high temperature is reported to reduce the effectiveness of some chemical controls (Palm, 1975; Paiva *et al.*, 1995) or increase others (Grafius, 1986; Yadwad and Kallapur, 1988; Zhu *et al.*, 2006). Humidity levels can also modify the efficacy of some pesticides (Imai *et al.*, 1995), as can the timing and the amount of rain following the application of pesticides (Suss *et al.*, 1994). On a simpler level, rain can affect the ability to apply the control measures at the time of most need; for example, when wet conditions encourage the development of pest or disease populations but prevent the application of sprays.

Pest and pathogen populations may themselves develop resistance to chemical or other control measures. The ability to overcome resistance may carry a fitness cost to the pest or pathogen which might be influenced by climate change. For example, *M. persicae* with esterase-based insecticide resistance are less likely to survive a cold, wet winter (Foster *et al.*, 1996, 1997).

The effectiveness of host-plant resistance genes can break down under changed climatic conditions. There are many examples of resistance breakdown at high temperatures (e.g. Eisbein and Haack, 1985; Kuginuki *et al.*, 1991; Ellis *et al.*, 1994; Huang *et al.*, 2006; Li *et al.*, 2006). Moreover, humidity can also influence this breakdown (Wang *et al.*, 2005).

Control by natural enemies has not been used to a great extent yet in fruit or field vegetable crops. However, natural enemies will have their own climate optima, although not necessarily the same as their hosts (Campbell *et al.*, 1974; Islam and Chapman, 2001). It is pertinent to ask whether climate change will affect the enemies to the same extent/in the same direction as the pests they are intended to control. For example, pea aphid (*Acyrthosiphon pisum*) is more resistant to infection by the entomopathogenic fungus, *Erynia neoaphidis*, at 28°C than at 18°C (Stacey and Fellowes, 2002). Similarly, experiments showed that the effects of temperature on the feeding rate of a ladybird predator, *Coccinella septempunctata*, and the reproduction rate of a prey species, *A. pisum*, are different. Above 10°C, the consumption rate of the ladybird exceeded the reproduction rate of the aphid, whereas below this temperature, aphid numbers built up more quickly than could be checked by the predator (Dunn, 1952). These examples indicate how difficult it will be to predict how climate change will impinge on the complex web of biotic interactions occurring in outdoor crops.

6.8 Conclusions

The potential impacts of climate change on the fruit and vegetable industries in the UK are complex, but it is clear that many of the current challenges facing these industries will be exacerbated. For perennial fruit crops, the lack of sufficient winter chilling, the reduced winter mortality of pests and pathogens and the increased risk of spring and autumn frost damage may limit productivity in some regions, while the longer growing season could reduce the winter soil water recharge period in the major fruit growing areas which are already under water stress (Environment Agency, 2009). Similarly, for field vegetables, changes in the mean temperature could affect production adversely through impacts on the crop and on pests and diseases, and increased incidences of periods of drought could be damaging to crops grown without irrigation. Nevertheless, higher temperatures, lower cloud cover and longer growing seasons may increase potential primary production, and these conditions are already providing opportunities to extend the production season for some crops and to grow non-hardy perennial crops such as peach and apricot that have hitherto been unsuited to the UK climate.

More frequent extreme summer rainfall events may help to reduce the intensity of some foliar pests and pathogens, but short-term and more prolonged soil flooding and associated root death during sensitive crop developmental stages may increase susceptibility to summer droughts and soilborne pathogens, as well as reducing crop yields and quality. Soil wetness and flooding are predicted to increase in winter throughout the UK, and wetter soils will reduce the accessibility of land and increase the risk of soil damage and erosion; an increased incidence of soil flooding will also reduce root survival and fruit tree stability.

In some ways, the need to anticipate and adapt to climate change is more urgent for perennial crops than for annual crops, and fruit growers need to become more aware of the potential impact of warmer winters and temporary soil flooding to help the industry adapt and continue to deliver good yields of quality fruit. For tree fruit, the reduction in chilling for perennial fruit tree crops has the potential to limit production. The soft fruit industry faces similar challenges. For both the soft fruit and vegetable industries, limited water availability for irrigation will further jeopardize production. However, the uptake of new approaches to irrigation scheduling and better water management will provide opportunities. For the vegetable industry, yield and quality are both important, but so is continuity of supply, and extreme weather events such as droughts and flooding that disrupt continuity will be a major threat in the future. For all crops, there may be potential 'genetic' solutions, but the identification of useful traits and their incorporation into commercially viable varieties takes a good deal of time. It is therefore imperative, from all points of view, that the fruit and vegetable industries plan now so that they are able to adapt to climate change and continue to deliver good yields of quality produce in a changing climate.

References

ADAS (2012) Diffuse Pollution: Best Practice Advisory Programme for Soft Fruit in the South East Region. Report 26477 produced by ADAS on behalf of the Environment Agency, Bristol, UK.

Amador, M.L., Sancho, S. and Rubio-Cabetas, M.J. (2009) Biochemical and molecular aspects involved in waterlogging tolerance in *Prunus* rootstocks. *Acta Horticulturae* 814, 715–720.

Andersen, P.C., Lombard, P.B. and Westwood, M.N. (1984) Leaf conductance, growth, and survival of willow and deciduous fruit tree species under flooded soil conditions. *Journal of the American Society of Horticultural Science* 109, 132–138.

Anon. (2012) Agriculture in the United Kingdom 2012. National Statistics. Produced by Department for Environment, Food and Rural Affairs, Department of Agriculture and Rural Development (Northern Ireland), The Scottish Government, Rural and Environment Research and Analysis Directorate, Welsh Assembly Government, The Department for Rural Affairs and Heritage (https://www.gov.uk/government/collections/agriculture-in-the-united-kingdom, accessed 16 December 2013).

Atkinson, C.J. and Lucas, A.S. (1996) The response of flowering date and cropping of *Pyrus communis* cv. Concorde to autumn warming. *Journal of Horticultural Science* 71, 427–434.

Atkinson, C.J. and Taylor, L. (1994) The influence of autumn temperature on flowering time and cropping of *Pyrus communis* cv. Conference. *Journal of Horticultural Science* 69, 1067–1075.

Atkinson, C.J., Sunley, R.J., Jones, H.G., Brennan, R. and Darby, P. (2004) Desk study on winter chill in fruit. Department of Environment, Food and Rural Affairs CTC 0206, London.

Atkinson, C.J., Brennan, R.M. and Jones, H.G. (2013) Declining chilling and its impact on temperate perennial crops. *Environmental and Experimental Botany* 91, 48–62.

Bailey-Serres, J. and Voesenek, L.A.C.J. (2008) Flooding stress: acclimations and genetic diversity. *Annual Review of Plant Biology* 59, 313–339.

Berrie, A.M. and Xu, X. (2003) Managing apple scab (*Venturia inaequalis*) and powdery mildew (*Podosphaera leucotricha*) using Adem™. *International Journal of Pest Management* 49, 243–249.

Blanke, M.M. and Cooke, D.T. (2004) Effects of flooding and drought on stomatal activity, transpiration, photosynthesis, water potential and water channel activity in strawberry stolons and leaves. *Plant Growth Regulation* 42, 153–160.

Bowden, J., Cochrane, J., Emmett, B.J., Minall, T.E. and Sherlock, P.L. (1983) A survey of cutworm attacks in England and Wales, and a descriptive population model for *Agrotis segetum* (Lepidoptera: Noctuidae). *Annals of Applied Biology* 102, 29–47.

Bradford, K.J. and Hsiao, T.C. (1982) Stomatal behaviour and water relations of waterlogged tomato plants. *Plant Physiology* 70, 1508–1513.

Butts, R.A. and McEwen, F.L. (1981) Seasonal populations of the diamondback moth, *Plutella xylostella* (Lepidoptera: Plutellidae), in relation to day-degree accumulation. *Canadian Entomologist* 113, 127–131.

Calabria, G., Máca, J., Bächli, G., Serra, L. and Pascual, M. (2012) First records of the potential pest species *Drosophila suzukii* (Diptera: Drosophilidae) in Europe. *Journal of Applied Entomology* 136, 139–147.

Campbell, A., Frazer, B.D., Gilbert, N., Gutierrez, A.P. and Mackauer, M. (1974) Temperature requirements of some aphids and their parasites. *Journal of Applied Ecology* 11, 431–438.

Campoy, J.A., Ruiz, D. and Egea, J. (2011) Dormancy in temperate fruit trees in a global warming context: a review. *Scientia Horticulturae* 130, 357–372.

CCRA (2012) The UK Climate Change Risk Assessment 2011: Evidence Report. Defra, London.

Chandler, W.H. (1960) Some studies of the rest in apple trees. *Proceedings of the American Society of Horticultural Science* 76, 1–10.

Chapman, J.W., Reynolds, D.R., Smith, A.D., Riley, J.R., Pedgley, D.E. and Woiwod, I.P. (2002) High altitude migration of the diamondback moth *Plutella xylostella* to the UK: a study using radar, aerial netting, and ground trapping. *Ecological Entomology* 27, 641–650.

Clarkson, J.P., Kennedy, R. and Phelps, K. (2000) The effect of temperature and water potential on the production of conidia by sclerotia of *Botrytis squamosa*. *Plant Pathology* 49, 119–128.

Clarkson, J.P., Phelps, K., Whipps, J.M., Young, C.S., Smith, J.A. and Watling, M. (2007) Forecasting Sclerotinia disease on lettuce: a predictive model for carpogenic germination of *Sclerotinia sclerotiorum* sclerotia. *Phytopathology* 97, 621–631.

Coakley, S.M., Scherm, H. and Chakraborty, S. (1999) Climate change and plant disease management. *Annual Review of Phytopathology* 37, 399–426.

Collier, R., Fellows, J., Adams, S., Semenov, M. and Thomas, B. (2008) Vulnerability of horticultural crop production to extreme weather events. *Aspects of Applied Biology* 88, 3–13.

Collier, R.H., Finch, S., Phelps, K. and Thompson, A.R. (1990) Possible impact of global warming on cabbage root fly (*Delia radicum*) activity in the UK. *Annals of Applied Biology* 118, 261–271.

Cramer, C. (2000) Breeding and genetics of fusarium basal rot resistance in onion. *Euphytica* 115, 159–166.

De Visser, C.L.M. (1998) Development of a downy mildew advisory model based on Downcast. *European Journal of Plant Pathology* 104, 933–943.

Defra (2007) Opportunities for reducing water use in agriculture. Final Report for Defra project WU0101. Defra, London.

Defra (2009a) Scoping study on the potential impact of environmental factors associated with climate change on major UK crops. Final Report DEFRA Project AC0309. Defra, London.

Defra (2009b) Identification and knowledge transfer of novel and emerging technology with the potential to improve water use efficiency within English and Welsh agriculture. Defra Project WU0123. Defra, London.

Defra (2010) Climate change impacts and adaptation – a risk based approach. Final Report DEFRA Project AC0310. Defra, London.

Defra (2012) Defra Basic Horticultural Statistics 2012. Defra, London.

Dichio, B., Xiloyannis, C., Celano, G. and Vicinanza, L. (2004) Performance of new selections of *Prunus* rootstocks, resistant to root knot nematodes, in waterlogging conditions. *Acta Horticulturae* 658, 403–406.

Dodds, P.A.A., Taylor, J.M., Else, M.A., Atkinson, C.J. and Davies, W.J. (2007) Partial rootzone drying increases antioxidant activity in strawberries, Proceedings of the 1st International Symposium on Human Health Effects of Fruits and Vegetables. *Acta Horticulturae* 744, 295–302.

Drew, M.C. (1990) Sensing soil oxygen. *Plant, Cell and Environment* 13, 681–693.

Drew, M.C. (1997) Oxygen deficiency and root metabolism: injury and acclimation under hypoxia and anoxia. *Annual Review of Plant Physiology and Plant Molecular Biology* 48, 223–250.

Dunn, J.A. (1952) The effect of temperature on the pea aphid–ladybird relationship. Second report of the National Vegetable Research Station, Wellesbourne, UK.

Easterling, W.E., Aggarwal, P.K., Batima, P., Brander, K.M., Erda, L., Howden, S.M., et al. (2007) Food, fibre and forest products. In: Parry, M.L., Canziani, O.F., Palutikof, J.P., van der Linden, P.J. and Hanson, C.E. (eds) *Climate Change 2007: Impacts, Adaptation and Vulnerability*. Contribution of Working Group II to the Fourth Assessment Report of the Intergovernmental Panel on Climate Change. Cambridge University Press, Cambridge, UK, pp. 273–313.

Eisbein, K. and Haack, I. (1985) Changes in the resistance behaviour of spinach towards a strain of cucumber mosaic virus under the influence of higher temperatures. *Archiv fur Phytopathologie und Pflanzenschutz* 21, 411–413.

Ellis, P.R., Pink, D.A.C. and Ramsey, A.D. (1994) Inheritance of resistance to lettuce root aphid in the lettuce cultivars 'Avoncrisp' and 'Lakeland'. *Annals of Applied Biology* 124, 141–151.

Else, M.A. and Atkinson, C.J. (2010) Climate change impacts on UK top and soft fruit production. *Outlook on Agriculture* 39, 257–262.

Else, M.A., Davies, W.J., Malone, M. and Jackson, M.B. (1995) A negative hydraulic message from oxygen-deficient roots of tomato plants. *Plant Physiology* 109, 1017–1024.

Else, M.A., Tiekstra, A.E., Davies, W.J., Croker, S. and Jackson, M.B. (1996) Stomatal closure in flooded tomato plants involves abscisic acid and a chemically unidentified anti-transpirant in xylem sap. *Plant Physiology* 112, 239–247.

Else, M.A., Coupland, D., Dutton, L. and Jackson, M.B. (2001) Decreased root hydraulic conductivity reduces leaf water potential, initiates stomatal closure and slows leaf expansion in flooded plants of castor oil (*Ricinus communis*) despite diminished delivery of ABA from the roots to shoots in xylem sap. *Physiologia Plantarum* 111, 46–54.

Else, M.A., Taylor, J.M. and Atkinson, C.J. (2006) Anti-transpirant activity in xylem sap from flooded tomato (*Lycopersicon esculentum* Mill.) plants is not due to pH-mediated redistributions or root- or shoot-sourced ABA. *Journal of Experimental Botany* 57, 3349–3357.

Else, M.A., Dodds, P.A.A. and Wood, M. (2011) The effects of quality composts on field-grown strawberry. Final Report for WRAP project OAV023-006 (http://www.wrap.org.uk/sites/files/wrap/OAV023-006_EMRStrawberry_Final Report_101125_FORMATTED.pdf, accessed 16 December 2013).

Environment Agency (2009) *Water for People and the Environment, Water Resources Strategy for England and Wales*. Environment Agency, Bristol, UK.

Esbjerg, P. (1988) Behaviour of 1st- and 2nd-instar cutworms (*Agrotis segetum* Schiff.) (Lep., Noctuidae): the influence of soil moisture. *Journal of Applied Entomology* 105, 295–302.

Evans, N., Baierl, A., Semenov, M.A., Gladders, P. and Fitt, B.D.L. (2008) Range and severity of a plant disease increased by global warming. *Journal of the Royal Society Interface* 5, 525–531.

Finch, S. and Collier, R.H. (1985) Laboratory studies on aestivation in the cabbage root fly (*Delia radicum*). *Entomologia Experimentalis et Applicata* 38, 137–143.

Foster, S.P., Harrington, R., Devonshire, A.L., Denholm, I., Devine, G.J., Kenward, M.G., et al. (1996) Comparative survival of insecticide-susceptible and resistant peach-potato aphids, *Myzus persicae* (Sulzer) (Hemiptera: Aphididae), in low temperature field trials. *Bulletin of Entomological Research* 86, 17–27.

Foster, S.P., Harrington, R., Devonshire, A.L., Denholm, I., Clark, S.J and Mugglestone, M.A. (1997) Evidence for a possible fitness trade-off between insecticide resistance and the low temperature movement that is essential for survival of UK populations of *Myzus persicae* (Hemiptera: Aphididae). *Bulletin of Entomological Research* 87, 573–579.

Gambrell, R.P., DeLaune, R.D. and Patrick, W.H. Jr (1991) Redox processes in soils following oxygen depletion. In: Jackson, M.B., Davies, D.D. and Lambers, H. (eds) *Plant Life Under Oxygen Deprivation. Ecology, Physiology and Biochemistry*. SPB Academic, The Hague, the Netherland, pp. 101–117.

Garrett, K.A., Dendy, S.P., Frank, E.E., Rouse, M.N. and Travers, S.E. (2006) Climate change effects on plant disease: genomes to ecosystems. *Annual Review of Phytopathology* 44, 489–509.

Gibberd, M.R., Gray, J.D., Cocks, P.S. and Colmer, T.D. (2001) Waterlogging tolerance among a diverse range of *Trifolium* accessions is related to root porosity, lateral root formation and 'aerotropic rooting'. *Annals of Botany* 88, 579–589.

Gibbs, J. and Greenway, H. (2003) Mechanisms of anoxia tolerance in plants. I. Growth, survival and anaerobic catabolism. *Functional Plant Biology* 30, 1–47.

Gilles, T., Phelps, K., Clarkson, J.P. and Kennedy, R. (2004) Development of MILLIONCAST, an improved model for predicting downy mildew sporulation on onions. *Plant Disease* 88, 695–702.

Grafius, E. (1986) Effects of temperature on pyrethroid toxicity to Colorado potato beetle (Coleoptera: Chrysomelidae). *Journal of Economic Entomology* 79, 588–591.

Guerriero, R., Viti, R., Iacona, C. and Bartolini, S. (2010) Is apricot germplasm capable of withstanding warmer winters? This is what we learned from last winter. *Acta Horticulturae* 862, 265–272.

Harrington, R., Hullé, M. and Plantegenest, M. (2007) Monitoring and forecasting. In: van Emden, H.F. and Harrington, R. (eds) *Aphids as Crop Pests*. CAB International, Wallingford, UK, pp. 515–536.

Harrison, J.R. and Barlow, C.A. (1972) Population-growth of the pea aphid, *Acyrthosiphon pisum* (Homoptera: Aphididae) after exposure to extreme temperatures. *Annals of the Entomological Society of America* 65, 1011–1015.

Harvell, C.D., Mitchell, C.E., Ward, J.R., Altizer, S., Dobson, A.P., Ostfeld, R.S., *et al.* (2002) Climate warming and disease risk for terrestrial and marine biota. *Science* 296, 2158–2162.

Hellmann, J.J., Byers, J.E., Bierwagen, B.G. and Dukes, J.S. (2008) Five potential consequences of climate change for invasive species. *Conservation Biology* 22, 534–543.

Hong, C.X., Fitt, B.D.L. and Welham, S.J. (1996) Effects of wetness period and temperature on development of dark pod spot (*Alternaria brassicae*) on oilseed rape (*Brassica napus*). *Plant Pathology* 45, 1077–1089.

Hook, D.D. and Brown, C.L. (1973) Root adaptations and relative flood tolerance of five hardwood species. *Forest Science* 19, 225–229.

Horvath, D.P., Anderson, J.V., Chao, W.S. and Foley, M.E. (2003) Knowing when to grow: signals regulating bud dormancy. *Trends in Plant Science* 8, 534–540.

Huang, Y.J., Evans, N., Li, Z.Q., Eckert, M., Chevre, A.M., Renard, M., *et al.* (2006) Temperature and leaf wetness duration affect phenotypic expression of *Rlm6*-mediated resistance to *Leptosphaeria maculans* in *Brassica napus*. *New Phytologist* 170, 129–141.

Hullé, M., Coeur d'Acier, A., Bankhead-Dronnet, S. and Harrington, R. (2010) Aphids in the face of global changes. *Comptes Rendus Biologies* 333, 497–503.

Imai, T., Tsuchiya, S. and Fujimori, T. (1995) Humidity effects on activity of insecticidal soap for the green peach aphid, *Myzus persicae* (Sulzer) (Hemiptera, Aphididae). *Applied Entomology and Zoology* 30, 185–188.

Islam, S.S. and Chapman, R.B. (2001) Effect of temperature on predation by Tasmanian lacewing larvae. In: Zydenbos S.M. (ed.) *New Zealand Plant Protection*. New Zealand Plant Protection Society, Rotorua, New Zealand, pp. 244–247.

Jackson, M.B., Gales, K. and Campbell, D.J. (1978) Effect of waterlogged soil conditions on the production of ethylene and on water relationships. *Journal of Experimental Botany* 29, 183–193.

Jacobs, J.N., Jacobs, G. and Cook, N.C. (2002) Chilling period influences the progression of bud dormancy more than does chilling temperature in apple and pear shoots. *Journal of Horticultural Science* 77, 333–339.

Jurik, M., Mucha, V. and Vorosova, T. (1987) The effects of temperature and some other factors on retention of non-persistent viruses by aphids. *Biologia (Bratislava)* 42, 315–318.

Karapatzak, E.K., Wagstaffe, A., Hadley, P. and Battey, N.H. (2012) High-temperature-induced reductions in cropping in everbearing strawberries (*Fragaria ananassa*) are associated with reduced pollen performance. *Annals of Applied Biology* 161, 255–265.

King, J., Tiffin, D., Drakes, D. and Smith, K. (2006) Water use in agriculture: establishing a baseline. Final Report for Defra WU0102. Defra, London.

Kitaya, Y., Yabuki, K. and Kiyota, M. (1984) Studies on the control of the gasceous environment in the rhizosphere. 2. Effect of carbon dioxide in the rhizosphere on growth of cucumber. *Journal of Agricultural Meterology* 40, 119–124.

Knox, J.W. and Wade, S. (2012) Commentary: Assessing climate risks to UK agriculture. *Nature Climate Change* 2, 378 (doi:10.1038/nclimate1538).

Knox, J.W., Weatherhead, E.K., Rodriguez-Diaz, J.A. and Kay, M.G. (2009) Developing a strategy to improve irrigation efficiency in a temperate climate: a case study in England. *Outlook on Agriculture* 38, 303–309.

Knox, J.W., Morris, J. and Hess, T.M. (2010a) Identifying the future risks to UK agricultural crop production – putting climate change in context. *Outlook on Agriculture* 39, 249–256.

Knox, J.W., Rodriguez-Diaz, J.A., Weatherhead, E.K. and Kay, M.G. (2010b) Development of a water strategy for horticulture in England and Wales. *Journal of Horticultural Science and Biotechnology* 85, 89–93.

Knox J.W., Hurford, A., Hargreaves, L. and Wall, E. (2012) Climate change risk assessment for the agriculture sector. Final Report for Defra project GA0204. Defra, London.

Knox J.W., Daccache, A., Hess, T.M., Else, M.A., Kay, M., Burton, M., et al. (2013) Benchmarking agricultural water use and productivity in key commodity crops. Final Report for Defra project WU0122.

Kozlowski, T.T. (1997) Responses of woody plants to flooding and salinity. *Tree Physiology* Monograph 1, 1–29.

Kramer, P.J. (1940) Causes of decreased absorption of water by plants in poorly aerated media. *American Journal of Botany* 27, 216–220.

Kramer, P.J. (1969) *Plant and Soil Water Relationships: A Modern Synthesis*. McGraw Hill, London.

Kreuzwieser, J., Papadopoulou, E. and Rennenberg, H. (2004) Interaction of flooding with carbon metabolism of forest trees. *Plant Biology* 6, 299–306.

Kuginuki, Y., Yoshikawa, H. and Yui, S. (1991) Degradation of clubroot resistance in Chinese cabbage. Effect of temperature and daylength. *Cruciferae Newsletter* 4–5, 140–141.

Li, H., Smyth, F., Barbetti, M.J. and Sivasithamparam, K. (2006) Relationship between *Brassica napus* seedling and adult plant responses to *Leptosphaeria maculans* is determined by plant growth stage at inoculation and temperature regime. *Field Crop Research* 96, 428–437.

Lincoln, D.E., Fajer, E.D. and Johnson, R.H. (1993) Plant–insect herbivore interactions in elevated CO_2 environments. *Trends in Ecology and Evolution* 8, 64–68.

Luedeling, E. (2012) Climate change impacts on winter chill for temperate fruit and nut production: a review. *Scientia Horticulturae* 44, 219–229.

Malik, A.I., Colmer, T.D., Lambers, H., Setter, T.L. and Schortemeyer, M. (2003) Aerenchyma formation and radial O_2 loss along adventitious roots of wheat with only the apical root portion exposed to O_2-deficiency. *Plant Cell and Environment* 26, 1713–1722.

Mancuso, S. and Boselli, M. (2002) Characterisation of the oxygen fluxes in the division, elongation and mature zones of *Vitis* roots: influence of oxygen availability. *Planta* 214, 767–774.

Mees, G.C. and Weatherley, P.E. (1957) The mechanism of water absorption by roots II. The role of hydrostatic pressure gradients across the cortex. *Proceedings of the Royal Society Series B* 14, 381–391.

Minchinton, E.J., Auer, D.P.F., Thomson, F.M., Trapnell, L.N., Petkowski, J.E., Galea, V., et al. (2013) Evaluation of the efficacy and economics of irrigation management, plant resistance and Brassicaspot™ models for management of white blister on Brassica crops. *Australasian Plant Pathology* 42, 169–178.

Nascimento, W.M., Vieira, J.V. and Alvares, M.C. (2003) Physiological maturity of carrot seeds cv. Alvorada under tropical conditions. *Acta Horticulturae* 607, 49–51.

Paiva, E.A.S., Picanco, M.C., Corte, M.L. and Castro Gava, G.J.D. (1995) Variation in toxicity of deltamethrin and methamidophos to *Brevicoryne brassicae* (L., 1758) (Hemiptera: Aphididae) at different air temperatures. *Cientifica (Jaboticabal)* 23, 325–329.

Palm, G. (1975) Problems every year in the control of red spider mite (*Panonychus ulmi*). *Mitteilungen des Obstbauversuchringes des Alten Landes* 30, 303–306.

Parent, C., Capelli, N., Berger, A., Crèvecoeur, M. and Dat, J.F. (2008) An overview of plant responses to soil waterlogging. *Plant Stress* 2, 20–27.

Phelps, K., Collier, R.H., Reader, R.J. and Finch, S. (1993) Monte Carlo simulation method for forecasting the timing of pest insect attacks. *Crop Protection* 12, 335–342.

Pielaat, A., van den Bosch, F., Fitt, B.D.L. and Jeger, M.J. (2002) Simulation of vertical spread of plant diseases in a crop canopy by stem extension and splash dispersal. *Ecological Modelling* 151, 195–212.

Pistelli, L., Iacona, C., Miano, D., Cirilli, M., Colao, M.C., Mensuali-Sodi, A., et al. (2012) Novel *Prunus* rootstock somaclonal variants with divergent ability to tolerate waterlogging. *Tree Physiology* 32, 355–368.

Pitcairn, M.J., Zalom, F.G. and Rice, R.E. (1992) Degree-day forecasting of generation time of cydia-pomonella (Lepidoptera, Tortricidae) populations in California. *Environmental Entomology* 21, 441–446.

Ponnamperuma, F.N. (1972) The chemistry of submerged soils. *Advances in Agronomy* 24, 29–96.

Rodriguez-Gamir, J., Ancillo, G., Gonzalez-Mas, M.C., Primo-Millo, E., Iglesias, D.J. and Forner-Giner, M.A. (2011) Root signalling and modulation of stomatal closure in flooded citrus seedlings. *Plant Physiology and Biochemistry* 49, 636–645.

Rom, C. and Brown, S.A. (1979) Water tolerance of apples on clonal rootstocks and peaches on seedling rootstocks. *Compact Fruit Tree* 12, 30–33.

Rowe, R.N. and Beardsell, D.V. (1973) Waterlogging of fruit trees. *Horticultural Abstracts* 43, 533–548.

Rutto, K.L., Mizutani, F. and Kadoya, K. (2002) Effect of root-zone flooding on mycorrhizal and non-mycorrhizal peach (*Prunus persica* Batsch) seedlings. *Scientia Horticulturae* 94, 285–295.

Scaife, A., Cox, E.F. and Morris, G.E.L. (1987) The relationship between shoot weight, plant density and time during the propagation of four vegetable species. *Annals of Botany* 59, 325–334.

Schildwacht, P.M. (1989) Is a decreased water potential after withholding oxygen from the roots the cause of the decline of leaf-elongation rates in *Zea mays* L. and *Phaseolus vulgaris* L.? *Planta* 177, 178–184.

Semenov, M.A. (2007) Development of high-resolution UKCIP02-based climate change scenarios in the UK. *Agricultural and Forest Meteorology* 144, 127–138.

Semenov, M.A., Mitchell, R.A.C., Whitmore, A.P., Hawkesford, M.J., Parry, M.A.J. and Shewry, P.R. (2012) Commentary: Shortcomings in wheat yield predictions. *Nature Climate Change* 2, 380–382.

Setter, T. and Belford, B. (1990) Waterlogging: how it reduces plant growth and how plants can overcome its effects. *Western Australian Journal of Agriculture* 31, 51–55.

Smit, B. and Stachowiak, M. (1988) Effects of hypoxia and elevated carbon dioxide concentration on water flux through *Populus* roots. *Tree Physiology* 4, 153–165.

Sparks, T.H., Dennis, R.L.H., Croxton, P.J. and Cade, M. (2007) Increased migration of Lepidoptera linked to climate change. *European Journal of Entomology* 104, 139–143.

Stacey, D.A. and Fellowes, M.D.E. (2002) Influence of temperature on pea aphid *Acyrthosiphon pisum* (Hemiptera: Aphididae) resistance to natural enemy attack. *Bulletin of Entomological Research* 92, 351–357.

Stolzy, L.H. and Sojka, R.E. (1984) Effects of flooding on plant disease. In: Kozlowski, T.T. (ed.) *Flooding and Plant Growth*. Academic, Orlando, Florida, pp. 221–264.

Sunley, R.J., Jones, H.G., Atkinson, C.J. and Brennan, R.M. (2006) Phenology and yield modelling: the impacts of climate change on UK blackcurrant varieties. *Journal of Horticultural Science and Biotechnology* 81, 949–958.

Suss, A., Jahn, M. and Klementz, D. (1994) Model investigations on the effect of rain on the efficacy and residue levels of selected insecticides and fungicides. *Nachrichtenblatt des Deutschen Pflanzenschutzdienstes* 46, 97–104.

Tournaire-Roux, C., Sutka, M., Javot, H., Gout, E., Gerbeau, P., Luu, D.T., *et al.* (2003) Cytosolic pH regulates root water transport during anoxic stress through gating of aquaporins. *Nature* 425, 393–397.

Trought, M.C.T. and Drew, M.C. (1981) Alleviation of injury to young wheat plants in anaerobic solution culture in relation to the supply of nitrate and other inorganic nutrients. *Journal of Experimental Botany* 32, 509–522.

Wade, S. and Counsell, C. (2013) Climate and the demand for water for horticulture and agriculture: summary report. HR Wallingford report commissioned by Kent County Council/Environment Agency/UKWIR, April 2013. HR Wallingford, Wallingford, UK.

Wagstaffe, A. and Battey, N.H. (2006) Characterisation of the thermodormancy response in the everbearing strawberry 'Everest'. *Journal of Horticultural Science and Biotechnology* 81, 1086–1092.

Wang, C., Cai, X. and Zheng, Z. (2005) High humidity represses Cf-4/Avr4- and Cf-9/Avr9-dependent hypersensitive cell death and defense gene expression. *Planta* 222, 947–956.

Weatherhead, E.K. (2006) Survey of irrigation of outdoor crops in 2005: England and Wales. Cranfield University, Cranfield, UK.

Wilcox, W.F. and Mircetich, S.M. (1985) Effects of flooding duration on the development of *Phytophthora* root and crown rots of cherry. *Phytopathology* 75, 1451–1455.

Wurr, D.C.E., Fellows, J.R. and Suckling, R.F. (1988) Crop continuity and prediction of maturity in the crisp lettuce variety Saladin. *Journal of Agricultural Science Cambridge* 111, 481–486.

Wurr, D.C.E., Fellows, J.R. and Phelps, K. (1996) Investigating trends in vegetable crop response to increasing temperature associated with climate change. *Scientia Horticulturae* 66, 225–263.

Wurr, D.C.E., Fellows, J.R. and Fuller, M.P. (2004) Simulated effects of climate change on the production pattern of winter cauliflower. *Scientia Horticulturae* 101, 359–372.

Xiloyannis, C., Dichio, B., Tuzio, A.C., Kleinhentz, M., Salesses, G., Gomez-Aparisi, J., *et al.* (2007) Characterization and selection of *Prunus* rootstocks resistant to abiotic stresses: waterlogging, drought and iron chlorosis. *Acta Horticulturae* 732, 246–251.

Xu, X., Harris, D.C. and Berrie, A.M. (2000) Modeling infection of strawberry flowers by *Botrytis cinerea* using field data. *Phyopathology* 90, 1367–1374.

Xu, X.M. and Robinson, J. (2005) Modelling the effects of wetness duration and fruit maturity on infection of apple fruits of Cox's Orange Pippin and two clones of Gala by *Venturia inaequalis*. *Plant Pathology* 54, 347–356.

Yadwad, V.B. and Kallapur, V.L. (1988) Influence of temperature on knock-down and mortality to fenitrothion in the three lepidopteran species of insects. *Insect Science and its Application* 9, 531–534.

Zhu, Y.J., Sengonca, C. and Liu, B. (2006) Toxicity of biocide GCSC-BtA on arthropod pests under different temperature conditions. *Journal of Pest Science* 79, 89–94.

7 Intensive Livestock Systems for Dairy Cows

Robert J. Collier, Laun W. Hall and John F. Smith*

University of Arizona, Tucson, Arizona, USA

7.1 Introduction

The objectives of intensive livestock systems are to take advantage of scale effects to maximize profitability, to provide a uniform thermoneutral environment and consistent nutrition in order to maximize production output and to reduce the impacts of adverse environmental conditions. The use of intensive livestock systems is increasing, and will continue to do so for the immediate future because they are essential to achieving increases in animal productivity. However, the proper construction and management of these systems presents several challenges to producers, who must consider several factors including management of the microenvironments inside the facility, maximizing the efficiency of the labour, capital and nutrients required, as well as waste disposal in the form of waste water and manure.

There is now a strong scientific consensus that climate change is occurring and it is projected that the global average temperature will likely rise an additional 1.1–5.4°C over the next century (CCSP, 2008). These changes will have large and measurable impacts on animal productivity worldwide (Klinedinst et al., 1993) through a variety of routes including changes in food availability and quality, changes in pest and pathogen populations, alteration in immunity and both direct and indirect impacts on animal performance such as growth, reproduction and lactation. As productivity increases with continued genetic, nutritional and management improvement, the sensitivity of high-producing animals to heat stress increases. Currently, heat stress causes an estimated US$1.7 billion in losses in animal production each year (St-Pierre et al., 2003).

In modern dairy facilities, the objective is to achieve consistent high milk production, feed efficiency and reproductive efficiency while maintaining the health of the dairy cow. Heat stress reduces the intake, milk production, health and reproduction of dairy cows. Spain et al. (1998) showed that lactating cows under heat stress decreased feed intake by 6–16% as compared to thermal neutral conditions. Holter et al. (1996) reported heat stress depressed the intake of cows more than heifers. Other studies have reported similar results. In addition to a reduction in feed intake, there is also a 30–50% reduction in the efficiency of energy utilization for milk production (McDowell et al., 1969). In many parts of the world, milk production and reproductive performance decline drastically during periods of heat stress. Heat stress also has a negative impact on a dairy farm's future by reducing the peak milk production of cows that go through the transition period during periods of heat stress. The impact of reduced peak milk production often lingers into late autumn or early winter. Cows can be managed and cooled to minimize the impact of heat stress. The method used will vary depending on the severity of the climate and the ambient relative humidity. In modern dairy facilities, it is essential to minimize variation in the cow's core body temperature during periods of heat stress

*Deceased

to maximize milk production and reproductive performance.

Heat stress in intensive animal production is a worldwide issue with tremendous economic impact. Differences in core body temperature, metabolic rate and management requirements dictate differences in housing and management practices required to maintain animal productivity under adverse environmental conditions. The basal metabolism between different animals used in production varies. These animal types include poultry (layers and broilers – turkeys and chickens), swine, sheep, goats, beef cattle and dairy cattle.

Within a given species, responses to heat stress will differ based on variables such as age, hair or feather type, colour, breed, gender and stage in production. Environmentally induced hyperthermia occurs when an animal is unable to dissipate excess heat due to changes in ambient conditions. Animals undergo physiological changes induced by hyperthermia that are broadly classified as acclimation. This process is under endocrine regulation and is identified as homeorhetic in nature in that the metabolism of the animal is altered to support the requirement to acclimate to the stress. Over many generations under constant conditions of heat stress, these physiological changes become genetically fixed and we refer to these animals as being 'adapted' to their environment. Examples are cattle of Bos indicus breeds, which have adapted to conditions of high temperature and humidity as opposed to Bos taurus cattle, which have not.

Heat stress imposes dramatic challenges to the metabolism of high-producing animals and forces changes in many metabolic pathways as the animal attempts to maintain homeostasis while adjusting its productive output. The process of acclimation to thermal stress is homeorhetic and involves changes in tissue sensitivity to homeostatic regulators of the endocrine system, as well as changes in the secretion rate of some of these regulators too. Also, in some cases, acclimation involves the use of alternate metabolic pathways not yet fully described (Baumgard and Rhoads, 2012).

The maintenance of a homeothermic state within an organism has high metabolic costs. Environmental factors such as high temperature with or without increased levels of humidity, wind and solar exposure can impair the ability to maintain homeostasis and increases the need for metabolic fuel.

7.2 Seasonal Production Patterns

Pronounced seasonal patterns of milk yield and composition are evident in cattle. These seasonal patterns are largely induced by climatological variables, breed effects and management factors such as feed quality and reproductive management. The month of parturition is known to have a pronounced impact on subsequent milk yield and composition. The highest yields occur following January and February parturition, while the lowest yields occur following calvings in August and September. The seasonal pattern in milk yield is related to the direct and indirect effects of environment on milk production. The direct effects are related to the impacts of elevated temperature on milk yield; the indirect effects are due to the impact of the photoperiod and the negative consequence of heat stress, during late pregnancy, on maternal and fetal metabolism and the circulating plasma endocrine patterns, which are altered by the stress (Collier et al., 1982). There is also a seasonal pattern in milk protein that parallels the seasonal pattern in milk yield (Collier et al., 2010). Interestingly, the milk protein yield pattern appears to be affected more directly by temperature, as the nadir occurs during the hottest part of the summer (June and July), as opposed to the milk yield curve, which has its nadir in October. The protein yield pattern may reflect the need for the production of heat-shock proteins by mammary epithelial cells during periods of heat stress, which would reduce milk protein synthesis rates, while the milk yield curve displays both direct and carry-over effects related to the indirect effects on the pregnancy and the metabolic state of the cow.

The majority of studies published on the climatic effects on milk composition and yield have evaluated the effects of temperature. Dairy cattle are sensitive to heat stress because of the high metabolic heat production during feed intake associated with rumen fermentation and milk yield. Likewise, for the same reasons, dairy cattle are relatively resistant to cold stress. Heat stress in cattle is characterized by increased rectal temperature, elevated respiration rates and decreased feed intake, which subsequently decreases milk yield. The environmental temperature range from −5 to 23.9°C has little impact on milk yield, and composition and is referred to as the thermoneutral zone for the lactating dairy cow. However, temperatures above 23.9°C are known to decrease solids-not-fat (SNF), protein, lactose and the fat percentage of milk. Due to its involvement in the osmotic regulation of milk, the impact of temperature on lactose and the mineral content of milk is much smaller than the impact of temperature on protein and fat yields. Generally, in temperate regions, the fat content may average 0.4% lower and the protein content 0.2% lower in summer as compared to the winter months. An alternative approach to evaluating cooling needs in cattle is to use the temperature humidity index (THI). This combined measure of both ambient temperature and relative humidity has been shown to be more effective in evaluating the environmental effects on lactating cattle than temperature alone.

7.3 Estimating Thermal Loads on Animals Using the Temperature Humidity Index

From a practical perspective, dairy producers want to know which environmental temperatures will lead to losses in the productivity of dairy cows. Thus, the THI first developed by Thom (1958) and extended to cattle by Berry et al. (1964) is currently used to estimate the cooling requirements of dairy cattle in order to improve the efficiency of management strategies to alleviate heat stress. The Livestock Conservation Institute evaluated the biological responses to varying THI values and categorized them into mild, moderate and severe stress levels for cattle (Whittier, 1993; Armstrong, 1994). However, as pointed out by Berman (2005), the supporting data for these designations are not published. In addition, the index is based on a retrospective analysis of studies carried out at the University of Missouri in the 1950s and early 1960s on a total of 56 cows averaging 15.5 kg day^{-1}, (range 2.7–31.8 kg day^{-1}). In contrast, average production per cow in the USA is presently over 30 kg day^{-1}, with many cows producing above 50 kg day^{-1} at peak lactation. The sensitivity of cattle to thermal stress is increased when milk production is increased, thus reducing the 'threshold temperature' when milk loss begins to occur (Berman, 2005). This is due to the fact that metabolic heat output is increased as the production levels of the animal increase. For example, the heat production of cows producing 18.5 and 31.6 kg day^{-1} of milk has been shown to be 27.3 and 48.5%, respectively, higher than non-lactating cows (Purwanto et al., 1990). Research has shown that when milk production is increased from 35 to 45 kg day^{-1}, the threshold temperature for heat stress is reduced by 5°C (Berman, 2005). These and other data indicate that THI predictions on milk yield are currently underestimating the severity of heat stress on Holstein cattle. Radiant heat load and/or convection effects were not evaluated by Berry et al. (1964), and the majority of dairy cows are currently housed under a shade structure during heat stress months. Shade structures alleviate much of the radiant heat load, but there is still a significant radiant heat load on dairy cows housed under metal shades. In Israel, a typical shade structure is estimated to add 3°C to the effective ambient temperature surrounding the animals (Berman, 2005). The use of fans for cooling management systems causes varying convection levels under shade structures as well.

An additional factor in using THI values is the management time interval. In prior research, the milk yield response to a given THI was the average yield in the second week at a given environmental heat load; therefore, milk yield measurements were not recorded until 2 weeks after experiencing the environment (Berry et al., 1964). In order to avoid economic production losses, dairy producers need to be informed of the level of cooling to be implemented immediately when heat stress occurs. Research has indicated that the effects of a given temperature on milk production are maximal between 24 and 48 h following heat stress (Collier et al., 1981; Spiers et al., 2004) (Fig. 7.1).

It has also been reported that ambient weather conditions 2 days prior to milk yield measurement had the greatest correlation to decreases in milk production and dry matter intake (West et al., 2003). Research has shown that the total number of hours when THI is greater than 72 or 80 over a 4-day interval has the highest correlation with milk yield (Linvill and Pardue, 1992). Collectively, these findings indicate that current THI values for lactating dairy cows underestimate the impact of a given thermal load on animal productivity and have an inappropriate time interval associated with cooling management decisions. Avoiding a decline in milk production over a 48-h period will automatically prevent a decrease in lactation persistency 2 weeks later. Utilizing the THI in order to reduce milk production losses has been effective, however, the current THI is in need of updating on an appropriate timescale with data from higher-producing animals. Therefore, the THI was re-evaluated in controlled environment facilities using high-producing dairy cows and including radiant energy impacts on animal performance. The specific objectives were to determine the effects of minimum, maximum and average THI and the number of hours at a given THI on milk production of high-producing dairy cows.

Evaluation of the data on minimum THI indicates that milk yield losses become significant when minimum THI on any given day is 65 or greater. Average losses in milk

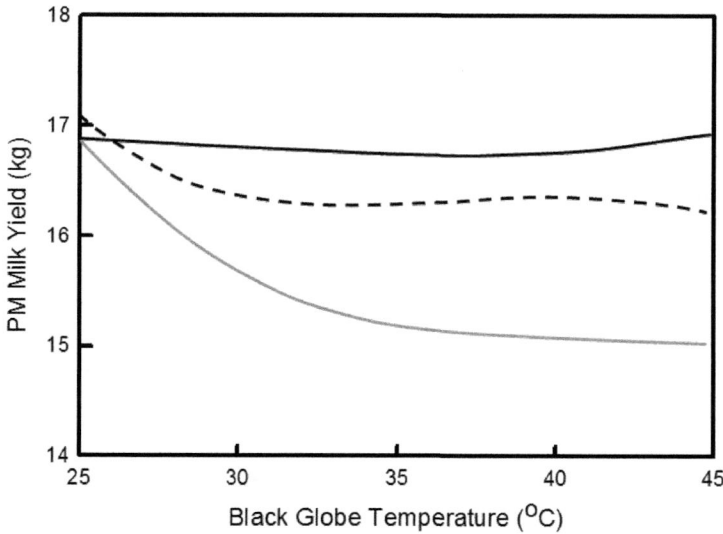

Fig. 7.1. Least squares regressions of average afternoon black globe temperature 2 days prior to (——), 1 day prior to (-----), or day of (——) a.m. milking on milk yield (from Collier et al., 1981). (Redrawn).

yield per day were 2.2 kg day^{-1} between a minimum THI of 65 and 73. This suggests that cooling of dairy cows should be initiated any time the minimum THI is 65 or above or when the average THI is 68 (Zimbleman et al., 2009).

Finally, in an evaluation of a large data set using over 1 million Italian Holstein lactation records, Bernabucci et al. (2010) indicated that milk yield losses began at an average THI of 68. These same investigators also demonstrated that death losses in dairy cows increased sharply above a minimum THI of 75.

7.4 Strategies to Reduce the Impact of Heat in Dairy Cattle Facilities

Since dairy facilities are being constructed throughout the world, producers must select carefully the appropriate facility for the local climate. Facility options vary depending on the severity of heat and cold stress. The capital costs to construct a dairy will vary considerably depending on the environmental conditions present in a specific locale. Currently, the alleviation of thermal stress conditions employs two main strategies: (i) reducing the heat load on dairy animals; and (ii) maximizing heat loss from dairy cows.

The primary tool used to reduce the heat load on animals is to provide shade to reduce the solar heat load. The secondary objective is met by the use of evaporative cooling and fans when appropriate. Evaporative cooling reduces the air temperature and increases the thermal gradient for cows to dissipate heat from their skin surface, while fans increase convective air movement across the skin surface and increase convective heat loss. A third possibility is to use conductive cooling, where the beds the cows lie on are cooled, allowing the cows to increase heat loss through the bedding material. It is envisioned that under several thermal stress conditions, all three routes of increasing heat loss will be utilized to some degree on dairy facilities.

7.5 Water Availability

Water availability is the first priority in reducing the impact of heat stress. Water should be made available to cows entering and leaving the parlour and in housing areas. Lactating dairy cows will consume 10% of their daily water intake if given the chance at the parlour. Often, decisions concerning water access are not made by determining summertime needs. Recommendations concerning access to water in housing areas vary greatly. Current recommendations suggest a range of 3–9 linear cm per cow (Smith et al., 2000). In cooler climates, the typical rule is one waterer every 61 cm of space for every 10–20 cows. In warmer climates, the recommendation is 9 cm of space for every cow in the pen. Typically, water is provided at each crossover in four-row and six-row freestall barns. In most cases, four- and six-row freestall barns have the same number of crossovers. Thus, water access in a six-row barn is reduced by 37.5% as compared to a four-row barn (Table 7.1). When overcrowding is considered (Table 7.2), water access is greatly reduced, and the magnitude of reduction is greater in six-row barns. Milk is 87% water, and water intake is critical for peak dry matter intake. When building six-row barns, or overcrowding either four-row or six-row barns, it is important to consider the amount of water space available based on actual stocking density rather than the number of stalls. In the summer, 9 linear cm of water per space per cow should be provided.

Many times, the distance between crossovers in freestall barns is increased to reduce the initial investment cost. However, producers will have to be careful to meet the summertime requirement for access to water when constructing freestall housing. A number of factors will impact the water space available per cow. Some of these factors include distance between crossovers, barn configuration, trough length and stall width. In four-row barns with a head-to-head configuration with 122 cm stalls, crossovers should be located every 30 m. Ideally, these crossovers would have a

Table 7.1. Average pen dimensions, stalls, cows and allotted space per animal.

					Per cow		
	Pen width	Pen length	Stall per pen	Cows per pen	Area	Feedline space	Linear water space
Barn style	m	m			m²	cm	cm
Four-row barn	11.9	73.2	100	100	8.5	73.7	9.1
Six-row barn	14.3	73.2	160	160	6.5	45.7	5.7
Two-row barn	11.9	73.2	100	100	8.5	73.7	9.1
Three-row barn	14.3	73.2	160	160	6.5	45.7	5.7

Table 7.2. Effect of stocking rate on space per cow for area, feed and water in four- and six-row barns.

	Area per cow m²		Linear feedlot space per cow cm		Linear water space per cow cm	
Stocking rate[a] (%)	Four-row barn	Six-row barn	Four-row barn	Six-row barn	Four-row barn	Six-row barn
100	8.5	6.5	73.7	45.7	9.1	5.7
110	7.8	5.9	66.0	40.6	8.3	5.2
120	7.1	5.4	61.0	38.1	7.6	4.8
130	6.6	5.0	55.9	35.6	7.0	4.4
140	6.1	4.6	53.3	33.0	6.5	4.2

Note: [a]Stocking density = cows/stalls.

minimum width of 4.3 m and have the longest possible water trough installed. If a tail-to-tail configuration is used, the width of the crossovers will need to be increased to accommodate two short water troughs versus one long trough. If stalls narrower than 122 cm are used, the distance between crossovers should be calculated to provide adequate summertime access to water. In freestall barns with inadequate water space, water troughs can be added on the outside walls to increase trough space. If stalls are located on the outside walls, this will require giving up some stalls to provide additional water troughs.

7.6 Shade

Providing shade in housing areas and the holding pen is the second step. Research from Florida, California, Australia and Arizona indicated that when high-producing cows were exposed to a THI greater than 80 but were provided shade, they produced an additional 1.8–4.1 kg of additional milk per day versus cows that did not have shade (Armstrong, 2000). Natural shading provided by trees is effective, but most often shades are constructed from solid steel or aluminium. Using more porous materials such as shade cloth or snow fence is not as effective as solid shades. Table 7.3 lists the effectiveness of different shade materials in descending order (Kelly and Bond, 1958; Bond et al., 1961). Providing 4.5 m² of solid shade per mature dairy cow is ideal to reduce solar radiation (Wiersma, 1982). Shades should be constructed at a height of a least 4.3 m, with a north–south orientation to prevent wet areas from developing under them.

7.7 Orientation of Naturally Ventilated Freestalls

The first naturally ventilated freestall barn design criteria to be considered should be the orientation of the structure. Barns with a north–south orientation have a greater solar radiation exposure than barns with an

Table 7.3. Shade material listed in descending order of effectiveness, relative to new corrugated aluminium as the standard (=1) (Bond et al., 1961).

Material	Description	Effectiveness
Hay	15 cm thick	1.203
Wood	Unpainted	1.060
Galvanized steel	Top white, bottom natural	1.053
Aluminium	Top white, bottom natural	1.049
Neoprene-coated nylon	White, both sides	1.037
Aluminium	Standard	1.000
Galvanized steel	Standard	0.992
Asbestos board	Natural colour	0.956
Shade cloth	90% solid	0.839
Shade cloth	80% solid	0.819
Slatted wood	5 cm solid – 5 cm open	0.589

east–west orientation (Figs 7.2 and 7.3). Sunlight can enter north–south oriented barns directly, both in the morning and in the afternoon. While the afternoon sun is the most detrimental, during hot summer weather, morning sun can also modify cow behaviour. Because cows seek shade during the summer, direct sunlight will reduce stall usage. Thus, use of stalls located on the east and west outside walls of north–south oriented barns is impacted greatly when in direct sunlight. It is also important to consider that with greater sidewall heights, afternoon sunlight can reach as far as the west half of the structure. Protection from direct sunlight is vital for effective heat stress abatement. A trial in California (Smith et al., 2001a,b) showed respiration rates were higher ($P < 0.05$) in north–south versus east–west orientated barns in the morning (56.4 versus 52.2), afternoon (77.4 versus 68.8) and daily averages (66.9 versus 60.5). Barns with an east–west orientation provided greater protection from direct sunlight than north–south orientations. When working with a north–south oriented barn, shades can be used on the west wall to reduce the amount of sunlight entering the building. These curtains should be lowered at about 1:00 pm each day and raised at about 8:00 pm. The use of automatic curtains which lower slowly in the afternoon, as required to provide shade, may be the best choice. It is important to note that the curtain provides protection from direct sun, but it also blocks natural airflow. Therefore, the curtain should only be lowered during the time when protection from direct sunlight is required. The use of a minimum of 90% shade cloth or reflective curtain material is recommended. North–south orientated barns can also be enclosed and ventilated mechanically to eliminate the impact of solar radiation on the cows. In this situation, the air entering the barn can be cooled using evaporative pads or high-pressure misters.

7.8 Taking Advantage of Natural Ventilation

Maintaining adequate air quality can be accomplished easily by taking advantage of natural ventilation techniques. Armstrong et al. (1999) reported that a 4/12 roof pitch, with an open ridge, resulted in lower afternoon cow respiration rate increases as compared to reduced roof pitch or covering the ridge. They also observed that eave heights of 4.3 m resulted in lower increases in cow respiration rates as compared to shorter eave heights. Designing freestall barns that allow for maximum natural airflow during the summer will reduce the effects of heat stress. Open sidewalls, open roof ridges, correct sidewall heights and the absence of buildings or natural features that reduce airflow increase natural airflow. During the winter months, it is necessary to allow adequate ventilation to maintain air

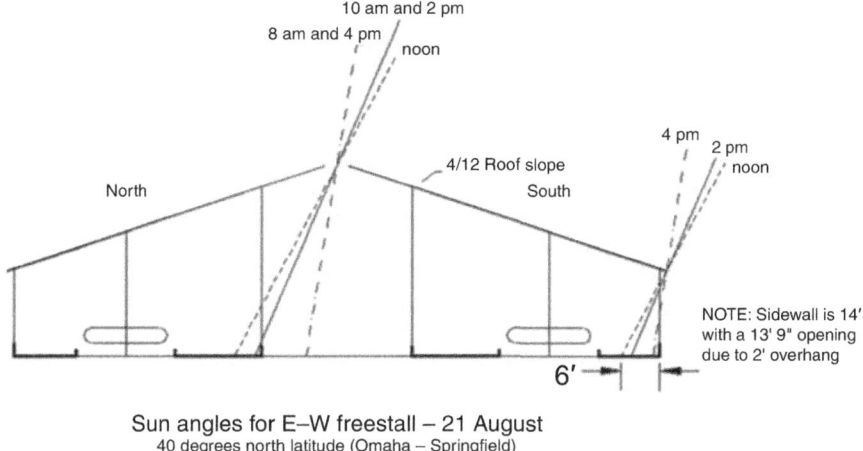

Fig. 7.2. Sun angles of an east–west oriented freestall barn (1 ft = 30.48 cm; 1 in = 2.54 cm).

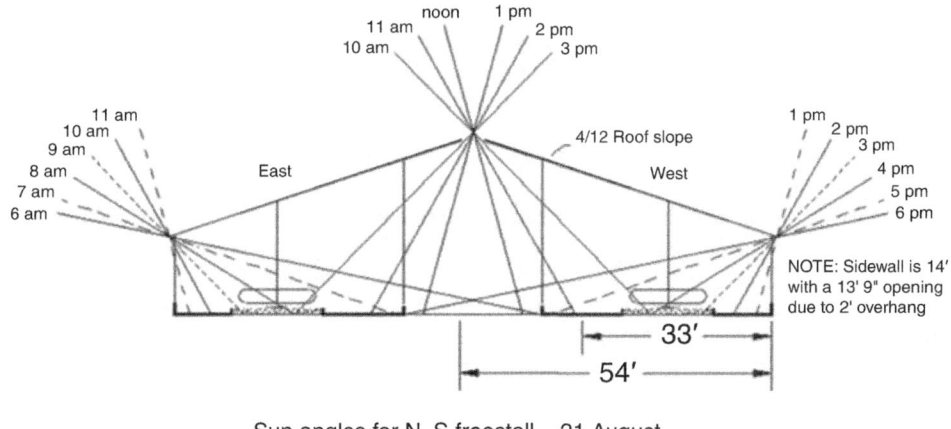

Fig. 7.3. Sun angles of a north–south oriented freestall barn (1 ft = 30.48 cm; 1 in = 2.54 cm).

quality while providing adequate protection from cold stress. In most cases, 5.08 cm of ridge row opening and 2.54 cm of eave opening is allowed for every 3.04 m of building width. The minimum distance between barns is equal to 1.5 times the building width and ideally would be 2 times the building width. Naturally ventilated four-row barns are ideally constructed with a minimum of 4.3 m sidewalls and a 4/12 roof pitch.

Another ventilation consideration is the width of the barn. Six-row barns are typically wider than four-row barns. This additional width reduces natural ventilation. Chastain et al. (1997) and Chastain (2000) indicated that summer ventilation rates were reduced by 37% in six-row barns as compared to four-row barns. In hot, humid climates, barn choice may increase heat stress, resulting in lower feed intake and milk production. A trial completed in north-west Iowa, USA (Smith et al., 2001a,b), indicated that respiration rates were higher ($P < 0.05$) in six-row versus four-row freestall barns in the morning (65.8 versus 60.5), afternoon (78.4 versus 73.8) and daily averages (72.1 versus 67.2).

7.9 Cow Cooling, Evaporating Water Off the Skin Surface

The use of low-pressure soaker and fan systems to wet and dry the cows effectively will increase heat loss from the cows. Dairy cows can be soaked in the holding pen, exit lanes and on feedlines. The goal should be to maximize the number of wet–dry cycles per hour. A study was conducted at Kansas State University to determine the effects of soak frequency and airflow on respiration rates, skin temperature and vaginal body temperature of heat-stressed dairy cattle (Brouk et al., 2005a,b). Sixteen heat-stressed lactating cows (eight primiparous and eight multiparous) were arranged in a replicated 8 × 8 Latin square design. The cattle were housed in freestall dairy barns and milked twice a day. During testing, the cattle were moved to a tie-stall barn for a 2-h period from either 1:00–3:00 pm or 3:00–5:00 pm on 8 different days in late August and early September. Afternoon temperatures ranged between 31.1 and 35.6°C, with a relative humidity between 75 and 85%. During the testing period, respiration rates were determined every 5 min by visual evaluation. Skin temperature of three sites was measured using an infrared thermometer, recorded every 5 min. Treatments (Table 7.4) were four different soaking frequencies with and without supplemental airflow. Soaking frequencies were control (no soaking), every 5 min, every 10 min or every 15 min. Supplemental airflow was either none or 0.33 m³ s⁻¹. Each wetting cycle provided similar amounts of water for all treatments. Initial data were collected for three initial 5-min periods prior to the start of the treatments.

Cows soaked every 5 min with supplemental airflow (5 + F) responded with the fastest and largest drop in body temperature and respiration rate, reducing the initial respiration rate by 47% at the end of 90 min of treatment (Figs 7.4 and 7.5). Soaking cows every 5 min without airflow (5) resulted in a similar response as soaking cows every 10 min with airflow (10 + F). Soaking cows every 15 min with airflow (15 + F) and soaking cows every 10 min without airflow (10) resulted in similar responses until the last 30 min of the study. Supplemental airflow without soaking (0 + F) resulted in little improvement over no soaking or airflow (0). Wetting had a greater effect on respiration rate and vaginal body temperature than airflow. However, the combination of wetting and airflow had the greatest effect on the respiration rate and vaginal body temperature. Respiration rates and vaginal body temperature were highly correlated (Fig. 7.5). When cooling heat-stressed dairy cattle, the most effective treatment included continuous supplemental airflow and wetting every 5 min.

These data suggest that different cooling strategies could be developed for different levels of heat stress. Under severe heat stress, soaking every 5 min with fan cooling will be the most effective. Under periods of moderate stress, soaking every 10 min with fan cooling may be adequate. Reducing soaking frequency when temperatures are lower could reduce water usage significantly. Data clearly indicate that the combination of soaking and fans is superior to either single treatment. If used singularly, soaking cows would have more impact than the use of fans only for cow cooling. Under periods of severe heat stress, soaking every 15 min with airflow is not adequate and soaking frequency must be increased. Cow cooling with soaking and supplemental airflow is very effective in reducing respiration rate. Many systems may be ineffective because they do not deliver adequate water to soak the cow and/or have an inadequate soaking frequency.

Table 7.4. Experimental treatments.

Treatment	Soaking frequency[a]	Supplemental airflow
0	None	None
0 + Fan	None	0.33 m³ s⁻¹
5	Every 5 min	None
5 + Fan	Every 5 min	0.33 m³ s⁻¹
10	Every 10 min	None
10 + Fan	Every 10 min	0.33 m³ s⁻¹
15	Every 15 min	None
15 + Fan	Every 15 min	0.33 m³ s⁻¹

Note: [a]1.3 l per headlock applied in 1 min; Fan = supplemental airflow.

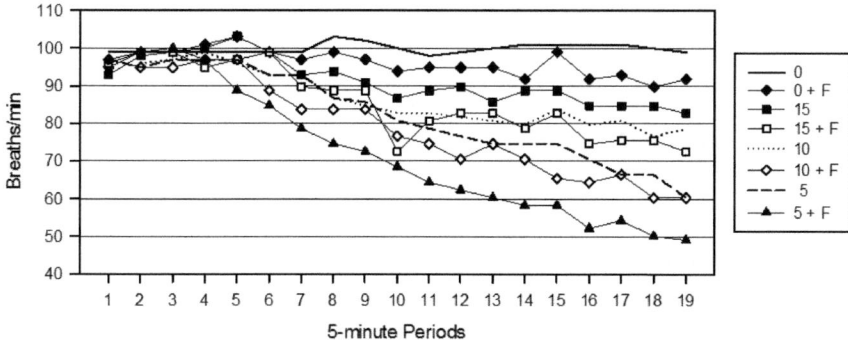

Fig. 7.4. Effect of sprinkling frequency and airflow on respiration rate of heat-stressed dairy cattle (from Brouk *et al.*, 2005a,b).

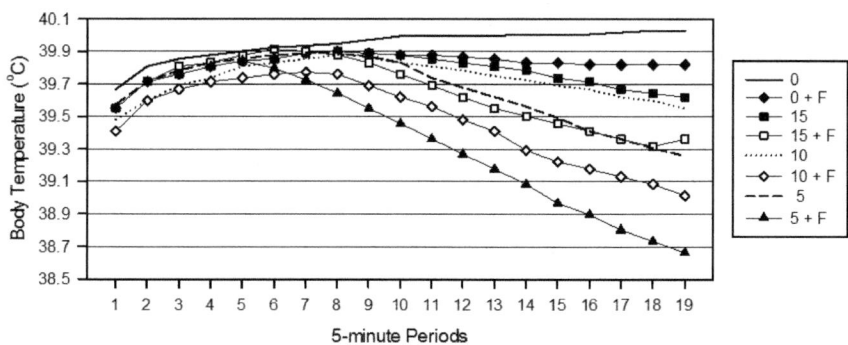

Fig. 7.5. Effect of sprinkling frequency and airflow on vaginal temperature of heat-stressed dairy cattle (from Brouk *et al.*, 2005a,b).

7.10 Providing a Cooler Environment for the Cow

Evaporative cooling can be used to cool the air around the cow. Producers have used tunnel or cross-ventilation with evaporative pads or high-pressure misters and combinations of fans and high-pressure sprayers to cool the air around the cow. This method works well in arid climates. As water is evaporated into the air, the temperature will drop and humidity will increase. As humidity increases, it becomes more difficult to change the environment in which the cow is housed. Producers often try to break the laws of physics by installing these systems in humid climates as the sole method of cooling cows. Another consideration in mechanically ventilated freestalls is the orientation of the airflow to the freestall. In a cross-ventilated facility, the airflow is perpendicular to the feed alley and parallel to the freestall. Airflow in a tunnel barn is parallel to the feedline and perpendicular to the stall. It is easier to baffle the airflow into a freestall in a cross-ventilated facility than in a tunnel-ventilated facility, because the baffles run parallel to the feedline. This allows the baffles to be constructed at a consistent height the entire length of the barn. Consequently, more of the cow's skin surface is impacted by the increased air speed in a cross-ventilated facility versus a tunnel-ventilated facility.

7.11 Using Multiple Methods to Cool Cows

In extremely hot, arid climates and hot, humid climates, it may be necessary to use a combination of soaking cows and evaporative cooling to change air temperature. Research completed in Saudi Arabia, Florida and Thailand would indicate that body temperature is lowered when feedline soaking is used in addition to cooling the air in the housing area with evaporation (Armstrong, 2000; Ortiz *et al.*, 2011; Brouk *et al.*, 2005a,b; Smith *et al.*, 2005, 2008). A combination approach to minimizing heat stress can be incorporated into a wide variety of facilities, including dry lots, desert barns, mechanically ventilated freestalls and naturally ventilated freestalls.

7.12 Considerations for Dry Lots and Desert Barns

Traditional dry lot dairies typically have shades and feedlines that are orientated north–south. The shade in the housing area is a significant distance from the feedline. The advantage of a north–south orientation is that cows move with the shade throughout the day, increasing the distribution of urine and faeces over more area. Providing shade over north–south orientated feedlines is difficult. Another disadvantage of this model is that it often makes cows choose between shade and cooling. There are a number of systems available in which the evaporative cooling follows the shade as it moves throughout the day. In the summer of 2011, a trial was completed in Arizona, USA, that evaluated the impact of the FlipFan® Dairy Cooling System (Schaefer Ventilation Equipment, Sauk Rapids, Minnesota) and fixed fans and misters on vaginal temperature and resting behaviour. Vaginal temperatures were lower and resting times higher when cows were cooled with the FlipFan® as compared to fixed fans and misters. Another approach in dry lot dairies is to use fixed evaporative cooling and incorporate a curtain on the prevailing wind side of the shade to overcome the movement of the shade and wind speed. This approach will concentrate cows over a smaller area, increasing the importance of grooming the bedding area. Feedline soakers are often incorporated into dry lot dairies. Wind speed can be a challenge when using feedline soakers in dry lot dairies.

Desert barns were developed to improve the ability to manage heat stress as compared to dry lots. The desert barn provides shade over the feedline and the housing area. Cows are not required to walk in the sunlight from the housing area to the feedline. Curtains are used on the sidewalls to minimize the impact of solar radiation and wind. The management requirements of the bedding area are increased as the area per cow is decreased. It is very easy to incorporate feedline soakers in combination with evaporative cooling in the desert barn model.

7.13 Mechanically Ventilated Freestalls

In recent years, there has been an increased interest in mechanically ventilated freestall facilities. As the air enters the facility, air temperature can be changed with evaporative pads or high-pressure misters. The evaporative cooling can be used easily in combination with feedline soakers. Most of these facilities rely only on mechanical ventilation. However, some producers are trying to implement a hybrid system that uses a combination of natural and mechanical ventilation. It can be difficult to make the transition from mechanical to natural ventilation consistently in these facilities. Orientation of the airflow to the cow (parallel versus perpendicular) and wind speed over the stalls is important in mechanically ventilated freestalls. The design of the ventilation system baffles and evaporative cooling in mechanically ventilated freestalls will determine how successful the facility will be over time.

7.14 Managing Four-Row Naturally Ventilated Freestall Barns

A series of three trials was completed from 1998 to 2000 to determine fan placement in four-row freestall barns (head-to-head configuration) with low-pressure sprinklers on the feedline (Meyer *et al.*, 1998; Brouk *et al.*, 1999, 2001). Based on these trials, fans should be mounted above the cows on the feedline and above head-to-head freestalls in a four-row freestall barn. If 91.4 cm fans are used, they should be located no more than 9.1 m apart. If 121.9 cm fans are used, they should be located no more than 12.2 m apart and operate when the temperature reaches 20°C. Fans should be mounted as low as possible, out of the reach of the cattle and in a manner that will not obstruct equipment movement. Fans should create airflow of a minimum velocity of 2.7 m s^{-1}. Feedline sprinklers should be used in addition to the fans. Feedline soaking systems should be installed to soak cows. There will be very little benefit from installing fans without soakers. Cows should be soaked every 5–15 min, depending on the severity of heat stress. The application rate per cycle should be 1.12 cm m^{-2} or 1.3–1.5 l per cow cycle, and soakers should operate when the temperature reaches 20°C. Using nozzles with check valves will reduce water usage and allow for consistent water delivery as the system cycles. Controllers are available that will increase soaking frequency as the temperature increases. In tail-to-tail four-row or six-row barns, additional fans may be needed over the stalls located near the sidewalls.

7.15 Holding Pen Cooling

The holding pen should be cooled with fans and sprinkler systems. An exit lane sprinkler system may be beneficial in hot climates. Holding pen time should not exceed 1 h. Soaker fans should move 28.3 m^3 min^{-1}, with a minimum velocity of 2.7 m s^{-1}. Most 76.2–91.4 cm fans will move between 283 and 340 m^3 min^{-1} per fan. If one fan is installed per 10 cows or 13.5 m^2, adequate ventilation will be provided. If the holding pen is less than 7.3 m wide with 2.4–3.0 m sidewall openings, fans may be installed on 1.8–2.4 m centres along the sidewalls. For holding pens wider than 7.3 m, fans are mounted parallel to the cow flow. Fans are spaced 1.8–2.4 m apart and in rows spaced either 6.1–9.1 m apart for 91.4 cm fans or 9.1–12.2 m apart for 121.9 cm fans (Harner *et al.*, 2000). In addition to the fans, a sprinkling system should deliver 1.25 l m^{-2} of area. Cycle times are generally set to soak cows every 5–15 min, depending on the temperature of the holding pen.

7.16 Bottlenecks to Cow Cooling

Often, producers do not plan to cool cows when they are building new dairy facilities. This creates serious problems in cooling cows. A list of potential obstacles to minimizing heat stress is given below:

1. Undersized water supply lines.
2. Failure to install supply lines to areas where cows will be soaked.
3. Undersized soaker lines.
4. Drift from soaker systems.
5. Lagoon capacity to handle runoff.
6. Ability to control runoff in areas where cows are soaked.
7. Incorrect fan placement.
8. Failure to install the electrical capacity for fans.
9. Fan maintenance.
10. Matching the cooling method/system with the climate.

7.17 Summary

Heat stress reduces intake, milk production, health and reproduction of dairy cows. The dairy cow can be managed and cooled to minimize the impact of heat stress. Listed below are the priorities for reducing heat stress in dairy facilities:

1. Improve water availability.
2. Provide shade in the housing areas and holding pen.
3. Reduce walking distance.

4. Reduce time in the holding pen.
5. Improve holding pen ventilation.
6. Add holding pen cooling and exit lane cooling.
7. Improve ventilation in cow housing areas (freestalls).
8. Cool close-up cows (3 weeks prior to calving).
9. Cool fresh cows and early lactation cows.
10. Cool mid and late lactation cows.

If these priorities are implemented, incremental improvements in milk production (assuming all else is equal) can be made over time to minimize the impact of heat stress.

References

Armstrong, D.V. (1994) Heat stress interactions with shade and cooling. *Journal of Dairy Science* 77, 2044–2050.
Armstrong, D.V. (2000) Methods to reduce heat stress for dairy cows. In: *Proceedings of Heart of America Dairy Management Conference.* Kansas State University, Missouri, Kansas, pp. 13–19.
Armstrong, D.V., Hillman, P.E., Meyer, M.J., Smith, J.F., Stokes, S.R. and Harner, J.P. III (1999) Heat stress management in freestall barns in the western US. *Proceedings of the 1999 Western Dairy Management Conference.* Department of Animal Sciences and Industry, Kansas State University, Manhattan, Kansas, pp. 87–95.
Baumgard, L.H. and Rhoads, R.P. (2012) Ruminant Nutrition Symposium: ruminant production and metabolic responses to heat stress. *Journal of Animal Science* 90, 1855–1865.
Berman, A.J. (2005) Estimates of heat stress relief needs for Holstein dairy cows. *Journal of Animal Science* 83, 1377–1384.
Bernabucci, U., Lacetera, N., Baumgard, L.H., Rhoads, R.P., Ronchi, B. and Nardone, A. (2010) Metabolic and hormonal acclimation to heat stress in domestic ruminants. *Animal* 4, 1167–1183.
Berry, I.L., Shanklin, M.D. and Johnson, H.D. (1964) Dairy shelter design based on milk production decline as affected by temperature and humidity. *Transactions of the American Society of Agricultural Engineers* 7, 329–331.
Bond, T.E., Kelly, C.F., Garrett, W.N. and Holm, L. (1961) Evaluation of materials for livestock shades applicable to other open-type structures. *California Agriculture* 15, 7–8.

Brouk, M.J., Smith, J.F., Harner, J.P. III, Pulkrabek, B.J., McCarty, D.T. and Shirley, J.E. (1999) Performance of Lactating Dairy Cattle Housed in a Four-row Freestall Barn Equipped with Three Different Cooling Systems. Dairy Day 1999 Publication, Kansas State University, Manhattan, Kansas.
Brouk, M.J., Smith, J.F., Harner, J.P. III and DeFrain, S.E. (2001) Drinking water requirements for lactating dairy cows. *KSU Dairy Day Report of Progress* 881, 35–39.
Brouk, M.J., Smith, J.F., Armstrong, D.V., VanBaale, M.J., Bray, D. and Harner, J.P. (2005a) Combining air cooling and feedline soaking for heat abatement of lactating dairy cattle housed in north central Florida. *Journal of Dairy Science* 88 (Suppl. 1); *Journal of Animal Science* 83 (Suppl. 1), 504.
Brouk, M.J., Cvetkovic, B., Smith, J.F. and Harner, J.P. (2005b) Utilizing data loggers and vaginal temperature data to evaluate heat stress of dairy cattle. *Journal of Dairy Science* 88 (Suppl. 1); *Journal of Animal Science* 83 (Suppl. 1), 505.
Chastain, J., Jacoboson, L., Beehler, J. and Martens, J. (1997) Improved lighting and ventilation systems for dairy facilities: its effects on herd health and milk production. In: Janni, K.A. (ed.) *Proceedings of the Fifth International Livestock Housing Conference.* American Society of Agricultural Engineers, St Joseph, Michigan, pp. 827–835.
Chastain, J.P. (2000) Designing and managing natural ventilation systems. In: *Dairy Housing and Equipment Systems: Managing and Planning for Profitability.* Proceedings from the Conference 'Dairy Housing and Equipment Systems: Managing and Planning for Profitability', Camp Hill, Pennsylvania, 1–3 February 2000. Natural Resource, Agriculture and Engineering Service (NRAES) Publication 129, pp. 147–163 (http://palspublishing.cals.cornell.edu/nra_order.taf?_function=detail&pr_id=36, accessed 18 December 2013).
Collier, R.J., Eley, R.M., Sharma, A.K., Pereira, R.J. and Buffington, D.E. (1981) Shade management in subtropical environment for milk yield and composition in Holstein and Jersey cows. *Journal of Dairy Science* 64, 844–849.
Collier, R.J., Doelger, S.G., Head, H.H., Thatcher, W.W. and Wilcox, C.J. (1982) Effects of heat stress during pregnancy on maternal hormone concentrations, calf birth weight and postpartum milk yield of Holstein cows. *Journal of Animal Science* 54, 309–319.
Collier, R.J., Romagnolo, D. and Baumgard, L.H. (2010) Lactation (f) galactopoiesis/seasonal

effects. In: Roginski, H., Fuquay, J.W. and Fox, P.F. (eds) *Encyclopedia of Dairy Sciences* (revised). Academic Press, New York, pp. 30–45.

Harner, J.P. III, Smith, J.F., Brouk, M.J. and Murphy, J.P. (2000) *Reducing Heat Stress in the Holding Pens*. Kansas State University Publication MF2468, Manhattan, Kansas.

Holter, J.B., West, J.W., McGillard, M.L. and Pell, A.N. (1996) Predicting ad libitum dry matter intake and yield of Jersey cows. *Journal of Dairy Science* 79, 912–921.

Kelly, C.F. and Bond, T.E. (1958) Effectiveness of artificial shade materials. *Agricultural Engineering* 39, 758–759.

Klinedinst, P.L., Wilhite, D.A., Hahn, G.L. and Hubbard, K.G. (1993) The potential effects of climate change on summer season dairy cattle milk production and reproduction. *Climatic Change* 23, 21–36.

Linvill, D.E. and Pardue, F.E. (1992) Heat stress and milk production in the South Carolina coastal plains. *Journal of Dairy Science* 75, 2598–2604.

McDowell, R.E., Moody, E.G., Van Soest, P.J., Lehman, R.P. and Ford, G.L. (1969) Effect of heat stress on energy and water utilization of lactating dairy cows. *Journal of Dairy Science* 52, 188–194.

Meyer, M.J., Smith, J.F., Harner, J.P. III, Shirley, J.E. and Titgemeyer, E.C. (1998) *Performance of Lactating Dairy Cattle in Three Different Cooling Systems*. Dairy Day Publication, Kansas State University, Manhattan, Kansas.

Ortiz, X.A., Smith, J.F., Bradford, B.J., Harner, J.P. and Oddy, A. (2011) Effect of complementation of cattle cooling systems with feedline soakers on lactating dairy cows in a desert environment. *Journal of Dairy Science* 94, 1026–1031.

Purwanto, B.P., Abo, Y., Sakamoto, R., Furumoto, F. and Yamamoto, S. (1990) Diurnal patterns of heat production and heart rate under thermoneutral conditions in Holstein Friesian cows differing in milk production. *Journal of Agricultural Science* 114, 139–142.

Smith, J.F., Harner, J.P. III, Brouk, M.J., Armstrong, D.V., Gamroth, M.J., Meyer, M., et al. (2000) *Relocation and Expansion Planning for Dairy Producers*. Kansas State University Publication MF2424, Manhattan, Kansas.

Smith, J.F., Brouk, M.J. and Harner, J.P. III (2001a) *Evaluation of Heat Stress in 4- and 6-Row Freestall Buildings Located in Northwest Iowa*. Dairy Day Publication, Kansas State University, Manhattan, Kansas.

Smith, J.F., Brouk, M.J. and Harner, J.P. III (2001b) *Influence of Freestall Building Orientation on Comfort of Lactating Dairy Cattle During Summer Heat Stress*. Dairy Day Publication, Kansas State University, Manhattan, Kansas.

Smith, J.F., Armstrong, D.V., Brouk, M.J., Wuthironarith, V. and Harner, J.P. III (2005) Impact of using feedline soakers in combination with tunnel ventilation and evaporative pads to minimize heat stress in lactating dairy cows located in Thailand. *Journal of Dairy Science* 88 (Suppl. 1); *Journal of Animal Science* 83 (Suppl. 1), 503.

Smith, J.F., Harner, J.P. III, Bradford, B.J., Overton, M.W. and Dhuyvetter, K.C. (2008) Opportunities with low profile cross ventilated freestall facilities. In: Harner, J. and Smith, J.F. (eds) *Dairy Housing of the Future*, 10–11 September 2008, Sioux Falls, South Dakota, Kansas State University, Manhattan, Kansas, pp. 1–20.

Spain, J.N., Spiers, D.E. and Snyder, B.L. (1998) The effects of strategically cooling dairy cows on milk production. *Journal of Animal Science* 76 (Suppl. 1), 103.

Spiers, D.E., Spain, J.N., Sampson, J.D. and Rhoads, R.P. (2004) Use of physiological parameters to predict milk yield and feed intake in heat-stressed dairy cows. *Journal of Thermal Biology* 29, 759–764.

St-Pierre, N.R., Cobanov, B. and Schnitkey, G. (2003) Economic losses from heat stress by US livestock industries. *Journal of Dairy Science* 86, E52–E77.

Thom, E.C. (1958) Cooling degree days. *Air Conditioning, Heating and Ventilating* 55, 65–69.

West, J.W., Mullinix, B.G. and Bernard, J.K. (2003) Effects of hot, humid weather on milk temperature, dry matter intake and milk yield in lactating dairy cows. *Journal of Dairy Science* 86, 232–242.

Whittier, J.C. (1993) *Hot Weather Livestock Stress*. University of Missouri, Extension Bulletin G2099, Mt Vernon, Virginia.

Wiersma, F. (1982) *Shades for Dairy Cattle*. University of Arizona Extension Service, WREP 51, Tucson, Arizona

Zimbleman, R.B, Rhoads, R.P., Baumgard, L.H. and Collier, R.J. (2009) Revised temperature humidity index (THI) for high producing dairy cows. *Journal of Dairy Science* 92, E-Suppl. 1, 347.

8 Climate Change and Integrated Crop–Livestock Systems in Temperate-Humid Regions of North and South America: Mitigation and Adaptation

Alan J. Franzluebbers

USDA – Agricultural Research Service, Raleigh, North Carolina, USA

8.1 Introduction

Integrated crop–livestock systems represent a diversity of agricultural approaches used for various reasons, but which can be characterized as mixing crop production aspects with animal production aspects. In contrast, specialized agricultural systems focus production on a single or a few species, whether for maize–soybean grain production in the Midwestern region of the USA, for small grain production in the Great Plains, for vegetable production in the Central Valley of California, for beef finishing in the High Plains of Texas, or for dairy production in the Upper Midwest and north-eastern USA.

Climate change can affect integrated crop–livestock systems in a similar way to specialized agricultural systems, but such mixed production systems offer some alternatives both to mitigate and to adapt to climate change, resulting in potentially less severe devastating effects on farm- and national-scale agricultural production outcomes. The attributes of integrated crop–livestock systems will be discussed in later sections.

What are the differences between integrated and specialized agricultural systems? As outlined in Table 8.1, operators of integrated agricultural systems must rely on similar climate, socio-economic, infrastructure and market conditions, but intentionally focus greater attention on utilizing natural capital to a greater extent and have a very high regard for minimizing negative environmental impacts and, through design, even enhancing some environmental impacts.

Some reasons for designing and implementing integrated crop–livestock systems are based on the production concerns of: (i) farms operating on marginal profit; (ii) economic vulnerability with specialized production; (iii) the high cost of fuel and nutrients; (iv) pests, which are more prolific in monocultures; and (v) yield decline could be overcome with rotation. Some environmental incentives for integrated crop–livestock systems are: (i) nutrient recycling could be improved in both crop and livestock systems; and (ii) the conservation of soil and water is more easily possible with sod-based management systems.

Agricultural systems with a diverse production of economic goods can avoid some of the negative impacts of climate change by chance and design, but some components of diverse crop–livestock systems will likely suffer from extreme events similar to simple agricultural systems. The extent of damage may be mitigated by diversity; for example, livestock can move to different locations on

Table 8.1. Characteristics of specialized and integrated agricultural systems.

Specialized agricultural systems, based on considerations of:	Integrated agricultural systems, based on considerations of:
o Climate	o Climate
o Socio-economics	o Socio-economics
o Infrastructure	o Infrastructure
o Markets	o Markets
	o Natural capital
➢ Leading to a focus typically on the most profitable system possible without high regard to other factors.	o Environmental impacts
➢ Most often a traditional system that fits the climate/infrastructure domain of a region without high regard to environmental factors.	➢ Leading to diverse agricultural enterprises to balance production and economic gains with minimal negative influence on the environment.
	➢ Typically, systems that rely on natural capital rather than purchased capital to maximize resource efficiency.

a farm to avoid floodwaters, seek shade or be fed supplements or even the residues of failed grain crops on the farm in the case of drought. Therefore, integrated crop–livestock systems can be viewed as less sensitive to climate change by absorbing negative impacts internally as resources to be regained in another component of the farming operation.

Temperate-humid climates are found in most continents, except Antarctica. Large areas predominate in North America (eastern USA), South America (southern Brazil, southern Chile, northern Argentina, Paraguay and Uruguay), Europe (many Western and central European countries), Asia (eastern China, Korea and Japan), Oceania (eastern Australia and New Zealand) and Africa (South Africa, Kenya and Madagascar). Agriculture in these areas is diverse, due to differences in cultural norms, technological adoption, wealth, market outlets, soil types and local climatic conditions. However, common climatic features are generally sufficient precipitation, negating the requirement for supplemental irrigation, a reasonably long summer growing period that allows for at least one cash crop to be grown each year, and a clear end to the growing season terminated by frost or near-frost-like conditions.

This chapter covers some of the agricultural characteristics of temperate-humid regions in North and South America. Topics covered are: (i) types of integrated crop–livestock systems; (ii) review of soil organic C sequestration and greenhouse gas emissions in integrated crop–livestock systems; (iii) how climate change could impact integrated crop–livestock systems; and (iv) prospects for developing greater resilience to climate extremes.

8.2 Types of Integrated Crop–Livestock Systems

Grasslands and agricultural cropping are capable of coexistence in ecologically-oriented approaches: for example, long-term sod-based rotations with maize, wheat, cotton, etc. (Katsvairo et al., 2006); short-term cover crops that can be grown for conservation cover, for grazing by ruminants or for hay harvest (Franzluebbers, 2007); temporally dynamic systems of intercropping or relay-cropping forage grasses with grain or fibre crops; and spatially dynamic systems of strip-cropping forages and grain or fibre crops, alley-cropping forages–shrubs–trees with grain or fibre crops, or silvopastoral systems with grazed forages as understorey for shade-timber-nut trees (Gliessman, 2007). Although not widespread in adoption in temperate-humid regions, these systems could help provide viable strategies to achieve the seemingly diametrically opposed world views for high agricultural production to meet the needs of a growing human population and environmental protection and restoration so that we can maintain the natural resource base supporting our very existence (Franzluebbers et al., 2011). The full value of grasslands as a forum for animal production to provide ecosystem services has been under-appreciated in many choices focused on highest production and profit, unfortunately at the disposal of people and environmental protection (Steiner and Franzluebbers, 2009).

8.2.1 Sod-based rotations

Sod-based rotations represent a diversity of proven systems to maintain yield, reduce pest incidence and help protect environmental quality in a wide range of climatic and edaphic conditions (Katsvairo et al., 2006; Allen et al., 2007; Franzluebbers, 2007; Sulc and Tracy, 2007). Rotations may be short (e.g. annual cover crops), intermediate (e.g. biennial legumes planted with a small-grain nurse crop and harvested for forage the following year, followed by planting of grain crops), or long (e.g. perennial forages planted following row crops and harvested for forage by grazing, haying or silage during several years thereafter) until rotated to cash crops again. Longer rotations generally enhance soil organic C sequestration, improve soil quality and foster deeper deposition of C.

Diverse crop rotations with perennial forage phases frequently increase yields of food and fibre crop phases and support higher, and often more stable, economic returns. A recent example in Iowa found greater maize grain yield during 8 years of evaluation under maize–soybean–oat–lucerne rotation (12.8 Mg ha^{-1}) than under the dominant rotation in the region of maize–soybean (12.2 Mg ha^{-1}) (Davis et al., 2012). Soybean yield was also greater in the more diverse rotation (3.8 versus 3.4 Mg ha^{-1}). Due to the high yield of all crops in the diverse 4-year rotation, net economic return was also greater (US$701 ha^{-1} year^{-1}) than in the maize–soybean rotation (US$690 ha^{-1} year^{-1}). These are significant advantages for livestock producers who diversify into crop production and for crop specialists who add pasture and livestock.

8.2.2 Grazing cover crops

Cover crops can provide a productive short-rotation opportunity for almost all temperate-humid regions, potentially improving long-term productivity by enhancing soil quality and reducing environmental threats by controlling erosion and nutrient losses, particularly during the vulnerable winter period. Despite extensive research conducted with cover crops (Hargrove, 1991; Tonitto et al., 2006; dos Santos et al., 2011; Djigal et al., 2012), and increasingly in combination with conservation tillage systems during the past two decades, there are still too few examples of how cover crops have been assimilated successfully into integrated crop–livestock systems.

Cover crops grazed by ruminant animals increase agronomic value, but whether environmental concerns with animal traffic can be overcome is an area of current research. On a Typic Kanhapludult in Georgia, grain yield was reduced by grazing cover crops in only 1 of 14 crop-tillage-year combinations during the first 3 years of a multiple-year trial (Franzluebbers and Stuedemann, 2007). Cover crop biomass, cattle performance and total gain were greater under no tillage than under conventional tillage. The integration of livestock with crops improved production from a whole-farm perspective by utilizing cover crop biomass as forage for cattle production. Large production gains by grazing of cover crops with cattle resulted in only small and variable effects on soil physical properties (Franzluebbers and Stuedemann, 2008a,b).

In Ohio, winter cover crops of annual ryegrass and oat/rye following no-till maize silage provided forage of excellent nutritive value to extend the grazing season during winter months. Yearling dairy heifer gains were substantial and achieved without detrimental effects on subsequent maize silage productivity, while cover crops increased labile soil organic C (Faé et al., 2009). Using conventional tillage in Illinois, soil organic C concentration and maize grain yield were greater in a maize/cover crop pasture rotation than in continuous maize without a cover crop (Tracy and Zhang, 2008).

8.2.3 Dual-purpose use of small grains

Winter wheat has been grown as a cash grain crop and used as forage in the southern Great Plains (Kansas, Oklahoma, Texas). Dual-purpose wheat is an economically important strategy in the region (Redmon et al., 1995) on an estimated 2.5 Mha (Taylor et al., 2010). Wheat is grazed during autumn, winter and spring, and grazing can be managed to have a minimal effect on subsequent grain yield, if soil moisture at planting and subsequent precipitation during the growing season are adequate. Wheat pasture is a valuable source of high-quality forage and stocker cattle can gain in excess of 1 kg day^{-1}. Although this system is widely practised, there is little research information available on how the system might affect soil organic C and greenhouse gas emissions; certainly an area of priority. In addition, research is critically needed on how this dual-purpose grain–grazing system managed under no tillage and in rotation with other suitable crops could help mitigate C emission by sequestering soil organic C and in adapting agriculture in this region to be more sustainable with changes in climate.

8.2.4 Grazing crop residues

Grazing livestock on crop residues after grain harvest represents one of the simplest and most economical methods for producers to integrate livestock into grain crop rotations. Crop residues are a vast feed resource available to ruminant livestock producers that can reduce feed costs effectively (Lawrence and Strohbehn, 1999). This practice was an integral part of many cattle operations in the western portion of the maize–soybean belt of the Midwestern USA, but the practice may be diminishing now that maize residues are being considered a cellulosic feedstock for biofuel production. Maize residues can provide 4–5 animal unit months of grazing per hectare. Observations suggested that grazing crop residues reduced soil surface cover by only 5–25% (Lesoing et al., 1996; Clark et al., 2004), so erosion control was not compromised. As with cover crop grazing, animal traffic on crop residues can cause soil compaction near the surface and increased soil surface roughness; however, grazing when soils are dry or frozen or performing tillage after the winter grazing period prevents significant detrimental effects on subsequent grain crop yields (Lesoing et al., 1996; Clark et al., 2004). Using a more intensive approach, providing feed supplementation to dairy heifers corralled on crop residues, resulted in improved capture of excreted N compared with spreading on to land similar amounts of manure collected from confined animals (Powell and Russelle, 2009).

8.2.5 Integrated crop–livestock–woodlands

Many agroforestry systems have the option for grazing of perennial forages, thereby providing great potential to increase system productivity through the intentional integration of trees, forages and livestock (Fike et al., 2004), as well as to capture greater amounts of C in woody biomass and in soil as organic matter. The spatial arrangements of components can be designed differently to optimize benefits to trees, forages, livestock and the environment. With wide tree spacing, annual cash crops can be grown to generate income when trees are too small for grazing animals to be present. In Mississippi, maize was grazed by cattle in such a system, rather than harvesting by machine (Glover Triplett, personal communication, 2007). During the initial research, accelerated tree growth from fertilizer applied to maize was observed, which could accelerate the tree rotation. With the increasing maturity of the trees, and as annual crops become shaded, perennial forages become better suited as the profitable herbaceous understorey.

8.2.6 Pasture cropping

Warm-humid regions have climatic advantages for using annual–perennial forage rotations in time that occupy the same space (Franzluebbers, 2007). In the south-eastern USA, it is common for beef producers to overseed dormant warm-season bermudagrass hay or pastureland with cool-season annual grasses (e.g. annual ryegrass, rye) and/or legumes such as crimson clover. Bermudagrass grows primarily from May through September in the region, while overseeded cool-season species extend the grazing season into cooler months. Pasture overseeding does not necessarily represent integration of cash crops and livestock, but it can be a component of an integrated approach.

Overseeding of bermudagrass with winter small grains is a technique to capitalize on the aggressive regrowth potential of the perennial sod, while capturing the seasonal opportunity to drill a small grain directly into the sod without tilling the soil. Historical research was conducted in Georgia during the 1960s and showed that rye grain yield was equivalent under sod-seeded conditions and clean-tilled conditions (Welch et al., 1967). Wheat grain production was low to moderate when sod-seeded, but could be enhanced with adequate application of N (Carreker et al., 1977). Small-grain silage harvest would help remove cool-season forage early enough prior to the onset of the early growing season of the warm-season perennial forage to avoid a reduction in first-cut hay production. In Australia, wheat and barley crops grown in mature lucerne stands were suppressed 16–26% compared with sole cropping (Harris et al., 2007). Crops did not respond to N input, whether or not grown with a companion, suggesting that fertility was not a limitation, but rather only water.

Another opportunity available to producers in areas with reliable precipitation in the summer is to seed summer cash crops directly into perennial cool-season forages. Supplemental irrigation would reduce the risk of stand establishment and crop failure in years with insufficient summer precipitation. Grain production of maize seeded directly into bermudagrass or tall fescue sods can vary from total failure to reasonable production, depending on the precipitation pattern and the control of perennial sod for successful establishment. In an evaluation in Georgia, maize grain yield was maximized with plough tillage of planting strips, because of the reduction in grass competition, but forage recovery was greatest with the least soil disturbance (Adams et al., 1970). When no-till planting maize into strip-killed tall fescue sod, maize grain yield from 1973 to 1976 in Georgia averaged 5.5 Mg ha^{-1} under non-irrigated conditions and 8.5 Mg ha^{-1} with supplemental irrigation (Box et al., 1980). Tall fescue forage yield when strip-killed averaged 6.0 and 4.7 Mg ha^{-1} under dryland and irrigated conditions, respectively. These studies with maize planted into tall fescue sod illustrate a multitude of opportunities available to producers, depending on irrigation availability and their need to balance grain and forage production. Living mulches are being developed for grain crop production in the Midwestern USA, and this practice could be used to produce forage (Singer et al., 2009). This practice may be especially applicable to producers with well-managed grazing systems, because they are often reluctant to destroy a productive perennial pasture sod in exchange for the rotational benefits to the following grain crop.

8.2.7 Regional integration

With increasing government regulations for nutrient management, owners of concentrated animal feeding operations (CAFOs) have been driven to develop business relationships with neighbouring farm operators for the distribution of excess manure to cropland or pasture. Planning approaches for integrating manure management among farms are being pursued. Examples also exist where crop farms grow feed for a neighbouring CAFO and receive manure back from the CAFO for application to their croplands (Russelle et al., 2007). Similar arrangements typify the large broiler chicken operations that have the need to spread floor litter (waste from broiler houses) on to the crop and pasturelands of neighbouring farms (Franzluebbers, 2007).

Therefore, among-farm integrated crop–livestock systems provide opportunities to capture some of the economic and environmental benefits of a fully integrated single-farm system. The crop farmer, in particular, has the opportunity to capture the benefits of crop rotation diversity and perennial species on sequestering soil organic C, of nutrient recycling on reducing N_2O emissions and of improved energy efficiency to reduce the C footprint from fossil fuel-based inputs. Development of among-farm integrated crop–livestock systems is enhanced by the small distances between farms, the existing basic trust between individuals and a willingness to begin slowly with modest exchanges (Files and Smith, 2001). Russelle et al. (2007) points out that an advantage of among-farm collaborations is that more people have a stake in assuring successful and mutually acceptable outcomes. This investment process may have particularly strong applications when designing systems to mitigate and adapt to climate change, because the climate change forces are beyond the control of each landowner and the response and outcomes to climate change are of value and of mutual interest to the community as a whole. It remains to be seen whether such collaborations can achieve the same range of production and environmental synergies as within-farm integration.

8.3 Soil Organic C Sequestration in Integrated Crop–Livestock Systems

Integrated crop–livestock systems rely on forages as part of the diversity of crop choices. These forages provide large benefit towards the balance of C in the soil. Forages have extensive, fibrous root systems that explore large volumes of soil, mostly in the surface 30 cm like most crops, but also penetrating deep into the profile with many perennial grasses. Perennial forages also extend the growing season compared with annual cash crops, thereby photosynthesizing, depositing rhizosphere C inputs and drying soil during longer periods of time than annual crops. It is this extended growth period, whether from perennial forages or cover cropping with annual forages following cash crops, that likely contributes greatly to soil C sequestration. Another key factor is that annual forages can be no-till planted and harvested and perennial forages remain without soil disturbance for several years. Lack of soil disturbance may be vital for integrated crop–livestock systems to enhance soil organic C accumulation rather than simply to maintain it.

In the long-term Morrow Plots in Indiana, high soil organic C was maintained only in a diverse crop rotation that included ley pasture compared with continuous cropping (Nafziger and Dunker, 2011). Soil organic C was always greater under a 3-year rotation of maize–oat/clover–hay (25 ± 2 g C kg^{-1} soil) than under monoculture maize (17 ± 3 g C kg^{-1} soil). High soil organic C was related positively with the achievement of greater maize grain yield over the lifetime of this study, initiated in 1876.

A couple of long-term experiments on pasture–crop rotations in Argentina and Uruguay have also shown the value of perennial forages in maintaining soil organic C. With conventional tillage cropping of a Typic Argiudoll in Argentina, soil organic C concentration declined progressively with continuous, annual cropping following the termination of perennial pasture (Studdert et al., 1997). Using a 6-year crop + 2-year pasture rotation, soil organic C also declined with time, but pasture phases were able to rebuild soil organic C for a short time. Using a 4-year crop + 4-year pasture rotation, soil organic C oscillated with peaks at the end of the pasture phase and troughs at the end of the cropping phase, and no temporal trend of decline over the long term. The authors concluded that 3–4 years of pasture were able to restore soil organic C to a sustainable level if cropping did not exceed 7 years.

An experiment initiated in 1964 on a Typic Argiudoll at La Estanzuela, Uruguay, showed that continuous cropping with no fertilizer caused gradual and steady decline in soil organic C (Garcia-Prechac et al., 2004). The application of fertilizer slowed soil organic C decline, but only crop–pasture systems with at least 2 years of perennial pasture in the rotation allowed soil organic C to stay at its initial level (Morón and Sawchik, 2002). Rotation with 3 years of cropping and 3 years of pasture was more beneficial on soil organic C concentration than 4 years of cropping and 2 years of pasture. These systems have been evaluated in the past with conventional tillage for all cropping, but recently the study has switched to no-tillage management of crops and soil organic C appears to have responded positively (Jorge Sawchik, personal communication, 2012). Further research on these systems is continuing to determine the long-term effects.

Cropping systems in Uruguay have also been investigated for about a decade under conventional tillage and no tillage. Continuous cropping systems under no tillage had greater soil organic C than under conventional tillage (Garcia-Prechac et al., 2004; Ernst and Siri-Prieto, 2009). In system-level comparisons for 12 years, soil organic C was similar between crop–pasture rotations under no tillage and continuous cropping under no tillage (Ernst and Siri-Prieto, 2009). Although soil organic C was similar, above-ground C input was greater under continuous cropping than under crop–pasture rotation, due to forage removal by animals, but below-ground C input was lower under continuous cropping than under crop–pasture rotation. In a detailed depth analysis, Salvo et al. (2010) found management effects on soil organic C at a depth of 0–3 cm only, not deeper: (i) soil organic C was greater under no tillage than under conventional tillage; (ii) soil organic C was greater under crop–pasture rotation than under continuous cropping; and (iii) soil organic C was greater under C_4 crops in the no-tillage crop–pasture rotations than under C_3 crops.

Rapid soil organic C accumulation with the establishment of forages has also been documented in Typic Kanhapludults in the south-eastern USA. With bermudagrass establishment, soil organic C accumulation during the first 5 years was dependent on the type of management, i.e. rate of accumulation was 1.40 Mg C ha^{-1} year^{-1} when grazed by cattle, 0.65 Mg C ha^{-1} year^{-1} when unharvested and 0.29 Mg C ha^{-1} year^{-1} when hay was removed (Franzluebbers et al., 2001). With tall fescue establishment, soil organic C accumulation during the first 8 years of management was 1.36 Mg C ha^{-1} year^{-1} when grazed by cattle and 0.69 Mg C ha^{-1} year^{-1} when hayed (Franzluebbers et al., 2012).

The rotation of pastures into a cropping phase often leads to a decline in soil organic C. In the long-term trial at Balcarce, Argentina, Studdert et al. (1997) reported mean soil organic C loss of 0.7 g C kg^{-1} soil year^{-1} during 5 years of cropping following a 4- to 5-year pasture phase and mean soil organic C loss of 0.3 g C kg^{-1} soil year^{-1} during 5 years of cropping following a 2-year pasture phase. These results were under conventional tillage in the cropping phase, but conversion of pasture to cropping using no tillage should not lead to such dramatic losses of soil organic C. For example, during the first 3 years on a Typic Kanhapludult in Georgia, stock of soil organic C declined with conversion of long-term pasture to conventional tillage cropping at 1.2 Mg C ha^{-1} year^{-1} compared with no tillage cropping (Franzluebbers and Stuedemann, 2008a). Soil organic C under no-tillage cropping at the end of 3 years was similar to that of continuation of perennial pasture. On a Vertic Argialboll in Nebraska, soil organic C was maintained at 90 Mg C ha^{-1} at a depth of 0–30 cm during 6 years of no-tillage maize production following chemical termination of bromegrass sod (Follett *et al.*, 2009). Therefore, conservation tillage systems combined with crop–pasture rotations may offer an important opportunity to improve agricultural soils further by avoiding losses of soil organic C associated with cropping, but also by including high-biomass forage crops (above- and below-ground) and cash crops in the rotation, possibly to enhance soil organic C levels beyond the levels obtained in long-term pastures. This might be possible, since opportunities for high soil fertility input with animal manures on cash-crop responsive annual crops can promote deep and robust rooting different from that of perennial crops, and significant soluble C and N inputs might enhance aggregation and soil biological diversity to promote sequestration of C.

In Brazil, some field trials have been established to compare sole cropping, sole pasture and integrated crop–livestock systems (Table 8.2). The value of no-tillage cropping with legume winter cover crops on soil organic C restoration was stressed (Freixo *et al.*, 2002), and such a system

Table 8.2. Summary of results in Brazil comparing soil organic C under different management systems.

Location	Soil type	Years of management	Soil organic C stock	Reference
Federal district	Typic Acrustox	13	0–30 cm: Native Cerrado (61 Mg ha^{-1}), continuous cropping (60 Mg ha^{-1}), integrated crop–livestock system (53 Mg ha^{-1}), continuous pasture (53 Mg ha^{-1})	Marchão *et al.*, 2009
Federal district	Oxisol	20	0–30 cm: Native Cerrado (97 Mg ha^{-1}), conventional tillage cropping (72 Mg ha^{-1}), no-tillage cropping (86 Mg ha^{-1})	Freitas *et al.*, 2000
Goiás	45 fields	Chrono-sequence of 0–12 years	0–20 cm: Native Cerrado (68 Mg ha^{-1}), continuous cropping (47 Mg ha^{-1}), continuous pasture (55 Mg ha^{-1})	Corbeels *et al.*, 2006
Goiás	Typic Hapludox	8	0–20 cm: Native Cerrado (35 Mg ha^{-1}), conventional tillage cropping (39 Mg ha^{-1}), no-tillage cropping (41 Mg ha^{-1})	Bayer *et al.*, 2006
Mato Grosso do Sul	Typic Hapludox	5	0–20 cm: Native Cerrado (54 Mg ha^{-1}), conventional tillage cropping (54 Mg ha^{-1}), no-tillage cropping (57 Mg ha^{-1})	Bayer *et al.*, 2006
Rio Grande do Sul	Paleudult	17	0–17.5 cm: Native grassland (39 Mg ha^{-1}), conventional tillage cropping (25 Mg ha^{-1}), no-tillage cropping (42 Mg ha^{-1})	Dieckow *et al.*, 2005
Rio Grande do Sul	Typic Hapludox	13	0–10 cm: Native forest (34 Mg ha^{-1}), conventional tillage cropping (21 Mg ha^{-1}), no-tillage cropping (25 Mg ha^{-1})	Freixo *et al.*, 2002

would lend itself for grazing by livestock. More recent research has been focused increasingly on grazing of cover crops in integrated crop–livestock systems in Brazil (Carvalho et al., 2010). This research has appeared mostly in Brazilian scientific journals in Portuguese, but upcoming special issues in English-language journals are forthcoming from a successful symposium on the topic in Porto Algere (Brazil) in 2012. The different types of results in the Cerrado make a clear interpretation of the effects of perennial grasses and integrated crop–livestock systems on soil organic C difficult. Further research in this important geographical region utilizing integrated crop–livestock techniques will be important to clarify responses.

8.4 Greenhouse Gas Emissions in Integrated Crop–Livestock Systems

The effect of integrated crop–livestock systems on greenhouse gas emissions (CO_2, CH_4 and N_2O) is largely unknown. There is a large spatial and temporal complexity of integrated crop–livestock system designs, which make projections of greenhouse gas emissions uncertain. For example, the crop canopy can be grazed at different times of the year, depending on the cover crop or forage phase, resulting in fluxes of CO_2 dependent on photosynthesis and heterotrophic decomposition of sloughing roots from grazed plants. Fertilization strategies and the dynamics of nutrient cycling in simple and complex rotations can affect the timing and quantity of inorganic N available for potential denitrification, since inorganic N is a large determinant of N_2O emission. The availability of soluble C and compacted surface soil could be enhanced with significant animal traffic in integrated crop–livestock systems. However, these variables in integrated crop–livestock systems need to be tested, since even under perennial pastures with nearly year-long grazing, soil bulk density is unaffected by moderate grazing (Franzluebbers and Stuedemann, 2010).

The impact of integrated crop–livestock systems on CH_4 and N_2O emissions is complicated by the multitude of competing properties affecting methanogenesis, methanotrophy, nitrification and denitrificaiton. Ammonium-N accumulation in soil can suppress methanotrophic activity but increase nitrification activity, whereas soluble C accumulation and low O_2 concentration in the soil (either through compacted soil or from high heterotrophic activity) can promote methanogenesis and denitrification. Both net CH_4 emission and uptake were observed during different periods of the year in an evaluation of a soil managed under different long-term no-tillage cropping systems in southern Brazil (Bayer et al., 2012). The greatest variation in CH_4 flux was observed after cover crop termination in the spring, when flux was related to ammonium-N and dissolved organic C concentrations in soil. Manure application to soil (mostly studied as a fertilizer amendment and rarely investigated under natural grazing environments) will often increase soluble C, provide abundant and readily available ammonium-N and lead to O_2 depletion due to large retention of water in soil and/or promotion of high soil heterotrophic activity – all prerequisites for N_2O emission (Petersen et al., 2006; Webb et al., 2010).

Soil organic C is often improved with perennial forages in rotation with crops, especially with deep-rooted perennial forages, in which ancillary soil benefits can accrue to mitigate and adapt agricultural systems significantly to climate changes. Some specific indirect effects of improved soil organic C include: (i) improved water infiltration can occur with concurrent reduction in nitrate leaching to avoid indirect N_2O emissions; (ii) reduced runoff losses to avoid CH_4 and N_2O emissions from wetland sediment deposition; and (iii) reduced need for supplemental irrigation and fertilizer N to avoid untimely N_2O emissions during saturated periods and limit soil organic C mineralization. Annual and perennial forage legumes further reduce purchased N requirement, which can also serve to reduce N_2O emissions.

The direct agronomic and environmental benefits from cover crops in annual cropping systems are numerous, yet the indirect benefits on climate change issues are equally important, including:

- reducing water, soil and nutrient runoff to control indirect emissions of CH_4 and N_2O;
- improving soil tilth, structure, water infiltration and nutrient cycling to enhance soil organic C sequestration and limit opportunities for N_2O emission;
- modifying soil moisture, by increasing uptake and reducing evaporation at different times of the year to enhance soil organic C sequestration and suppress flushes of N_2O emission;
- enhancing soil biological diversity to foster soil C sequestration;
- controlling weeds through competition, allelopathy and microclimatic alteration to foster a competitive aerobic soil environment;
- controlling insect and disease pressures ecologically to avoid productivity gaps;
- serving as a nutrient trap in high-fertility systems to avoid direct and indirect N_2O emissions;
- if leguminous, providing biologically fixed N to the cropping system to reduce N fertilizer input and limit N_2O emissions.

Trees integrated with pasture can improve certain ecosystem services compared with treeless pastures. For example, some tree species can remove nutrients from deeper in the soil profiles that would otherwise be lost to water bodies and lead to indirect N_2O emission (Nair et al., 2007).

The opportunity to maintain the integrity of high surface-soil organic matter in a perennial pasture without disturbing the soil while directly seeding a crop to produce grain would be valuable to avoid the loss of soil organic C, and potentially avoiding bursts of N_2O following termination. There is a need to understand altered nutrient cycling that could occur when one or more phases of the rotation are grazed by livestock.

Data from various crops, crop rotations and tillage systems can serve as a guideline as to what greenhouse gas emissions might be like under integrated crop–livestock systems. Emission of N_2O was related to the N fertilizer rate (170 kg N ha^{-1} applied to maize, 83 kg N ha^{-1} applied to wheat and none applied to soybean), type and amount of residues from the previous crop and residual N within a crop rotation (Table 8.3). With N_2O having 310 times the global warming potential of CO_2, the values of C equivalence reported in Table 8.2 range from 0.07 to 0.35 Mg C ha^{-1} year^{-1}, certainly levels of concern compared with the potential changes in soil organic C in agricultural systems.

From a meta-analysis across a diversity of studies in eastern Canada, N_2O emissions were greater following annual crops that were incorporated by tillage in the autumn (2.41 ± 1.79 kg N_2O-N ha^{-1}) and not incorporated (1.19 ± 0.79 kg N_2O-N ha^{-1}) than following perennial crops that were not incorporated (0.29 ± 0.39 kg N_2O-N ha^{-1}) (Gregorich et al., 2005). Emissions were also greater following annual crops than following perennial crops whether unfertilized (1.53 ± 1.00 versus 0.16 ± 0.21 kg N_2O-N ha^{-1}) or fertilized (2.82 ± 2.78 versus 0.62 ± 1.10 kg N_2O-N ha^{-1}).

Table 8.3. Influence of crop and crop rotation on N_2O emission from a clay loam soil in Woodslee, Ontario, Canada. (Source: Drury et al., 2008.)

Crop rotation	Nitrous oxide emission (kg N_2O-N ha^{-1})		
	Maize	Soybean	Wheat
Continuous maize	2.62 ± 1.82	0.84 ± 0.52	0.51 ± 0.15
Maize–soybean	1.34 ± 0.52	0.70 ± 0.43	–
Maize–soybean–wheat	1.64 ± 0.76	0.73 ± 0.24	0.72 ± 0.33

8.5 Integrated Crop–Livestock Systems as a Strategy for Climate Change Adaptation

Agricultural production is widely dependent on a stable climate. Although high agricultural production can occur under a wide range of climatic conditions, it is the expected climate that is a key factor in management decisions to select crop species and variety, select animal type and management style and integrate these selections into a whole-farm system. It is not reasonable to grow a maize crop with expected summer-dominated precipitation that shifts to winter-dominated precipitation. Adaptations of integrated crop–livestock systems to climate change can be accomplished reasonably through changes in the following:

- *crop* – crop selection, variety of crop and crop sequence to alter timing of water use, quantity of water needed, depth of rooting, nutrient-use efficiency, need for rotation and reliability of production;
- *cover crop* – species and mixtures of species to alter N and water availability to following cash crop, termination method to control weeds and water and establishment method to control soil disturbance;
- *tillage* – conservation tillage or no tillage to overcome excessive and/or deficit periods of water availability through reduced evaporation and to build soil organic matter to store water and nutrients;
- *fertilization* – slow-release N sources and split applications to avoid leaching losses, changing timing of fertilizer application to foster rapid uptake by a cover crop and allow cover crop to release nutrients through decomposition, and utilizing organic sources of nutrients from other farm components and/or regionally available waste streams/recyclable resources;
- *weed control* – smothering techniques with cover crops, crop sequencing to avoid weed emergence, seed spacing to close canopy more rapidly, relay planting to avoid open seasons and capture growing season opportunities;
- *livestock* – species, breed, class of animal, sale strategy, forage management plan, stocking rate, stocking density, grazing on both perennial and annual forages, distribution on the landscape defined by season, grazing method, grazing season and rotation sequence of livestock species on a pasture;
- *manure management* – in-field deposition, control with stocking density and frequency of herd movements, variable application rate to encourage excessive forage growth and limit more susceptible losses to cropping phases, and composting with bedding materials to stabilize nutrients to avoid losses;
- *seasonality of management* – contracting some planting or harvest operations to avoid delays in other key operations, stockpiling of forage for winter to avoid the need for hay harvest and feeding, and diversifying crops to allow for stubble grazing and manure application on to stubble fields at different times of the year;
- *management style* – flexibility in thinking, watching market signals, soil health monitoring, biodiversity assessment and community involvement.

Climate change issues are likely to be centred on overcoming the frequency and severity of droughts, managing land for excess precipitation, coping with fluctuating springtime temperature variations and dealing with heat stress. Some of these issues can be managed effectively by a variety of available technologies and techniques to maintain high soil organic matter content, utilizing water and nutrients throughout the year, and incorporating a diversity of crops and forages in the farm system design to hedge against single-commodity production and spread risk.

8.6 Prospects for Developing Greater Resilience to Climate Extremes

Forage and grazing lands historically have provided a sustainable and resilient land cover, rooted by a variety of vigorous grasses and forbs and serving as key biospheric engineering components mediating a plethora of essential ecosystem services, notably water cycling, nutrient cycling, gas exchange with the atmosphere, climate regulation, food and feed production and aesthetic experience. Robust, perennial grasses are considered a vital link in keeping the soil surface intact and maintaining the functional capabilities of extensive land areas, whether naturally occurring or human planted. Despite the perceived diametric uses of land, natural grassland cover and agricultural food production systems have inherent synergies that should and must be captured for achieving ecological sustainability. Integrated crop–livestock systems offer a unique opportunity to develop high-yielding agricultural systems, if appropriate conservation management approaches are implemented, while maintaining or improving environmental outcomes from these globally vital agricultural production areas in temperate-humid regions.

To characterize and analyse how integrated agricultural systems can become more sustainable, four basic categories need to be considered (Hanson and Franzluebbers, 2008): (i) social/political; (ii) economic; (iii) environmental; and (iv) technological. Within these categories, the following issues are of importance for designing sustainable systems.

- Systems must be flexible enough to respond to future challenges.
- Emerging social and political factors need to address the current issues of rising fuel costs, obesity, potential decreases in commodity subsidies, consumer awareness and demands to know how food is produced, and economic returns to land.
- Market-driven economies are impacted by policy, technology and environmental concerns.
- Previously mentioned factors will affect the scale of operation, while controlling management flexibility.
- Agricultural incentives control management decisions, sometimes to the detriment of the environment if not formulated holistically.
- Integrated agricultural systems need to be developed that are adapted to the environment, rather than trying to adapt environmental constraints to the agricultural systems.
- Enhanced technology increases the complexity of farming.
- Future agricultural systems must address emerging issues in land use, decline in workforce and societal support of farming, global competition, changing social values in both taste and convenience of food and increasing concerns for food safety and the environment.
- A set of robust principles should be defined and subsequently used to design adaptable integrated agricultural systems.

The landscape features of an area should be a key component of decision making for designing robust integrated crop–livestock systems for a farm, and for a group of farms in the same area if appropriate. In free-market economies, it would be difficult to design landscapes without the consideration of landowners, but regional planning is possible and government incentives can be offered to encourage criteria-based implementation of key conservation practices in congruence with integrated crop–livestock systems principles. Designing farms within the landscape can capture opportunities to mitigate and adapt to the consequences of climate change, whether from flooding, prolonged drought, excessive heat, early springtime, unseasonal temperature, etc. Trees, forages, cash crops and livestock can be designed spatially and temporally to anticipate the negative consequences of climate change, as well as the desired outcomes to overcome climate change. In the spirit of integrated crop–livestock systems, a diversity of approaches may be the best hedge against uncertain risks but, with experiences and experimentation, successful agricultural adoption strategies using integrated crop–livestock systems as a basis will likely be a part of such decisions. Obviously, further research and development are still needed to quantify and predict better the impacts of climate change in different regions of the world.

As stated for the case of integrated crop–livestock systems in the temperate-humid region of the USA: 'A transformation of agriculture in North America is needed to increase production, mitigate past environmental damage, protect biological diversity, reduce dependence on fossil fuels, provide healthier foods, and increase economic and cultural opportunities in rural America. When incorporated into diverse agricultural systems that include livestock, perennial grasses and legumes and a wide variety of annual forages offer enhanced agro-ecosystem resilience in the face of uncertain climate and market conditions' (Franzluebbers et al., 2011).

8.7 Concluding Remarks

A diversity of integrated crop–livestock systems exist, so specific responses to climate change will likely be diverse. Systems having a perennial forage component are likely to lead to an increase in soil organic C sequestration, a process that can improve soil quality and the resilience of a cropping system against climate perturbations to crop growth and yield. Forages offer an opportunity for grazing livestock to use high-quality cellulosic feedstuffs – an agronomic product not otherwise competing with human food production. In addition, livestock transformation of feedstuffs through processing in the rumen can lead to enhanced nutrient cycling and efficient resource use on the farm scale. Important opportunities in integrated crop–livestock systems exist to distribute animal manures spatially on the farm naturally through dung deposition with grazing and through the mechanical distribution of manure from concentrated feeding locations to nutrient-poor areas or environmentally sensitive areas elsewhere on the farm.

Although ruminant animals emit methane, the quantity of emission is roughly the same per animal whether in feedlot or on pasture. The age of the animal prior to slaughter is a key factor in total methane emission per product. Nitrous oxide emission is likely reduced in pasture-based systems compared with feedlot-based systems due to the lower intensity of manure deposition on a much larger landscape area. Reduction in N fertilizer inputs required for crops in rotation with forages will reduce nitrous oxide emissions. The large gain in soil organic C with the establishment and maintenance of perennial pastures is a key mitigation strategy offered by integrated crop–livestock systems, but is also a key adaptation strategy to overcome drought and partially control flooding by improving soil quality when forages are distributed appropriately across a landscape scale. The diversity of farming operations in integrated crop–livestock systems reduces the overall risk of failure, despite any one farming component being affected negatively, as in specialized agricultural systems. This diversity also offers resilience of the farming system against perturbations caused by extreme weather events and climate change.

References

Adams, W.E., Pallas, J.E. Jr and Dawson, R.N. (1970) Tillage methods for corn-sod systems in the Southern Piedmont. *Agronomy Journal* 62, 646–649.

Allen, V.G., Baker, M.T., Segarra, E. and Brown, C.P. (2007) Integrated irrigated crop–livestock systems in dry climates. *Agronomy Journal* 99, 346–360.

Bayer, C., Martin-Neto, L., Mielniczuk, J., Pavinato, A. and Dieckow, J. (2006) Carbon sequestration in two Brazilian Cerrado soils under no-till. *Soil and Tillage Research* 86, 237–245.

Bayer, C., Gomes, J., Vieira, F.C.B., Zanatta, J.A., Piccolo, M.C. and Dieckow, J. (2012) Methane emission from soil under long-term no-till cropping systems. *Soil and Tillage Research* 124, 1–7.

Box, J.E. Jr, Wilkinson, S.R., Dawson, R.N. and Kazachyn, J. (1980) Soil water effects on no-till corn production in strip and completely killed mulches. *Agronomy Journal* 72, 797–802.

Carreker, J.R., Wilkinson, S.R., Barnett, A.P. and Box, J.E. (1977) *Soil and Water Management Systems for Sloping Land.* ARS-S-160, USDA, Agricultural Research Service, Washington, DC.

Carvalho, P.C.F., Anghinoni, I., de Moraes, A., de Souza, E.D., Sulc, R.M., Reisdorfer-Lang, C., et al. (2010) Managing grazing animals to achieve nutrient cycling and soil improvement in no-till integrated systems. *Nutrient Cycling in Agroecosystems* 88, 259–273.

Clark, J.T., Russell, J.R., Karlen, D.L., Singleton, P.L., Busby, W.D. and Peterson, B.C. (2004) Soil surface property and soybean yield response to corn stover grazing. *Agronomy Journal* 96, 1364–1371.

Corbeels, M., Scopel, E., Cardoso, A., Bernoux, M., Douzet, J.M. and Siqueira Neto, M.S. (2006) Soil carbon storage potential of direct seeding mulch-based cropping systems in the Cerrados of Brazil. *Global Change Biology* 12, 1773–1787.

Davis, A.S., Hill, J.D., Chase, C.A., Johanns, A.M. and Liebman, M. (2012) Increasing cropping system diversity balances productivity, profitability and environmental health. *Public Library of Science One* 7, e47149. doi:10.1371/journal.pone.0047149.

Dieckow, J., Mielniczuk, J., Knicker, H., Bayer, C., Dick, D.P. and Köegel-Knabner, I. (2005) Soil C and N stocks as affected by cropping systems and nitrogen fertilisation in a southern Brazil Acrisol managed under no-tillage for 17 years. *Soil and Tillage Research* 81, 87–95.

Djigal, D., Saj, S., Rabary, B., Blanchart, E. and Villenave, C. (2012) Mulch type affects soil biological functioning and crop yield of conservation agriculture systems in a long-term experiment in Madagascar. *Soil and Tillage Research* 118, 11–21.

dos Santos, N.Z., Dieckow, J., Bayer, C., Molin, R., Favaretto, N., Pauletti, V. and Piva, J.T. (2011) Forages, cover crops and related shoot and root additions in no-till rotations to C sequestration in a subtropical Ferralsol. *Soil and Tillage Research* 111, 208–218.

Drury, C.F., Yang, X.M., Reynolds, W.D. and McLaughlin, N.B. (2008) Nitrous oxide and carbon dioxide emissions from monoculture and rotational cropping of corn, soybean and winter wheat. *Canadian Journal of Soil Science* 88, 163–174.

Ernst, O. and Siri-Prieto, G. (2009) Impact of perennial pasture and tillage systems on carbon input and soil quality indicators. *Soil and Tillage Research* 105, 260–268.

Faé, G.S., Sulc, R.M., Barker, D.J., Dick, R.P., Eastridge, M.L. and Lorenz, N. (2009) Integrating winter annual forages into a no-till corn silage system. *Agronomy Journal* 101, 1286–1296.

Fike, J.H., Buergler, A.L., Burger, J.A. and Kallenbach, R.L. (2004) Considerations for establishing and managing silvopastures. *Forage and Grazinglands* doi:10.1094/FG-2004-1209-01-RV (http://www.plantmanagementnetwork.org/pub/fg/review/2004/silvo/, accessed 27 November 2013).

Files, A.C. and Smith, S.N. (2001) Agricultural integration: systems in action. *Maine Agriculture Center Publication 002*, University of Maine, Orono, Maine (http://www.slideshare.net/Aliki85w/agricultural-integration-systems-in-action-the-university-of-maine, accessed 13 December 2013).

Follett, R.F., Varvel, G.E., Kimble, J.M. and Vogel, K.P. (2009) No-till corn after bromegrass: effect on soil carbon and soil aggregates. *Agronomy Journal* 101, 261–268.

Franzluebbers, A.J. (2007) Integrated crop–livestock systems in the southeastern USA. *Agronomy Journal* 99, 361–372.

Franzluebbers, A.J. and Stuedemann, J.A. (2007) Crop and cattle responses to tillage systems for integrated crop–livestock production in the Southern Piedmont, USA. *Renewable Agriculture and Food Systems* 22, 168–180.

Franzluebbers, A.J. and Stuedemann, J.A. (2008a) Early response of soil organic fractions to tillage and integrated crop–livestock production. *Soil Science Society of America Journal* 72, 613–625.

Franzluebbers, A.J. and Stuedemann, J.A. (2008b) Soil physical responses to cattle grazing cover crops under conventional, and no tillage in the Southern Piedmont, USA. *Soil and Tillage Research* 100, 141–153.

Franzluebbers, A.J. and Stuedemann, J.A. (2010) Surface soil changes during twelve years of pasture management in the Southern Piedmont, USA. *Soil Science Society of America Journal* 74, 2131–2141.

Franzluebbers, A.J., Stuedemann, J.A. and Wilkinson, S.R. (2001) Bermudagrass management in the Southern Piedmont USA: I. Soil and surface residue carbon and sulfur. *Soil Science Society of America Journal* 65, 834–841.

Franzluebbers, A.J., Sulc, R.M. and Russelle, M.P. (2011) Opportunities and challenges for integrating North-American crop and livestock systems. In: Lemaire, G., Hodgson, J. and Chabbi, A. (eds) *Grassland Productivity and Ecosystem Services*. CAB International, Wallingford, UK, pp. 208–218.

Franzluebbers, A.J., Endale, D.M., Buyer, J.S. and Stuedemann, J.A. (2012) Tall fescue management in the Piedmont: sequestration of soil organic carbon and total nitrogen. *Soil Science Society of America Journal* 76, 1016–1026.

Freitas, P.L., Blancaneaux, P., Gavinelli, E., Larré-Larrouy, M.C. and Feller, C. (2000) Nature and level of organic stock in clayey oxisols under different land use and management systems. *Pesquisa Agropecuária Brasileira* 35, 157–170.

Freixo, A.A., Machado, P.L.O.A., dos Santos, H.P., Silva, C.A. and Fadigas, F.S. (2002) Soil organic carbon and fractions of a Rhodic Ferralsol under the influence of tillage and crop rotation systems in southern Brazil. *Soil and Tillage Research* 64, 221–230.

García-Prechac, F., Ernst, O., Siri-Prieto, G. and Terra, J.A. (2004) Integrating no-till into crop–pasture rotations in Uruguay. *Soil and Tillage Research* 77, 1–13.

Gliessman, S.R. (2007) *Agroecology: The Ecology of Sustainable Food Systems.* CRC Press, Boca Raton, Florida.

Gregorich, E.G., Rochette, P., VandenBygaart, A.J. and Angers, D.A. (2005) Greenhouse gas contributions of agricultural soils and potential mitigation practices in Eastern Canada. *Soil and Tillage Research* 83, 53–72.

Hanson, J.D. and Franzluebbers, A.J. (2008) Principles of integrated agricultural systems. *Renewable Agriculture and Food Systems* 23, 263–264.

Hargrove, W.L. (ed.) (1991) *Cover Crops for Clean Water.* Proceedings of an International Conference, West Tennessee Experimental Station, 9–11 April 1991, Jackson, Tennessee. Soil and Water Conservation Society, Ankeny, Iowa, 198 pp.

Harris, R.H., Clune, T.S., Peoples, M.B., Swan, A.D., Bellotti, W.D., Chen, W., et al. (2007) The importance of in-crop lucerne suppression and nitrogen for cereal companion crops in south-eastern Australia. *Field Crops Research* 104, 31–43.

Katsvairo, T.W., Wright, D.L., Marois, J.J., Hartzog, D.L., Rich, J.R. and Wiatrak, P.J. (2006) Sod-livestock integration into the peanut–cotton rotation: a systems farming approach. *Agronomy Journal* 98, 1156–1171.

Lawrence, J.D. and Strohbehn, D.R. (1999) Understanding and managing costs in beef cow–calf herds. Iowa Beef Center, Iowa State University, Ames, Iowa (http://www2.econ.iastate.edu/faculty/Lawrence/Acrobat/IRMWhitePaper.pdf, accesssed 13 December 2013).

Lesoing, G., Shain, D., Klopfenstein, T., Gosey, J. and Schroeder, M. (1996) Effect of sorghum and cornstalk grazing on crop production. In: *Nebraska Beef Cattle Report*, University of Nebraska Cooperative Extension MP 66-A. University of Nebraska, Lincoln, Nebraska (http://digitalcommons.unl.edu/cgi/viewcontent.cgi?article=1475&context=animalscinbcr, accessed 13 December 2013).

Marchão, R.L., Becquer, T., Brunet, D., Balbino, L.C., Vilela, L. and Brossard, M. (2009) Carbon and nitrogen stocks in a Brazilian clayey Oxisol: 13-year effects of integrated crop–livestock management systems. *Soil and Tillage Research* 103, 442–450.

Morón, A. and Sawchik, J. (2002) Soil quality indicators in a long-term crop-pasture rotation experiment in Uruguay. In: *Proceedings of the 17th World Congress of Soil Science*, Symposium No 32, Paper 1327, Thailand. CD-ROM. International Union of Soil Sciences, Bangkok.

Nafziger, E.D. and Dunker, R.E. (2011) Soil organic carbon trends over 10 years in the Morrow Plots. *Agronomy Journal* 103, 261–267.

Nair, V.D., Nair, P.K.R., Kalmbacher, R.S. and Ezenwa, I.V. (2007) Reducing nutrient loss from farms through silvopastoral practices on coarse-textured soils of Florida, USA. *Ecological Engineering* 29, 192–199.

Petersen, S.O., Regina, K., Pöllinger, A., Rigler, E., Valli, L., Yamulkif, S., et al. (2006) Nitrous oxide emissions from organic and conventional crop rotations in five European countries. *Agriculture, Ecosystems and Environment* 112, 200–206.

Powell, J.M. and Russelle, M.P. (2009) Dairy heifer management impacts manure N collection and cycling through crops in Wisconsin, USA. *Agriculture, Ecosystems and Environment* 131, 170–177.

Redmon, L.A., Horn, G.W., Krenzer, E.G. Jr and Bernardo, D.J. (1995) A review of livestock grazing and wheat grain yield: boom or bust? *Agronomy Journal* 87, 137–147.

Russelle, M.P., Entz, M.H. and Franzluebbers, A.J. (2007) Reconsidering integrated crop–livestock systems in North America. *Agronomy Journal* 99, 325–334.

Salvo, L., Hernandez, J. and Ernst, O. (2010) Distribution of soil organic carbon in different size fractions, under pasture and crop rotations with conventional tillage and no-till systems. *Soil and Tillage Research* 109, 116–122.

Singer, J.W., Franzluebbers, A.J. and Karlen, D.L. (2009) Grass-based farming systems: soil conservation and environmental quality. In: Wedin, W.F. and Fales, S.L. (eds) *Grassland: Quietness and Strength for a New American Agriculture.* American Society of Agronomy, Crop Science Society of America, Soil Science Society of America, Madison, Wisconsin, pp. 121–136.

Steiner, J.L. and Franzluebbers, A.J. (2009) Farming with grass – for people, for profit, for production, for protection. *Journal of Soil and Water Conservation* 64, 75A–80A.

Studdert, G.A., Echeverria, H.E. and Casanovas, E.M. (1997) Crop–pasture rotation for sustaining the quality and productivity of a Typic Argiudoll. *Soil Science Society of America Journal* 61, 1466–1472.

Sulc, R.M. and Tracy, B.F. (2007) Integrated crop–livestock systems in the U.S. Corn Belt. *Agronomy Journal* 99, 335–345.

Taylor, K.W., Epplin, F.M., Prorsen, B.W., Fieser, B.G. and Horn, G.W. (2010) Optimal grazing termination date for dual-purpose winter wheat production. *Journal of Agricultural and Applied Economics* 42, 87–103.

Tonitto, C., David, M.B. and Drinkwater, L.E. (2006) Replacing bare fallows with cover crops in fertilizer-intensive cropping systems: a meta-analysis of crop yield and N dynamics. *Agriculture, Ecosystems and Environment* 112, 58–72.

Tracy, B.F. and Zhang, Y. (2008) Soil compaction, corn yield response, and soil nutrient pool dynamics within an integrated crop–livestock system in Illinois. *Crop Science* 48, 1211–1218.

Webb, J., Pain, B., Bittman, S. and Morgan, J. (2010) The impacts of manure application methods on emissions of ammonia, nitrous oxide and on crop response – a review. *Agriculture, Ecosystems and Environment* 137, 39–46.

Welch, L.F., Wilkinson, S.R. and Hillsman, G.A. (1967) Rye seeded for grain in coastal bermudagrass. *Agronomy Journal* 59, 467–472.

9 Land Managed for Multiple Services

Richard Aspinall

James Hutton Institute, Craigiebuckler, Aberdeen, UK

9.1 Introduction

Land systems and their dynamics are managed to provide a variety of ecosystem services used by individuals and society. This chapter uses examples from agricultural land use to examine the nature and sources of the multiple goods and services produced by land systems as a set of services that are directly influenced by land management. First, the global-scale and scope of land change and pressures on land are reviewed briefly; this reveals the extent of the impacts of management on the global system. Second, the specification of land systems as coupled human–environment systems incorporating natural, technological, financial and human capital and flows is emphasized. This system's formulation supports the analysis and understanding of land dynamics at multiple spatial, temporal and organizational scales, and helps to reveal the full range of inputs of different forms of capital, and also the importance of human–environment relationships in the production of ecosystem goods and services. Third, the nature of different goods and services produced from land, and the sources of these goods as they relate to agricultural land systems and land management are reviewed, using examples based on farming in Scotland. This provides an understanding of ecosystem services that are based on land and land management, and emphasizes the roles of human activities in the delivery of ecosystem services. It (i) discusses the human processes by which ecosystem services are realized in land systems, (ii) increases understanding of the role of human activity in the delivery of ecosystem services, including the costs, benefits and trade-offs of services of different types of value, and (iii) identifies the need to include consideration of the delivery of multiple services at multiple scales from land through land management. The case study results demonstrate the need to move beyond the understanding of ecosystem services as solely a function of biodiversity and natural capital and to understand land management activities by people as central to the provision of ecosystem services and multiple benefits.

9.2 Land Change: Global Scale, Scope and Impact

Land use and land-use change and the consequences of land-use management activities are global in their scope and scale of impact (Foley *et al.*, 2005; Rockström *et al.*, 2009). Of the 510 million km² total surface area of the Earth, about 72% is ocean and 28%, about 140 million km², is land (Millennium Ecosystem Assessment, 2005). Cropland and pastures/rangelands together occupy about 43 million km² (Ramankutty *et al.*, 2008), or about 40% of the ice-free land area (Ramankutty and Foley, 1999); deserts, tundra, forests, mountains and urban areas cover the rest. Human land use and management practices and activities have been transformational, not only in replacing other habitats (Vitousek *et al.*, 1997b) but also through their intensity of resource use and impact (Monfreda et al., 2008). About 50% of the global nitrogen flux is now associated with human activity through the use of synthetic fertilizers and fossil fuel combustion (Matson et al., 1997; Vitousek et al., 1997a). The water footprint of humanity is estimated at 9087 Gm^3 year^{-1} (Hoekstra and Mekonnen, 2012), with 26% of total terrestrial evapotranspiration and 54% of geographically and temporally accessible runoff being used for human activities (Postel et al., 1996); agriculture accounts for 85% of consumptive water use (Gleick, 2003). About 15.6 Pg C year^{-1}, or 24%, of total potential net primary productivity of the land-based systems of the planet is appropriated for human use (Haberl et al., 2007), of which 53% is through harvest and 40% through land-use-induced productivity changes (Haberl et al., 2007). The total primary energy consumption of all human activities is about 520 EJ year^{-1} (Krausmann et al., 2009).

These demands placed on land and other natural resources, combined with global population growth and changing climate, lead to international and national concerns over food, water and energy security. This nexus of issues dominates international and government policy (Beddington, 2010) and is attracting an increasing level of scientific and political enquiry (Godfray et al., 2010). Many food, water and energy security issues are associated with inequalities in resource distribution and access (Ingram et al., 2010) (see Ingram, Chapter 16, this volume), but other reasons are related to the total amount of food, water and energy available, to drivers of land change and to the management of resources. Climate change is a significant global driver, with regional and local implications for land use (IPCC, 2007). The monitoring of trends and changes in weather and climate, developing scenarios for climate change and the use of climate models to project scientific understanding of climate to develop scenarios have together produced consensus on possible future trajectories of climate (IPCC, 2007). There are expected increases in mean summer temperature (in both the northern and southern hemispheres) and complex patterns of change in precipitation (IPCC, 2007). These changes, and associated changes in agroclimatic and other weather conditions (Rivington et al., 2013), will have considerable impacts on land use (Matthews et al., 2008) and are considered further in this chapter in relation to the case study presented for Scotland.

Meta-studies of land change identify the proximate and underlying drivers of change (Geist and Lambin, 2002, 2004). Underlying drivers include economic, policy, institutional, technological, cultural, demographic and environmental factors. This same set of factors form the basis for the different capitals of the sustainable livelihoods framework (Brocklesby and Fisher, 2003), and are also reflected in political and economic concerns over trade, policy and markets that dominate discussion of global change (Stern, 2007). Connections between the underlying and proximate drivers, global changes and local concerns, solutions and actions can be recognized in land-use issues and land management responses to global change. Although land use and land-use change are global-scale phenomena and have global impact, land use and land management are subject to the influences of a wide variety of processes operating at a range of scales to influence the dynamics of the land system (Lambin et al., 2003). These influences include global-scale political and economic factors, as well as other underlying factors. However, the international land systems science community and land managers recognize that global forces become the main determinants of land-use change primarily through their ability to amplify or attenuate local factors (Lambin et al., 2001; Lambin and Geist, 2006b) and via global teleconnections (Seto et al., 2012). Local factors, including land management practices, are, therefore, of primary importance, not only in driving land changes that are local, regional and global in scope and impact, but also in resolving land-use issues and delivery of ecosystem services.

9.3 The Land System as a Coupled Human–Environment System

The land system can be represented as a coupled human–environment system (Global Land Project, 2005; Foresight Land Use Futures Project, 2010). A systems approach and structure can represent not only the driving factors of land system change (Lambin et al., 2001) but also the structures and processes that link human and environment subsystems. It necessarily embeds the different spatial, temporal and organizational scales of environment (Levin, 1992; Chave and Levin, 2003) and human (Gibson et al., 2000) system processes. The environment subsystem includes the earth system with climatic, biological, biogeochemical, hydrological and other subsystems, linked most directly to the human subsystem through land cover. The human subsystem includes economic, institutional, demographic, technological and political factors, and links most directly to the environment system through land use. Ecosystem goods and services are defined as the benefits society receives from environment systems (Millennium Ecosystem Assessment, 2005). The interaction of land use and land cover is recognized in research by international programmes such as the Land Use and Cover Change (LUCC) (Lambin and Geist, 2006a) and the NASA Land-Cover/Land-Use Change (LCLUC) (Gutman et al., 2004) programmes and in the main elements of recent interdisciplinary research that has defined land systems and land change science, including observation and monitoring, impacts and modelling (Turner et al., 2007). Processes in the environment subsystem are driven largely by biodiversity, energy and material flows, while processes in the human subsystem are driven by flows of finance, ideas, innovation and human activity executed by decisions. Land, as a coupled human and environment system, reflects fully both the human and environment subsystems and their interaction. A coupled systems model thus emphasizes the systems-level interactions, not only of factors that affect land but also of the different activities and processes that have an impact on food, water and energy, as well as other goods and services provided by land systems. Land use, and the multiple services produced from land, are not, therefore, solely the product of environment systems; active land management is central to the delivery of ecosystem services.

9.4 Ecosystem Services

Terrestrial ecosystems are the primary source of goods and services that support human well-being and society (Costanza et al., 1997; Millennium Ecosystem Assessment, 2005). Global-scale environmental changes have raised the profile of multiple use of land and natural resources and the importance of ecosystem goods and services (Daily and Matson, 2008), leading to reviews of the state of global ecosystems and the goods and services provided (Millennium Ecosystem Assessment, 2005). National-scale reviews are also being carried out, such as the UK National Ecosystem Assessment (UK National Ecosystem Assessment, 2011).

There are numerous taxonomies of ecosystem goods and services. The Millennium Ecosystem Assessment (MEA) has popularized a categorization into four main types: supporting, regulating, provisioning and cultural services (Table 9.1). This has been continued in other assessments including the UK National Ecosystem Assessment (UK NEA; UK National Ecosystem Assessment, 2011). Other classifications are aligned closely to this typology, including the Common International Classification of Ecosystem Services (EEA, 2010).

The MEA and UK NEA are both founded on a conceptual model that emphasizes biodiversity and natural capital (Millennium Ecosystem Assessment, 2005; Mace et al., 2011). In this approach, human and technological capitals are considered to be drivers of change that are exogenous to the process-based foundation of ecosystem services in biodiversity and natural capital. The Intergovernmental Science-Policy Platform on Biodiversity and Ecosystem Services (IPBES) will continue the process of linking biodiversity and ecosystem services (Turnhout

Table 9.1. Definitions of the four main categories of ecosystem goods and services. (Sources: Millennium Ecosystem Assessment, 2005; EEA, 2010; UK National Ecosystem Assessment, 2011.)

Ecosystem service	Definition and examples
Supporting	The underpinning system functions necessary for the production of all other ecosystem goods and services. Supporting services refer to the products and results of biogeochemical, physical, chemical and biological functions that operate in environmental systems. Biodiversity is a key part of this foundational system function. Examples of supporting services include soil formation, photosynthesis, primary production, nutrient cycling and water cycling.
Regulating	The goods, services and benefits obtained from environmental regulation functions provided by ecosystem processes. Examples of regulating services include air quality, climate regulation, flood and hazard regulation, erosion regulation, regulation of diseases and pests, regulation of water quality (water purification) and waste treatment.
Provisioning	The goods and other products obtained from ecosystems. Examples of provisioning services include food, fibre, fuel, fresh water, genetic and ornamental resources, biochemicals, natural medicines and pharmaceuticals. Since the environment also provides sources of energy, renewable and fossil sources of energy can also be considered among provisioning services.
Cultural	The non-material benefits people obtain from ecosystems through education, exercise and recreation, spiritual enrichment, cognitive development, reflection and aesthetic experience. The UK NEA provides a rich suite of definitions of cultural services that expand the range of cultural services defined by the MEA (Church *et al.*, 2011).

et al., 2012; Pe'er *et al.*, 2013). Without diminishing the importance and role of biodiversity, the full range of ecosystem goods and services obtained from land systems are not produced solely by biodiversity and natural capital, and there is recognition that evaluation of ecosystem goods and services against land systems, with explicit attention to land management, is needed (Global Land Project, 2005; Carpenter *et al.*, 2009). Physical, chemical and biogeochemical processes, and geodiversity (Gray, 2012), also contribute ecosystem goods and services; the Common International Classification of Ecosystem Services explicitly recognizes abiotic services (EEA, 2010).

Additionally, although valuation of services in the UK NEA includes concepts of health values and shared (social) values, quantitative valuation is primarily economic, with a focus on financial benefit (Mace *et al.*, 2011). Currently, most valuation protocols either omit consideration of the costs of generating services or preclude other accounting approaches that examine input–output relationships and changes in capital stocks and flows that generate values. The cost of recovering ecosystem goods and services is important, not only because energy, labour (human activity), money and other resources are expended to recover goods and services as part of land use and land management, but also because investment in natural capital through costs is necessary to ensure sustainability of goods and services. Goods and services can be properly considered technically as flows derived from *funds*, rather than from *stocks*; funds require investment to ensure their maintenance (Georgescu-Roegen, 1975; Giampietro *et al.*, 2009).

The remainder of this chapter considers the production of food from land systems and identifies some of the dominant human and economic processes that produce these services. The goods and services are described using examples from Scotland, a country with a detailed synthesis of ecosystem services as a result of the UK

National Ecosystem Assessment (Aspinall et al., 2011). Strictly, the 'food products' produced by agriculture are the raw materials for a longer chain of production that provides food to people. They are, however, the first stage in this chain, and without agricultural production there would be no food. Agriculture is the most immediate link between environmental and ecological systems (ecosystems) and food supply, and, as described above, the global scale, scope and impact of agriculture on the planetary environment is of significance for global sustainability. This case study considers the expenditures on as well as the income (the nominal economic 'value' that is reported in national statistics as the economic contribution of agriculture, and in ecosystem assessments such as the UK National Ecosystem Assessment) for agriculture in Scotland from 1940 to 2010. Additionally, the energy investment is computed to help evaluate the input of other capital resources to agriculture. This analysis gives an estimate of the relative change in land management intensity over time for food production by agriculture in Scotland. The energy output of food materials produced through agricultural production is also calculated to provide a simple input–output analysis in energy terms. Although the case study is Scotland, the trends and relationships that are identified can be considered to apply more broadly to similar agricultural types in other areas, particularly in northern Europe, where grain, grassland and livestock production under temperate climatic conditions prevail.

9.5 Data and Methods

Data for agricultural production, land use and economic costs and benefits from 1940 to the present were extracted from the annual series of Economic Reports for Scottish Agriculture (Rural and Environment Research and Analysis Directorate, 2010) and the Agricultural (June) Census (Rural and Environment Research and Analysis Directorate, 1939–present). These data are the basis for national assessments, and were the basis for assessments of production and economic value in the Scotland chapter of the UK National Ecosystem Assessment (Aspinall et al., 2011). Data for energy use in agriculture for each year from 1940 to the present were calculated using the methods described by Leach (Leach, 1976), with data from the annual series Digest of UK Energy Statistics (Department for Energy and Climate Change, 2010). Energy associated with pesticides and fertilizers was based on the recorded national usage statistics for pesticides (SASA, 2009) and fertilizers (Cooke, 1982; British Survey of Fertiliser Practice, 2011) rather than on the value of farming expenditure, as in Leach. This is possible because the analysis is of the aggregate farm production for the country rather than individual farm types, and usage statistics for fertilizers and pesticides are available for this national level of aggregation. The use of energy as a metric allows the comparison of different inputs and outputs. Annual economic values for inputs and outputs were computed to 2010 equivalent value using GDP deflators. It should be noted that aggregate statistics mask differences between types of farming and regions in Scotland, but in the context of evaluating the impacts of land management for food production as an ecosystem service, the process of accounting in ecosystem assessment and valuation at a national scale, the analysis is constructive.

9.6 Results

9.6.1 Provisioning services

The NEA reports the changes in provisioning services associated with Scottish agriculture from 1940 to 2010 (Aspinall et al., 2011). Figure 9.1a and b shows the changes in the production of crops and meat in Scotland from 1940 to 2010. The production of barley and wheat has increased since the 1950s and 1960s (from about 250,000 t year^{-1} to about 1.8 million t year^{-1} since the 1980s for barley, and from about 100,000 t year^{-1} in the 1940s–1950s to over 800,000 t year^{-1} for wheat), while the production of oats has

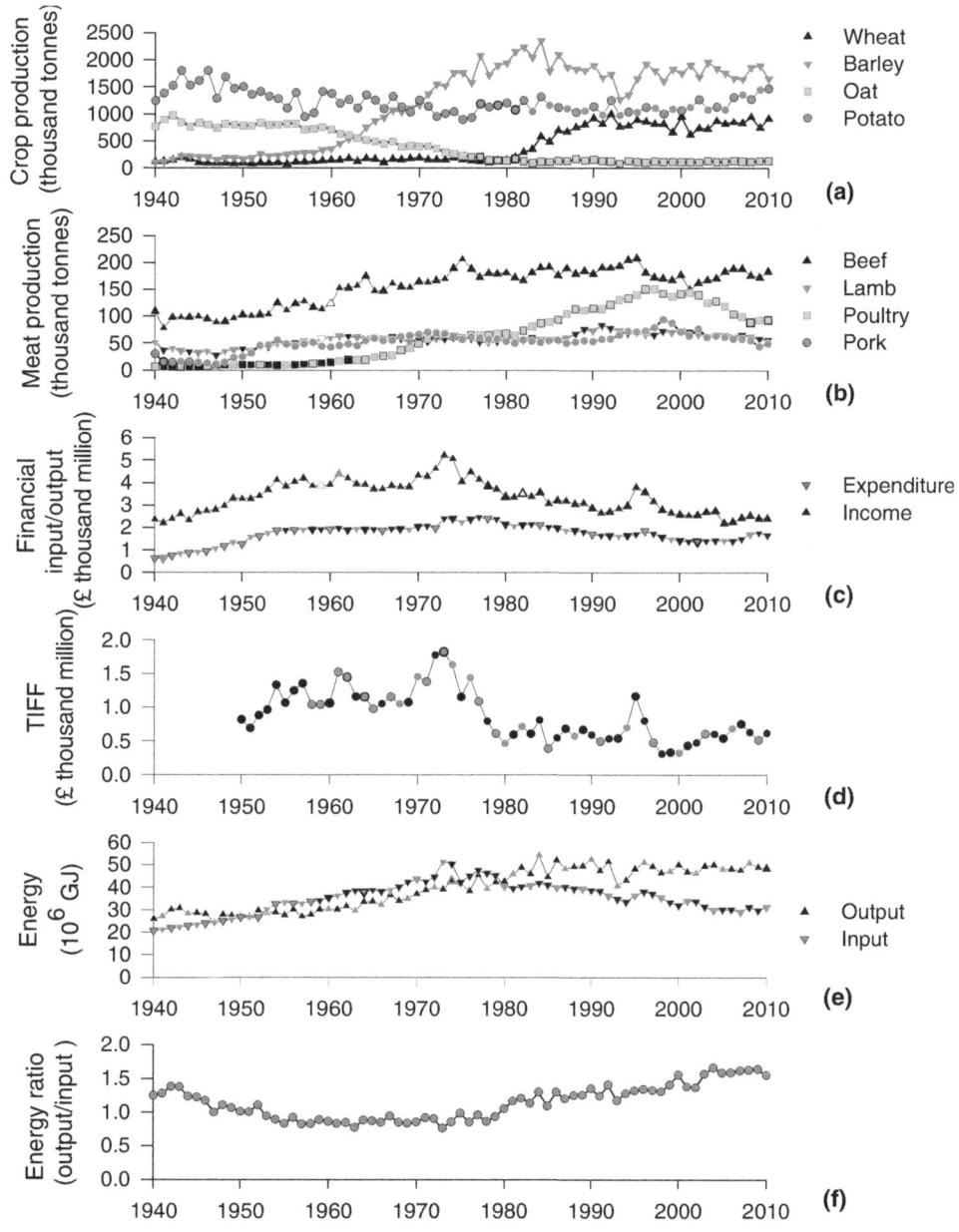

Fig. 9.1. Trends in production, financial performance and energy input and output for agriculture in Scotland 1940–2010: (a) production of crops (values in thousand tonnes); (b) production of meat (values in thousand tonnes); (c) annual income and expenditure (in thousand million £ at 2010 prices); (d) total income from farming (TIFF) (in thousand million £ at 2010 prices); (e) total energy inputs and gross outputs (in Gigajoules, GJ); (f) energy ratio. (Data on production and finances from the annual Agricultural (June) Census of Agriculture (Rural and Environment Research and Analysis Directorate, 1939–present) and annual Economic Reports on Scottish Agriculture (Rural and Environment Research and Analysis Directorate, 2010). Energy calculated using methods given in Leach (1976) using data from the annual series of publications in the *Digest of UK Energy Statistics* (Department for Energy and Climate Change, 2010).)

declined since the 1950s (from about 800,000 t year^{-1} during the 1940s and 1950s to about 120,000 t year^{-1} from the 1980s to the present). Part of the increase in barley and wheat is due to an increase in the area planted, mainly on land that previously grew oats, and part is due to an increase in yield as technologies have changed. Although the total area of land under cereals decreased from about 530,000 ha in the 1940s to about 430,000 ha in the 2000s, the total production of wheat, barley and oats together increased from about 1 million t year^{-1} in the 1940s and 1950s to about 3 million t year^{-1} in the 2000s. The production of potatoes has remained between 1 million and 1.5 million t year^{-1} since the 1950s, the area planted having decreased from 85,000 ha in the 1940s to 30,000 ha by the 2000s, while yields have increased from 17–18 t ha^{-1} to over 45–47 t ha^{-1} over the same period.

Beef meat production increased from about 95,000 t year^{-1} in the 1940s to about 180,000 t year^{-1} by the 1970s, and has remained at this level since. Lamb meat production similarly increased from about 40,000 t year^{-1} in the 1940s and 1950s to about 60,000 t year^{-1} during the 1960s–1980s and 65,000–70,000 t year^{-1} in the 1990s and 2000s. Chicken meat increased from about 10,000 t year^{-1} in the 1950s to a peak of over 150,000 t year^{-1} in the late 1990s; chicken production is currently about 100,000 t year^{-1}. Pork meat production increased to a peak of over 90,000 t in 1998 and has decreased to about 50,000 t year^{-1} in 2010.

9.6.2 Economic value

Table 9.2 shows the average income from and expenditure on farming and the total income from farming (TIFF – data from 1950 onwards) in Scotland for each decade from the 1940s to the 2000s. The absolute values of farm incomes, expenditures and TIFF have risen markedly since the 1940s, although they have declined in real terms when measured at 2010 prices (Table 9.2). The annual values of input, expenditure and TIFF for agriculture in Scotland at 2010 prices are shown in Fig. 9.1c and d. Analysis of financial values at 2010 prices shows that the relative value of income generally increased between the 1940s and 1973/74, after which it showed a steady decrease, although there was a short-term increase in the mid-1990s as world prices for grains increased due to shortages. Expenditure has generally increased from the 1940s to the present in real terms. The differences between income and expenditure show a general increase from the 1940s to 1974 but a decline since, such that, by 2010, the difference between income and expenditure was the lowest that it had been in real terms for the past 70 years. There are numerous

Table 9.2. Average values of gross input (expenditure), gross output (income) and total income from farming (in £ millions) for each decade from the 1940s to the 2000s. (Data sources: annual Agricultural (June) Census of Agriculture (Rural and Environment Research and Analysis Directorate, 1939–present) and annual Economic Reports on Scottish Agriculture (Rural and Environment Research and Analysis Directorate, 2010)).

	Gross input (expenditure)	Gross output (income)	Total income from farming	Gross input (expenditure)	Gross output (income)	Total income from farming
	Current prices			2010 prices		
1940s	36.2	76.8		929.7	2646.8	
1950s	82.0	172.8	50.5	1766.2	3759.4	1044.0
1960s	120.5	244.5	72.2	1929.5	3929.6	1165.4
1970s	334.9	595.1	171.8	2277.4	4276.8	1316.8
1980s	781.9	1212.4	237.7	2024.1	3131.0	607.4
1990s	1050.0	1884.5	372.1	1678.7	3001.4	598.4
2000s	1230.4	2048.1	465.0	1490.5	2404.4	561.2

reasons for these changes, which reflect complex interactions between policy changes and world market prices (Ritson and Harvey, 1997). For example, from 1947, the farm sector in the UK was supported through a policy system of Guaranteed Prices and Deficiency Payments, which gave foreign supplies access to domestic markets at world prices but which also maintained domestic food prices at (relatively cheap) world levels for consumers (Angus et al., 2009). In 1973, the UK joined the European Community and the Common Agricultural Policy (CAP); deficiency payments ceased and prices increased through the support policies of CAP. At the same time, although independently, there was a commodity price boom in world markets (Cooper et al., 1975), which increased input costs and strongly influenced both the overall economics of agriculture and the economic value of food as an ecosystem service. The relative changes in farm incomes and expenditures (Fig. 9.1c and d) are, therefore, a product of the combined influences of changes in UK and European policies and of world market prices, as well as farming's response to those changes and signals.

9.6.3 Energy input and output

The energy input and output and the energy ratio (output/input) for Scottish agriculture from 1940 to 2010 are shown in Fig. 9.1e and f. Total energy input grew to about 55×10^6 GJ year^{-1} by the mid-1970s, more than double the input for the late 1940s. Since the late 1970s, total energy input has decreased and is now about 36×10^6 GJ year^{-1}. Energy output increased more slowly, from 23×10^6 GJ year^{-1} in the late 1940s to about 45–50×10^6 GJ year^{-1} by the mid-1990s, remaining at about this level to the present. The energy ratio exceeds 1 during 1940–1945 and again from the mid-1980s to the present; from 1945 to the early 1980s, the energy ratio was less than 1. Increases in energy input between the 1940s and the 1970s were associated particularly with the increased use of fuel, nitrogen fertilizer and pesticides; decreases since the 1970s are generally associated with increased resource-use efficiency.

9.6.4 Trends in production, energetics and economics

Figure 9.2 shows the indicators of energy use and economic production from 1940 to 2010, 1950 being used as the reference year. Indices for TIFF and for energy input and output per hectare follow the trends described above for Fig. 9.1d and e, respectively. The index of person-years employed in agriculture shows the well-recognized decline in the agricultural workforce, which in Scotland is now less than one-fifth of the number that it was in the 1950s. The energy input and output per person both increase markedly over the whole 70-year period. The increases in energy input and output per person reflect that while the energy backing for each agricultural worker has increased by a factor of about 6 since 1950, due to a combination of technological change and decline in the workforce, the energy output per person has increased by a factor of about 10 over the same period. These patterns are consistent with technological innovation and employment in agriculture in the UK in general over the past 70 years (Burgess and Morris, 2009), and with changes in the industrial agriculture of many other countries.

9.7 Discussion

The quantity of food, as a provisioning ecosystem service, delivered by agriculture in Scotland has increased considerably over the past 70 years. Analysis of energy inputs and outputs associated with these services provides a mechanism for illuminating the consequences of human, technological and economic changes associated with the retrieval of ecosystem services. The trends in the generation of provisioning services from agriculture since 1940, illustrated in Figs 9.1 and 9.2, reflect changes in mechanization, area and varieties of crops grown, fertilizer and pesticide use, disease control, animal

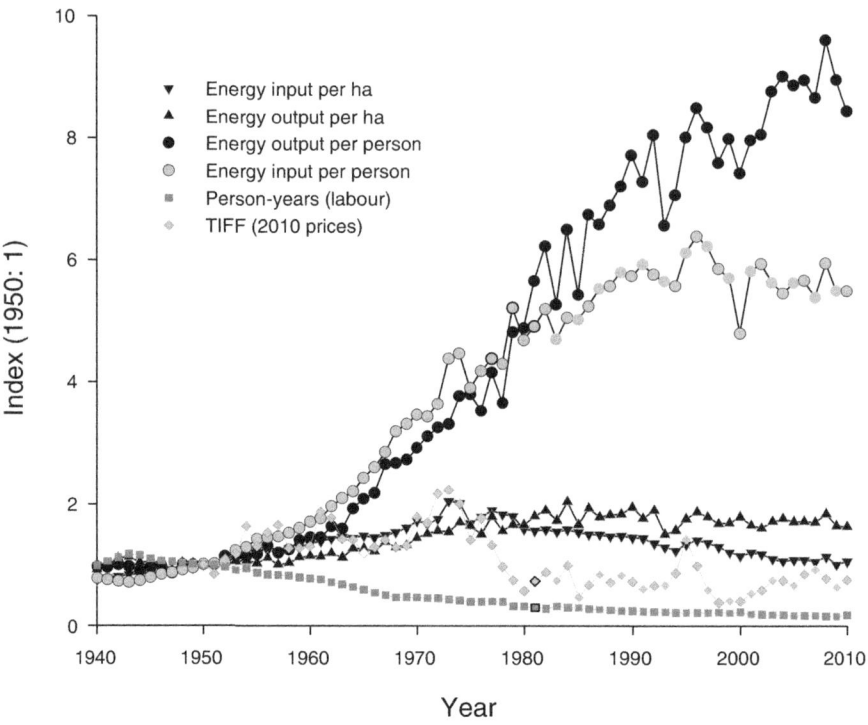

Fig. 9.2. Indices for energy input and output per hectare, energy input and output per person in the farming labour force, person-years and TIFF (in 2010 prices) for Scottish Agriculture 1940–2010. 1950 is used as the base year. (Data sources as for Fig. 9.1.)

breeding and numbers, plant and animal husbandry, land management, employment in agriculture and many other aspects of agricultural activity, including European and national policies, as well as world prices and the operation of markets and trade agreements. Figure 9.1e shows that the energy costs of producing food materials from agriculture increased from the 1940s to the 1970s, but have fallen since the 1980s. There is also a general picture of increased resource-use efficiency in agriculture over the past 30 years, and a marked absolute increase in the production of food as a provisioning service. This increased efficiency in energy use is associated with a decrease in total energy input, while energy output has remained at about $45–50 \times 10^6$ GJ year^{-1} since the 1980s.

However, considering economic inputs as well as outputs for agriculture provides a rather different picture of the economic value of agricultural provisioning services in Scotland over the past 70 years than is found in a focus on the absolute economic value of outputs. Although the current absolute value of agricultural output, at about £2300 million, is the highest it has been over the past 70 years, it has never been lower, both in real terms and in relation to the economy as a whole. Peak income, at 2010 prices over the whole period, was in the early 1970s. Additionally, at the same time as the relative value of output has decreased, the level of expenditure needed to maintain agricultural activity has continued to increase, despite increased efficiencies in resource use. This has produced falling gross valued added (income minus expenditure) since the mid-1970s.

Changes in the economics and energetics of food production in Scotland as the central provisioning ecosystem service provided by agriculture has four lessons and implications for the economic valuation of ecosystem services, both in the particular case of agricultural production and more generally in terms of the market values of goods and services.

1. Costs as well as income should be included in economic accounting for valuation of ecosystem services. As demonstrated by the case of food from agriculture, the recovery of many provisioning services requires considerable financial and other inputs. Expenditures are a major consideration in land management decisions alongside income and value of products, as they are in any business. Understanding costs is also informative about the nature and impacts of land management activities and decision making, and gives a more complete understanding of the human systems contribution to production and the recovery of ecosystem services.

2. For food as a provisioning service, the contributions of agriculture are too narrowly specified to provide a full picture of benefits, including of financial benefits. The value and role of the provisioning services of food from agriculture do not end at the farm gate, which is where the value currently allocated to food as a provisioning service is calculated. Food provision from agriculture also includes providing the raw ingredients for the modern food and drink industry (Scottish Executive, 2007). In the specific case of Scotland, but also more widely, the use of agricultural products beyond the farm is of significant economic value. About 33% of barley and about 60% of the wheat grown in Scotland is used in malting and distilling (Scottish Executive, 2007), and this adds greatly to the value of food as a provisioning service. The food and drink industry in Scotland is the leading industry by value of exports for Scotland, valued at about £4 billion (thousand million) in 2010/11; distilling of spirits accounts for about £3.3 billion, or over 80%, of this total. However, none of this value is attributed to food production in ecosystem services by current economic valuation methodologies, since it is value added beyond the production of 'food' directly from ecosystems.

3. There are values beyond economic value. Without agriculture there would be little or no food, a productive food supply system clearly being of relevance to food security issues and making food important beyond economic value. In addition, there are social, health and other cultural ecosystem service values (Church et al., 2011) that are influenced by food and agricultural land management. None of these is considered in the current narrow economic valuation of food as a provisioning service.

4. Agricultural land management has direct and indirect influences on ecosystem services beyond food production. In Scotland, agriculture directly influences about 75% of the land surface, through management schemes and activities for crop and livestock production (Rural and Environment Research and Analysis Directorate, 2010). Many other ecosystem services, of all the four main categories of services (Table 9.1), are thus inevitably affected by agricultural land management, because it occupies such a large proportion of the nation's land surface. Regulating services such as water quality and quantity, flood management, erosion, soil condition and pest and disease regulation are all affected directly and indirectly by land management and by agriculture.

The relationships between provisioning services from agriculture and natural capital are not only complex but also are multi-scale in space, time and organizational structure (Gibson et al., 2000). The increases in many provisioning services from agriculture since 1940 for Scotland, as elsewhere, are associated with increases in technological inputs, measured here in energy terms, and are not associated directly with natural capital. It is well recognized that technological changes have altered the quantity and quality of agricultural provisioning services at the same time as the capacity of natural capital has declined in Scotland (Aspinall et al., 2011). Despite the importance of technological changes, some aspects of agricultural output are clearly associated with natural capital. For example, insect pollination is considered to have contributed about £43 million in value to Scottish agriculture in 2010 (Aspinall et al., 2011); it should also be noted that insect pollination is a 'free' service, with no input costs, provided by a healthy environment. However, many of the benefits and changes in provisioning services are not founded on natural capital and biodiversity, as implicitly suggested by the conceptual framework of the NEA. A more holistic ecosystem assessment should include the analysis of human and technological capital as well as the consideration of natural capital.

As human and technological processes and innovation have become increasingly integral to the functioning of ecosystems, and may exceed natural processes in impact and importance (Rockström et al., 2009), and as recognition of ecosystem services as a set of benefits obtained from land systems increases, understanding the multi-scale relationships of land management, land systems and ecosystem services has become more important (Carpenter et al., 2009). Agricultural land management to produce food is connected to many other ecosystem services across a range of scales in space, time and organization. For example, agricultural field operations that produce crops influence important environmental properties and functions at both smaller and larger spatial scales. For example, soils provide supporting services (soil formation, nutrient cycling) and regulating services (erosion control, water quality), yet their physical microstructure and chemical and biological functions and processes are influenced strongly by field-scale land management (Allison et al., 2005, Castellanos-Navarrete et al., 2012; Tian et al., 2012). Similarly, field-scale operations also influence river catchment-scale ecosystems and associated services such as water quality and quantity, flood regulation, erosion regulation through hydrological and sediment transport processes (Friberg, 2010). For example, in the north-east of Scotland, the composition and relative proportion of different habitats, types of land cover and associated land uses within catchment areas, and especially the proportion of arable agriculture, was found to have a strong correlation with river water chemistry (Wright et al., 1991). Increased understanding of the links between river catchment condition, ecosystem services and agricultural management are changing the science of catchment management to include policy and planning issues so that multi-scale benefits of land management can be achieved (Sharpley et al., 2010).

In relation to climate change, land use can act as both a source and a sink of greenhouse gases. In 2011, agriculture contributed about 16% of Scotland's greenhouse gases (GHGs) (2% of CO_2, 52.7% CH_4 and 86.6% N_2O), while land-use change produced a net reduction in GHGs equivalent to 66.5% of total agricultural emissions (Salisbury et al., 2013). Reducing GHG emissions, while also managing land for the delivery of multiple other services, is thus of increasing importance. There are a variety of potential mitigation measures to reduce GHG emissions within agriculture (Smith et al., 2008; MacLeod et al., 2010), including changes in the management of cropland, grazing land, cultivated organic soils, livestock and manure; use of set-aside, land-use change and agroforestry; and reducing direct (fuel, heating, electricity) and indirect (fertilizers) energy use. The importance of achieving these measures in relation to the provision of goods and services beyond the reduction of GHG emissions has not been addressed directly, although they should also improve overall environmental quality, biodiversity and functioning of environmental systems (Matson et al., 1997; Lowe et al., 2009).

The challenge presented by climate change for developing systems of husbandry and agricultural practice that mitigate climate change by reduction in GHG emissions, and that deliver multiple ecosystem goods and services, is, however, made complex, not by changes in the raw climatic variables of temperature and rainfall but by changes in the associated complex seasonal dynamics of weather as it relates to the dynamics of land systems and land management. Thus, in Scotland, a warmer and wetter climate, as projected, might be expected to be beneficial for agriculture, improving the flexibility of land for agriculture by removing growing temperature limitations (Brown et al., 2008). However, the important changes in climate are less direct and are measured more usefully through a variety of agro-meteorological characteristics that relate to different land management activities and crop and livestock production requirements (Matthews et al., 2008; Rivington et al., 2013), which themselves change over the agricultural seasons. Projected climate changes for Scotland are not simply warmer and wetter, but describe an agroclimatic regime that is very different to the present climate, introducing new seasonal patterns that will necessitate substantially changed land management. Projected changes include later field access in spring, a lengthened growing season, increased soil moisture deficits and the likelihood of drought in summer, more extreme – and potentially damaging – rainfall later in the growing season and an end to air frosts earlier in the year (Rivington et al., 2013). These changes will have consequences for field management, choice and the performance of crops and varieties, harvest operations and diseases and pests, all of which will require management responses with consequent impacts on ecosystem services beyond food production.

Linking mitigation and adaptation to climate change, with a desire to achieve multiple benefits from land, will require considerable policy, planning and management activity involving farmers and other land managers, policy makers, scientists and others including a variety of governance organizations (Curry, 2001; Dwyer and Hodge, 2001). Taking a whole-systems approach to land (Foresight Land Use Futures Project, 2010) provides a mechanism for integrating human and environment subsystems and information, and is necessary not only for mitigation and adaptation related to climate change but also to ensure the multiple benefits required from land are achieved.

9.8 Conclusions

Recognition of the importance of all categories of ecosystem services, and that their derivation from land systems, as coupled human and environment systems, is influenced over large areas by agricultural land management for food production, offers a strong rationale and set of goals for developing strategies for agricultural and other land management to produce multiple benefits. Process-level understanding of the interactions of human, technological and natural capital in land systems, and at multiple scales, is necessary for evaluating the status, condition and trends of ecosystem services, and their sustainability. This understanding also provides a knowledge base necessary for developing strategies to produce multiple benefits. The multi-scale nature of ecosystems, land use and land management, and their interactions, as well as the diversity of interrelated ecosystem services, makes this scientifically challenging.

In the example of agricultural land in Scotland from 1940 to 2010 that is presented in this chapter, it is evident that land management for food production, as a core provisioning service, is a major influence on environment over a large area and has an influence on many ecosystem services other than food production. Many of the impacts of the global-scale and scope of land changes, reviewed early in the chapter, are the aggregate evidence of local-scale land management activities. The example also illustrates the importance of considering human activity and land management in conjunction with environmental system processes. Significant financial, technological and human capitals are expended to ensure food production, and many other ecosystem services are similarly dependent on human systems. Examination of ecosystem services within land systems as coupled environment and human systems supports analysis and understanding at multiple spatial, temporal and organizational scales, and helps to reveal the role of different inputs to the production of ecosystem goods and services.

Human processes, including land management, as well as biodiversity and natural capital, are central to producing the multiple benefits of ecosystem services from land. Responses of land use to climate change are, similarly, most appropriately considered in the context of a whole-systems understanding of land as a coupled human-environment system.

Note

The views expressed in this chapter are the author's own.

References

Allison, V.J., Miller, R.M., Jastrow, J.D., Matamala, R. and Zak, D.R. (2005) Changes in soil microbial community structure in a tallgrass prairie chronosequence. *Soil Science Society of America Journal* 69, 1412–1421.

Angus, A., Burgess, P.J. and Lingard, J. (2009) Agriculture and land use: demand for and supply of agricultural commodities, characteristics of the farming and food industries, and implications for land use in the UK. *Land Use Policy* 26S, S230–S242.

Aspinall, R.J., Green, D., Spray, C., Shimmield, T. and Wilson, J. (2011) Status and changes in the UK ecosystems and their services to society: Scotland. In: UK National Ecosystem Assessment (ed.) *The UK National Ecosystem Assessment Technical Report.* UNEP-WCMC, Cambridge, UK, pp. 895–977.

Beddington, J. (2010) Food security: contributions from science to a new and greener revolution. *Philosophical Transactions of the Royal Society B-Biological Sciences* 365, 61–71.

British Survey of Fertiliser Practice (2011) *Fertiliser Use on Farm Crops for Crop Year 2010.* Defra, York, UK.

Brocklesby, M.A. and Fisher, E. (2003) Community development in sustainable livelihoods approaches – an introduction. *Community Development Journal* 38, 185–198.

Brown, I., Towers, W., Rivington, M. and Black, H.I.J. (2008) Influence of climate change on agricultural land-use potential: adapting and updating the land capability system for Scotland. *Climate Research* 37, 43–57.

Burgess, P.J. and Morris, J. (2009) Agricultural technology and land use futures: the UK case. *Land Use Policy* 26S, S222–S229.

Carpenter, S.R., Mooney, H.A., Agard, J., Capistrano, D., DeFries, R.S., Díaz, S., et al. (2009) Science for managing ecosystem services: beyond the Millennium Ecosystem Assessment. *Proceedings of the National Academy of Sciences* 106, 1305–1312.

Castellanos-Navarrete, A., Rodriguez-Aragones, C., De Goede, R.G.M., Kooistra, M.J., Sayre, K.D., Brussard, L. and Pulleman, L.L. (2012) Earthworm activity and soil structural changes under conservation agriculture in central Mexico. *Soil and Tillage Research* 123, 61–70.

Chave, J. and Levin, S. (2003) Scale and scaling in ecological and economic systems. *Environmental and Resource Economics* 26, 527–557.

Church, A., Burgess, J. and Ravenscroft, N. (2011) Cultural services. In: UK National Ecosystem Assessment (ed.) *The UK National Ecosystem Assessment Technical Report.* UNEP-WCMC, Cambridge, UK, pp. 633–691.

Cooke, G. (1982) *Fertilizing for Maximum Yield.* Granada, London.

Cooper, R.N., Lawrence, R.Z., Bosworth, B. and Houthakker, H.S. (1975) The 1972-75 commodity boom. *Brookings Papers on Economic Activity* 1975, 671–723.

Costanza, R., d'Arge, R., deGroot, R., Farber, S., Grasso, M., Hannon, B., et al. (1997) The value of the world's ecosystem services and natural capital. *Nature* 387, 253–260.

Curry, N. (2001) Recreation and conservation in the countryside: policy challenges. In: Smout, T.C. (ed.) *Nature, Landscape and People since the Second World War.* Tuckwell Press, East Linton, UK, pp. 161–180.

Daily, G.C. and Matson, P.A. (2008) Ecosystem services: from theory to implementation. *Proceedings of the National Academy of Sciences* 105, 9455–9456.

Department for Energy and Climate Change (2010) *Digest of UK Energy Statistics 2010.* TSO, London.

Dwyer, J. and Hodge, I. (2001) The challenge of change: demands and expectations for farmed land. In: Smout, T.C. (ed.) *Nature, Landscape and People since the Second World War.* Tuckwell Press, East Linton, UK, pp. 117–134.

EEA (2010) Proposal for a Common International Classification of Ecosystem Goods and Services (CICES) for integrated environmental and economic accounting. Paper prepared for the EEA by Centre for Environmental Management, University of Nottingham, UK, Background document, ESA/STAT/AC.217, UNCEEA/5/7/Bk, Fifth Meeting of the UN Committee of Experts on Environmental-Economic Accounting, New York, 23–25 June 2010.

Foley, J.A., DeFries, R., Asner, G.P., Barford, C., Bonan, G., Carpenter, S.R., et al. (2005) Global consequences of land use. *Science* 309, 570–574.

Foresight Land Use Futures Project (2010) Land use futures: making the most of land in the 21st century. Final Project Report, London.

Friberg, N. (2010) Ecological consequences of river channel management. In: Ferrier, R.C. and Jenkins, A. (eds) *Handbook of Catchment Management*. John Wiley and Sons, Chichester, UK, pp. 77–105.

Geist, H.J. and Lambin, E.F. (2002) Proximate causes and underlying driving forces of tropical deforestation. *Bioscience* 52, 143–150.

Geist, H.J. and Lambin, E.F. (2004) Dynamic causal patterns of desertification. *Bioscience* 54, 817–829.

Georgescu-Roegen, N. (1975) Energy and economic myths. *Southern Economic Journal* 41, 347–381.

Giampietro, M., Mayumi, K. and Ramos-Martin, J. (2009) Multi-scale integrated analysis of societal and ecosystem metabolism (MuSIASEM): theoretical concepts and basic rationale. *Energy* 34, 313–322.

Gibson, C.C., Ostrom, E. and Ahn, T.K. (2000) The concept of scale and the human dimensions of global change: a survey. *Ecological Economics* 32, 217–239.

Gleick, P.H. (2003) Water use. *Annual Review of Environment and Resources* 28, 275–314.

Global Land Project (2005) Science plan and implementation strategy. *IGBP Report No 53/ IHDP Report No 19*. IGBP Secretariat, Stockholm.

Godfray, H.C.J., Beddington, J.R., Crute, I.R., Haddad, L., Lawrence, D., Muir, J.F., et al. (2010) Food security: the challenge of feeding 9 billion people. *Science* 327, 812–818.

Gray, M. (2012) Valuing geodiversity in an 'Ecosystem Services' context. *Scottish Geographical Journal* 128, 177–194.

Gutman, G., Janetos, A.C., Justice, C.O., Moran, E.F., Mustard, J.F., Rindfuss, R.R., et al. (eds) (2004) *Land Change Science. Observing, Monitoring and Understanding Trajectories of Change on the Earth's Surface,* Kluwer Academic Publishers, Dordrecht/Boston/London.

Haberl, H., Erb, K.H., Krausmann, F., Gaube, V., Bondeau, A., Plutzar, C., et al. (2007) Quantifying and mapping the human appropriation of net primary production in earth's terrestrial ecosystems. *Proceedings of the National Academy of Sciences* 104, 12942–12945.

Hoekstra, A.Y. and Mekonnen, M.M. (2012) The water footprint of humanity. *Proceedings of the National Academy of Sciences* 109, 3232–3237.

Ingram, J., Ericksen, P. and Liverman, D. (eds) (2010) *Food Security and Global Environmental Change*. Routledge, Oxford, UK.

IPCC (2007) *Climate Change 2007: Synthesis Report. Contribution of Working Groups I, II and III to the Fourth Assessment Report of the Intergovernmental Panel on Climate Change*. IPCC, Geneva, Switzerland.

Krausmann, F., Gingrich, S., Eisenmenger, N., Erb, K.-H., Haberl, H. and Fischer-Kowalski, M. (2009) Growth in global materials use, GDP and population during the 20th century. *Ecological Economics* 68, 2696–2705.

Lambin, E.F. and Geist, H. (eds) (2006a) *Land-Use and Land-Cover Change*. Springer-Verlag, Berlin.

Lambin, E.F. and Geist, H. (eds) (2006b) *Land-Use and Land-Cover Change: Local Processes and Global Impacts*. Springer-Verlag, Berlin.

Lambin, E.F., Turner, B.L., Geist, H.J., Agbola, S.B., Angelsen, A., Bruce, J.W., et al. (2001) The causes of land-use and land-cover change: moving beyond the myths. *Global Environmental Change – Human and Policy Dimensions* 11, 261–269.

Lambin, E.F., Geist, H.J. and Lepers, E. (2003) Dynamics of land-use and land-cover change in tropical regions. *Annual Review of Environment and Resources* 28, 205–241.

Leach, G. (1976) *Energy and Food Production*. IPC Science and Technology Press Ltd, Guildford, UK.

Levin, S.A. (1992) The problem of pattern and scale in ecology. *Ecology* 73, 1943–1967.

Lowe, P., Woods, A., Liddon, A. and Phillipson, J. (2009) Strategic land use for ecosystem services. In: Winter, M. and Lobley, M. (eds) *What is Land for?: The Food, Fuel and Climate Debate*. Earthscan, London.

Mace, G.M., Bateman, I.J., Albon, S., Balmford, A., Brown, C., Church, A., et al. (2011) Conceptual framework and methodology. In: UK National Ecosystem Assessment (ed.) *The UK National Ecosystem Assessment Technical Report*. UNEP-WCMC, Cambridge, UK, pp. 11–26.

MacLeod, M., Moran, D., Eory, V., Rees, R.M., Barnes, A., Topp, C.F.E., et al. (2010) Developing greenhouse gas marginal abatement cost curves for agricultural emissions from crops and soils in the UK. *Agricultural Systems* 103, 198–209.

Matson, P.A., Parton, W.J., Power, A.G. and Swift, M.J. (1997) Agricultural intensification and ecosystem properties. *Science* 277, 504–509.

Matthews, K.B., Rivington, M., Buchan, K., Miller, D. and Bellocchi, G. (2008) Characterising the agro-meteorological implications of climate

change scenarios for land management stakeholders. *Climate Research* 37, 59–75.
Millennium Ecosystem Assessment (2005) *Ecosystems and Human Well-being: Synthesis*. Island Press, Washington, DC.
Monfreda, C., Ramankutty, N. and Foley, J.A. (2008) Farming the planet: 2. Geographic distribution of crop areas, yields, physiological types, and net primary production in the year 2000. *Global Biogeochemical Cycles* 22, GB1022 (http://onlinelibrary.wiley.com/doi/10.1029/2007GB002947/abstract, accessed 3 December 2013).
Pe'er, G., McNeely, J.A., Dieterich, M., Jonsson, B.-G., Selva, N., Fitzgerald, J.M., et al. (2013) IPBES: opportunities and challenges for SCB and other learned societies. *Conservation Biology* 27, 1–3.
Postel, S.L., Daily, G.C. and Ehrlich, P.R. (1996) Human appropriation of renewable fresh water. *Science* 271, 785–788.
Ramankutty, N. and Foley, J.A. (1999) Estimating historical changes in global land cover: croplands from 1700 to 1992. *Global Biogeochemical Cycles* 13, 997–1027.
Ramankutty, N., Evan, A.T., Monfreda, C. and Foley, J.A. (2008) Farming the planet: 1. Geographic distribution of global agricultural lands in the year 2000. *Global Biogeochemical Cycles,* 22, GB1003, doi:10.1029/2007GB002952.
Ritson, C. and Harvey, D.R. (eds) (1997) *The Common Agricultural Policy*. CAB International, Wallingford, UK.
Rivington, M., Matthews, K.B., Buchan, K., Miller, D.G., Bellocchi, G. and Russell, G. (2013) Climate change impacts and adaptation scope for agriculture indicated by agro-meteorological metrics. *Agricultural Systems* 114, 15–31.
Rockström, J., Steffen, W., Noone, K., Persson, Å., Chapin, F.S. III, Lambin, E., et al. (2009) A safe operating space for humanity. *Nature* 461, 472–475.
Rural and Environment Research and Analysis Directorate (1939–present) Agricultural Census (June). Series of Annual Census Data, Scottish Government, Edinburgh, UK.
Rural and Environment Research and Analysis Directorate (2010) Economic Report on Scottish Agriculture, 2010 edn. Scottish Government, Edinburgh, UK.
Salisbury, E., Claxton, R., Goodwin, J., Thistlethwaite, G., MacCarthy, J., Pang, Y., et al. (2013) *Greenhouse Gas Inventories for England, Scotland, Wales and Northern Ireland: 1990–2011*. Report to the Department of Energy and Climate Change, The Scottish Government, The Welsh Government and The Northern Ireland Department of the Environment. National Atmospheric Emissions Inventory.
SASA (2009) *Pesticide Usage in Scotland. Arable Crops 2008*. SASA, Edinburgh, UK.
Scottish Executive (2007) *Scottish Primary Food and Drink Produce Processed in Scotland*. DTZ, Edinburgh, UK.
Seto, K.C., Reenberg, A., Boone, C.G., Fragkias, M., Haase, D., Langanke, T., et al. (2012) Urban land teleconnections and sustainability. *Proceedings of the National Academy of Sciences* 109, 7687–7692.
Sharpley, A., Matlock, M., Heathwaite, L. and Simpson, T. (2010) Managing agricultural catchments to sustain production and water quality. In: Ferrier, R.C. and Jenkins, A. (eds) *Handbook of Catchment Management*. John Wiley and Sons, Chichester, UK, pp. 107–132.
Smith, P., Martino, D., Cai, Z., Gwary, D., Janzen, H., Kumar, P., et al. (2008) Greenhouse gas mitigation in agriculture. *Philosophical Transactions of the Royal Society B-Biological Sciences* 363, 789–813.
Stern, N. (2007) *The Economics of Climate Change. The Stern Review*. Cambridge University Press, Cambridge, UK.
Tian, J., Fan, M.S., Guo, J.H., Marschner, P., Li, X.L. and Kuzyakov, Y. (2012) Effects of land use intensity on dissolved organic carbon properties and microbial community structure. *European Journal of Soil Biology* 52, 67–72.
Turner, B.L., Lambin, E.F. and Reenberg, A. (2007) The emergence of land change science for global environmental change and sustainability. *Proceedings of the National Academy of Sciences* 104, 20666–20671.
Turnhout, E., Bloomfield, R., Hulme, M., Vogel, J. and Wynne, B. (2012) Conservation policy: listen to the voices of experience. *Nature* 488, 454–455.
UK National Ecosystem Assessment (2011) *The UK National Ecosystem Assessment Technical Report*. UNEP-WCMC, Cambridge, UK.
Vitousek, P.M., Aber, J.D., Howarth, R.W., Likens, G.E., Matson, P.A., Schindler, D.W., et al. (1997a) Human alteration of the global nitrogen cycle: sources and consequences. *Ecological Applications* 7, 737–750.
Vitousek, P.M., Mooney, H.A., Lubchenco, J. and Melillo, J.M. (1997b) Human domination of Earth's ecosystems. *Science* 277, 494–499.
Wright, G.G., Edwards, A.C., Morrice, J.G. and Pugh, K. (1991) North east Scotland river catchment nitrate loading in relation to agricultural intensity. *Chemical Ecology* 5, 263–281.

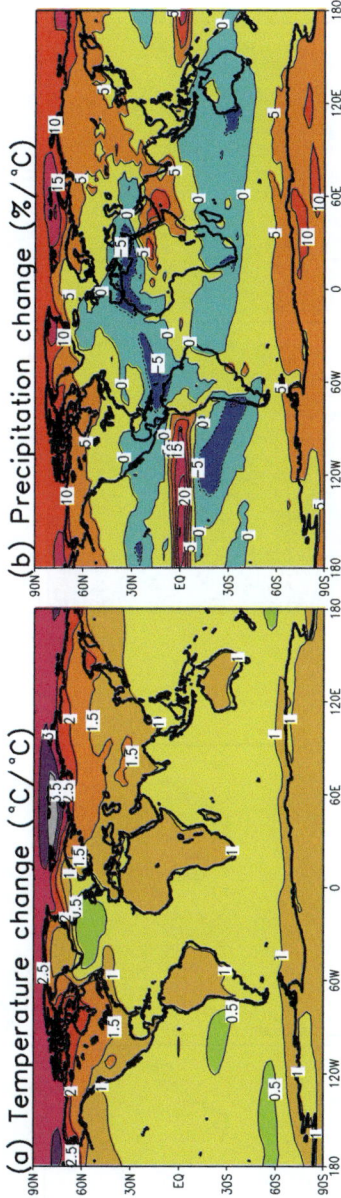

Plate 1. Normalized annual mean temperature (a) and precipitation (b) changes from 25 global climate models from CMIP5, run under the RCP4.5 scenario for the 21st century. The figure shows regional changes per 1°C of global mean warming. For example, for each 1°C rise in the global mean temperature, the warming over Canada amounts to 1–2°C, precipitation in the Caribbean region decreases by 5%, and so on. (Figure courtesy of Jouni Räisänen, University of Helsinki.)

2a

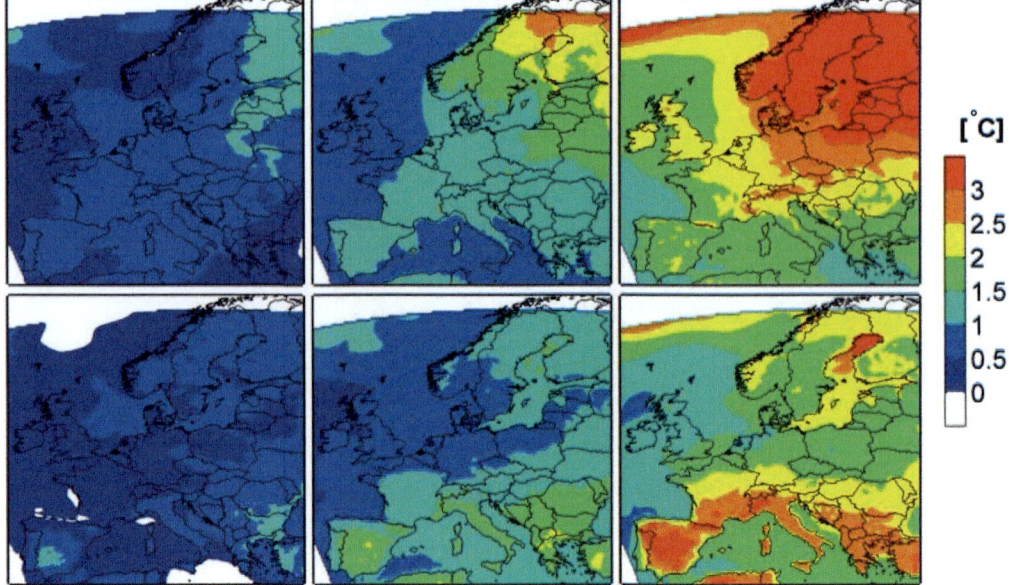

2b

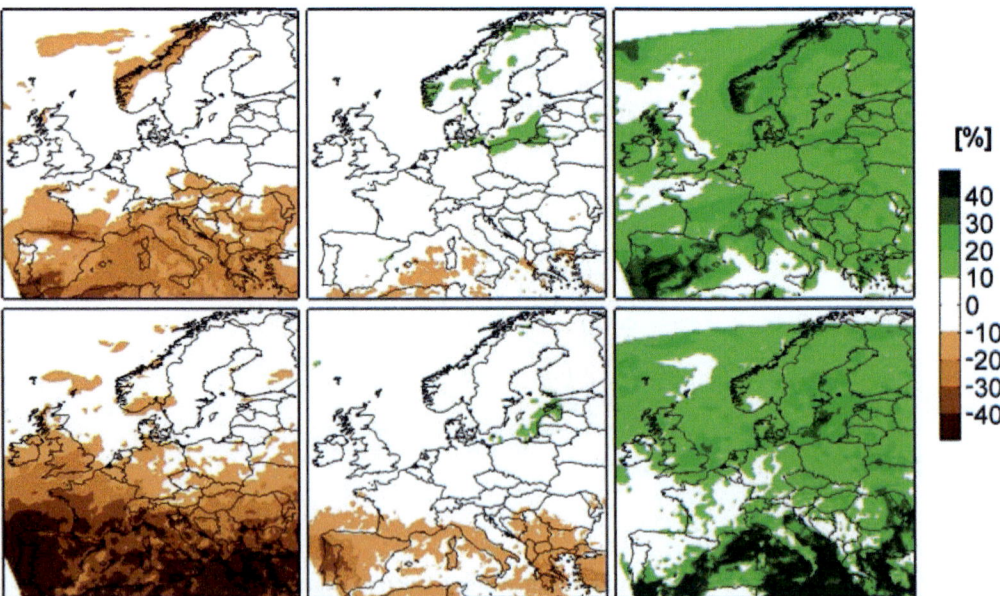

Plate 2. Projected regional temperature and precipitation changes according to 21 regional climate models for the period 2021–2050 relative to 1961–1990. Top two rows: temperature (winter and summer). Bottom two rows: precipitation (winter and summer). The three columns are from the models that project the next-to-smallest, median and next-to-largest changes (the 2nd, 11th and 20th of the 21 models after they were ranked by magnitude of change). (After Kjellström et al., 2013.)

3a

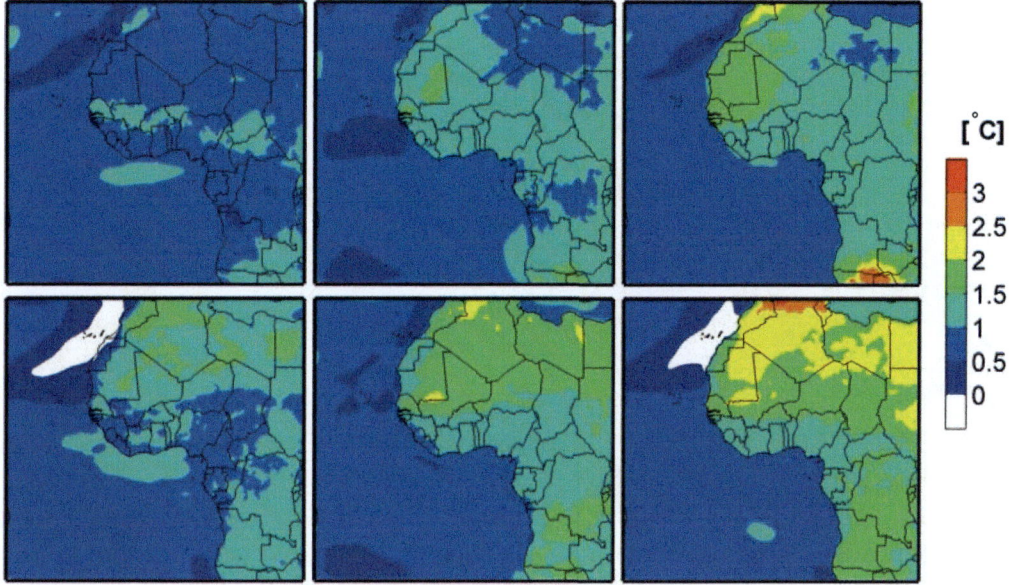

3b

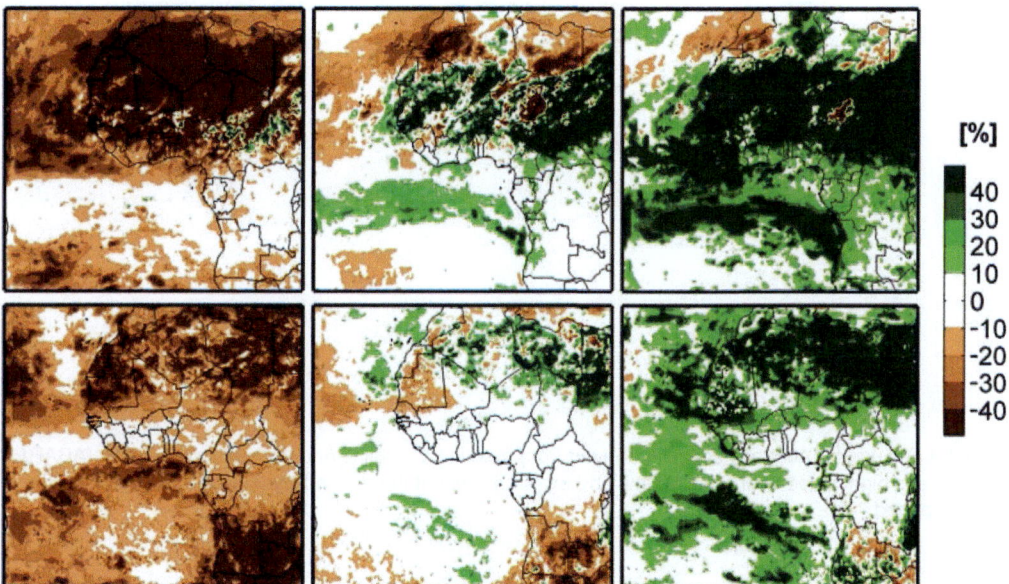

Plate 3. Projected regional temperature and precipitation changes according to 11 regional climate models for the period 2021–2050 relative to 1961–1990. Top two rows: temperature (northern hemisphere winter and summer). Bottom two rows: precipitation (northern hemisphere winter and summer). The three columns are from the models that project the next-to-smallest, median and next-to-largest changes (the 2nd, 6th and 10th of the 11 models after they were ranked by magnitude of change). (Figure courtesy of Fredrik Boberg, DMI.)

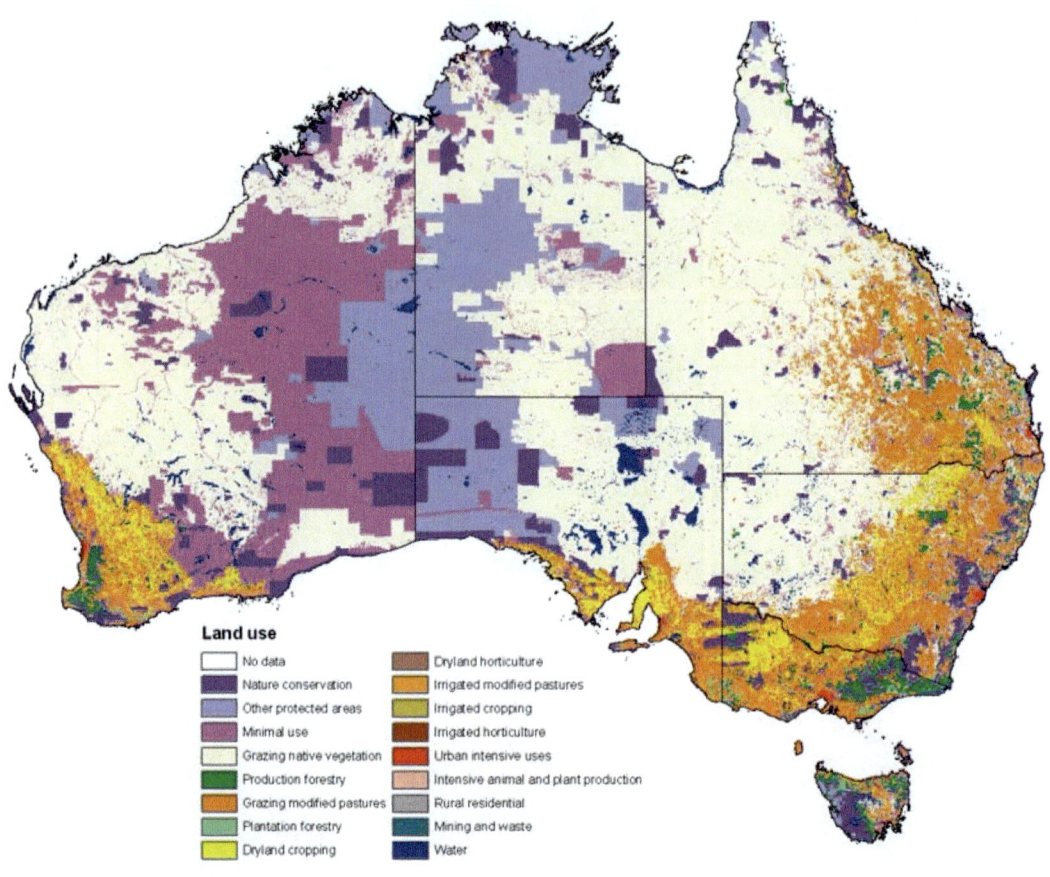

Plate 4. Land use for Australia 2005/06 derived from modelled, satellite and agricultural statistics (ABARES, 2011). Areas highlighted in orange, yellow and brown represent the dryland agricultural boundary for the 2005/06 period.

Plate 5. Modern intensive tree fruit growing systems are reliant on irrigation to deliver the yields and quality demanded by growers, retailers and consumers.
Plate 6. Prunus avium trees in an orchard at East Malling Research (UK) were flooded for 3 weeks during the winter of 2002.

Plate 7. A prolonged period of soil anoxia in the dormant season resulted in extensive root death, which was evident the following spring when newly emerged leaves desiccated and died.

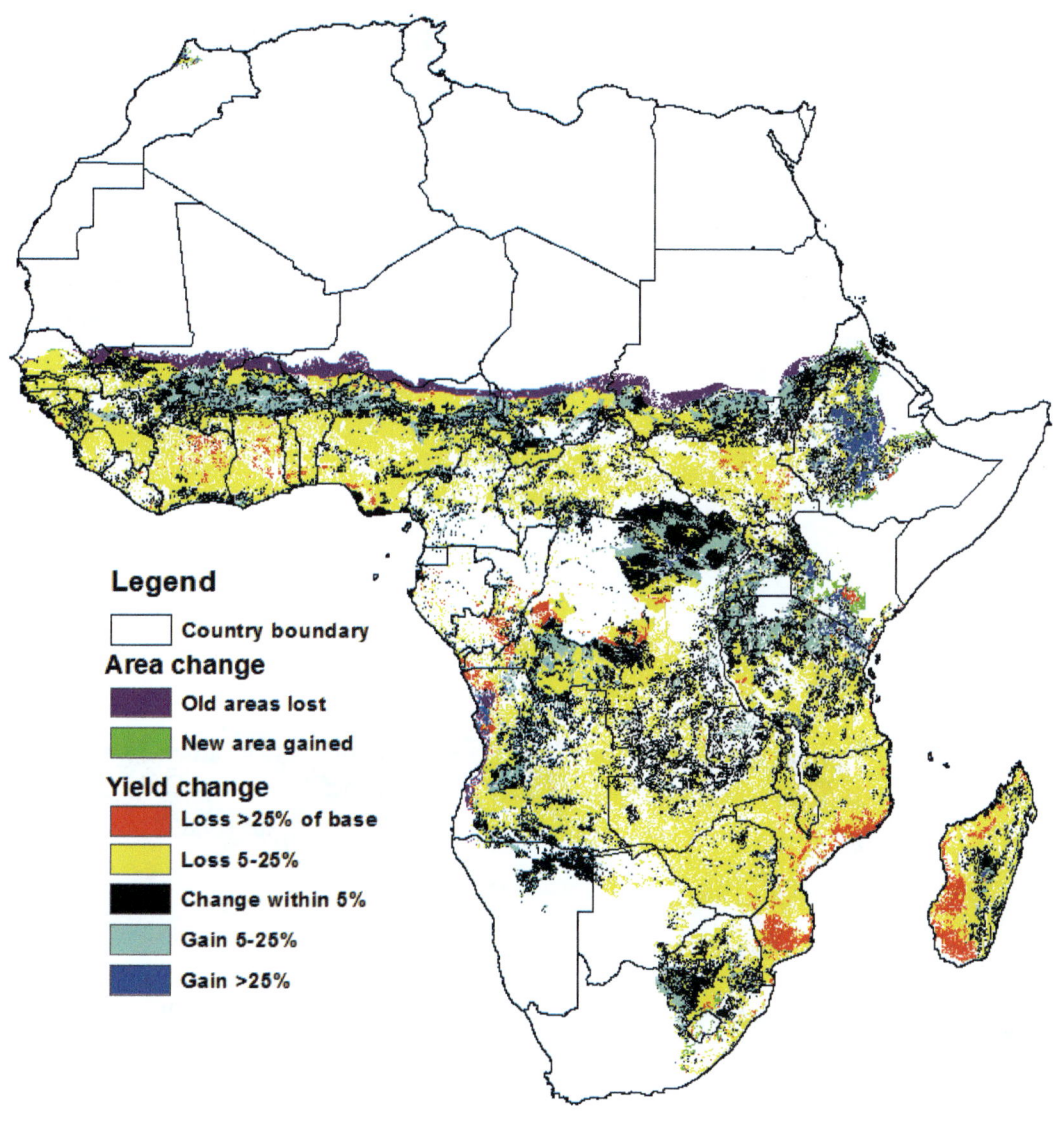

Plate 8. Potential changes in area and yield of maize (%) under low-input conditions (<5 kg N ha-1) in Africa by the 2050s under MIROC3.2 GCM and A1B emission scenario.

9

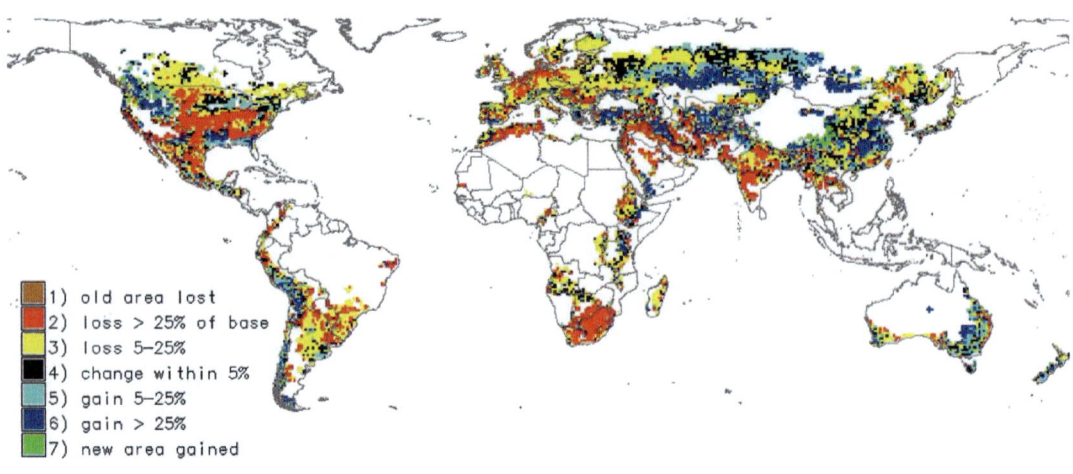

10

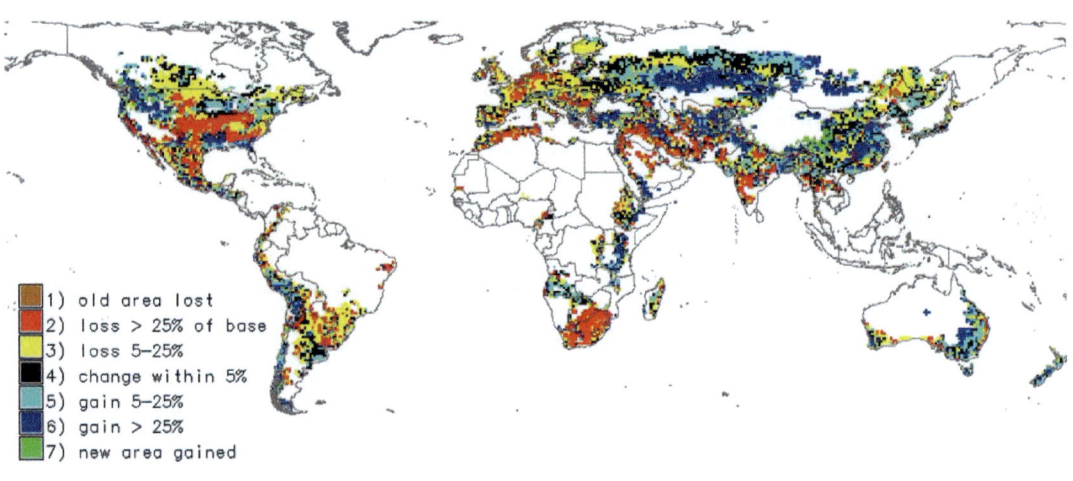

Plate 9. Effect of climate change (MIR-A1) on yield for conventional wheat cultivars – preliminary results
Plate 10. Effect of climate change (MIR-A1) on yield for drought-tolerant wheat cultivars – preliminary results

10 Adaptation of Mixed Crop–Livestock Systems in Asia

Fujiang Hou

State Key Laboratory of Grassland Agro-Ecosystem, China College of Pastoral Agriculture Science and Technology, Lanzhou University, China

10.1 Introduction

The mixed farming system combining crop and livestock production, which usually is based on the interaction of arable crops such as forage crop, grain crop and oil crop, rangeland, woodland and livestock, is the dominant agricultural system of the world. It produces about half of the world's food (Herrero et al., 2010) and makes the largest contribution to the food supply of humans. The production system uses 90% of the total cropland, feeds 70% of sheep and goats and produces 88.5% of beef, 88% of milk, 61% of pork and 26% of poultry meat (Seré and Steinfeld, 1996; Blackburn, 1998). Approximately 84% of the total agricultural population is involved in the operation of mixed farming systems in developing countries (Blackburn, 1998). As one of the biggest developing areas, the situation in Asia is similar (Hou et al., 2009).

10.2 The Current Situation of Mixed Crop–Livestock Systems in Asia

Farming system evolution is the outcome of social, abiotic and biotic factors and their interactions (Ren, 1985). Various mixed crop–livestock systems exist due to the diversity of culture, environment, plants, animals and microbes, economic activities and the rich history of agricultural production in different countries. In terms of the interactions between livestock production and other components and ecoregions of farming systems, especially between plant and livestock, five types of mixed crop–livestock systems have been identified in Asia: farming systems based on rangeland; farming systems based on grain crops; farming systems based on crop/pasture rotations; agrosilvopastoral systems; and farming systems based on ponds (Fig. 10.1).

10.2.1 Farming systems based on rangeland

This type of production system is operated in the arid area (annual mean precipitation below 250 mm) of north-west China, central Asia and west Asia, of which the dominant landscape is the Gobi desert; some of the semi-arid area (between 250 mm and 500 mm annual mean precipitation), of which the dominant vegetation is steppe; the Qinghai-Tibetan Plateau and northern Russia, of which the dominant vegetation is tundra, alpine steppe or alpine meadow (Fig. 10.1). There are about 1900×10^4 km^2 of rangeland, which occupies 45% of the total land area in Asia. At a regional level, typical landscapes are coupled agroecosystems being made up of mountain, desert and oasis. Rivers originating in the mountain areas integrate three ecosystems of mountain, oasis and desert by supplying water and carrying the ingredients for life, while the desert supplies the existent

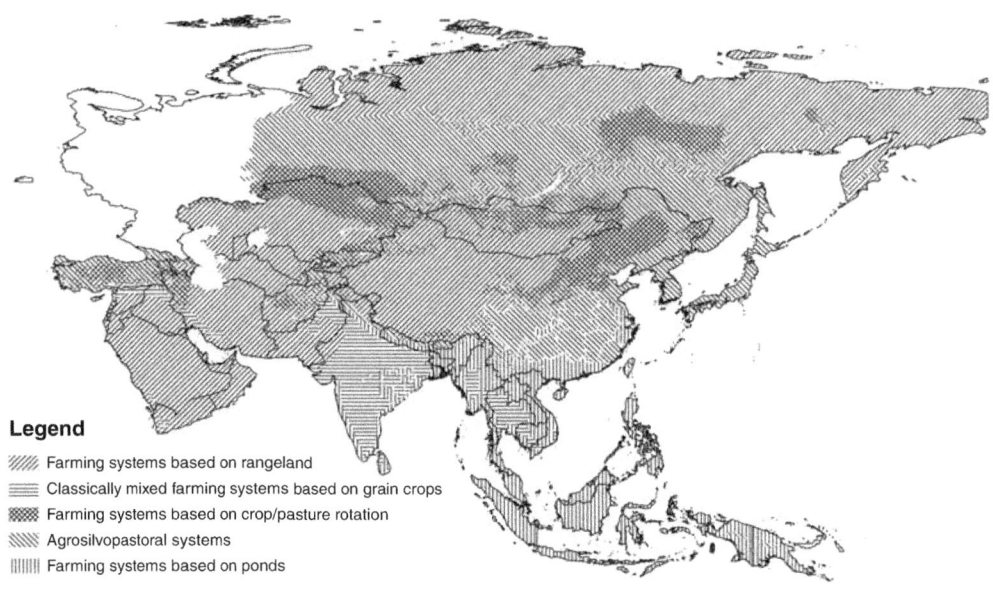

Legend
- Farming systems based on rangeland
- Classically mixed farming systems based on grain crops
- Farming systems based on crop/pasture rotation
- Agrosilvopastoral systems
- Farming systems based on ponds

Fig. 10.1. Sketch map of mixed crop–livestock systems in Asia.

background of oasis and mountain (Hou and Li, 2001). Cropland appeared over 2000 years ago, first in natural oases, and expanded rapidly through the cultivation of the rangeland (including saline meadows, which are distributed sporadically in desert region) and the establishment of irrigation facilities both in desert and mountain regions (Hou and Li, 2001). Mountain, desert and oasis account for 43%, 53% and 4%, respectively, of the total land area in the Xinjiang Uygur Autonomous Region of China (Hou, 2007), and most of the croplands are located in oases. This kind of spatial pattern is common to many arid regions and some semi-arid regions of the world. On the whole, as a result of drought, high elevation and cold, there is over 95% of rangeland in desert, tundra and alpine areas, and the forage crop area is less than 10% of the cropland in oasis areas (Ren et al., 1995; Hou, 2000). Farming systems are supported by water from rivers rising in mountain areas (Ren et al., 1999).

Mixed farming systems based on rangeland feed about 35% of the sheep, horses and donkeys in the whole of China and produce approximately 60% of the wool and cashmere and 33% of the total milk and mutton produced in China (Nan, 2005). In arid areas of China, the main crops are cotton (*Gossypium hirsutum* L.), wheat (*Triticum aestivum* L.) and maize (*Zea mays* L.), which account for 31%, 20% and 14% of total croplands, respectively. Lucerne (*Medicago sativa* L.) originates from Iran, has been planted for over 2000 years and is the dominant forage crop in this kind of farming system. The main livestock in arid areas are sheep, goats, cattle and camels. In semi-arid areas of China, maize is planted in about one-third of the croplands, while the planted area of soybean (*Glycine max* (L.) Merr.) and wheat is 13% and 7%, respectively. The main livestock are sheep, dairy cattle, goats and beef cattle. In the Qinghai-Tibetan Plateau, the main crops are rapeseed (*Brassica napus* L.), hulless barley (*Hordeum vulgare* L. var. nudum Hook.f.) and wheat, for which the planted areas occupy 26%, 22% and 20%, respectively. The main livestock are yak (*Bos grunniens*) and Tibetan sheep. The sown pasture area accounts for only about 0.2% of rangeland in the Qinghai-Tibetan Plateau,

1% in the arid area and 0.2% in the semi-arid area (Hou et al., 2008). In the tundra area of eastern Russia, rye (*Secale cereale* L.), oat (*Avena sativa* L.), triticale (*Triticale hexaploide* Lart.) and sugarbeet (*Beta vulgaris* L.) are planted as forage crops in small areas, and reindeer (*Rangifer tarandus*) is one of the dominant livestock.

In this type of agricultural system, crop, rangeland and livestock interact with each other in the following five ways: (i) livestock graze rangeland throughout the year; (ii) livestock often graze fallow cropland and stubble cropland after harvesting the crop; (iii) livestock supply draft power and manure for crop production; (iv) crop residues and forage crops are provided to livestock mostly in the winter and spring; and (v) in an abundant rainfall year, herbage is harvested in the rangeland and then made into hay to feed animals in the winter and spring, which is one of the prevalent utilization ways in native meadow. There is a net flow of nutrient elements from rangeland to cropland in two ways: first, livestock graze rangeland during the daytime and stay overnight on fallow cropland; or, second, more prevalent in Asia, livestock excrement is collected, after the animals have grazed rangeland during the day and have stayed overnight in pens, which is then applied to cropland. This extensive type of agricultural system has a high ecological efficiency as a result of low inputs. A high ratio of rangeland to cropland such as in the farming–pastoral ecotone in northern China (e.g. Z.B. Nan, 2008, unpublished results) leads to intensive fertilization of cropland.

10.2.2 Classically mixed farming systems based on grain crops

This kind of farming system is located in the plains and oases of temperate and subtropical Asia, where crop production is possible owing to favourable conditions of water (rainfall or irrigation), temperature and soil (Fig. 10.1). It is one of the most dominant regions for maize, wheat, cotton and soybean production in the world because of the high yields of maize, cotton and wheat and the third highest yield of soybean, which is next only to North (USA and Canada) and South America (Brazil and Argentina). With abundant and high-quality grain and straw resources, this type of agricultural system seldom grows forage crops but feeds 34% of cattle, 47% of goats, 26% of sheep, 42% of donkeys and generates 58% of beef and 50% of milk production in China (Hou et al., 2008) as well as nearly 80% of buffalo in India (A.K. Roy, 2013, unpublished). Interaction between crop production and livestock production occurs mainly in four ways (Wang and Zhou, 2007; Hou et al., 2009): (i) crop residues and grain are fed to livestock throughout the year; (ii) livestock supply manure and draft power for some crop production in the extensive systems of the developing regions, although there is an increasing level of mechanization in intensive crop production systems; (iii) livestock graze fallow cropland, stubble cropland and sparse rangeland; (iv) they also sometimes graze small grain crops such as wheat, barley and rye, which in these areas have been prevalent as multi-purpose crops (ground cover, energy, grain, forage, and so on) for a long time. The incorporation of small grain crops into grazing systems can overcome the feed gap of early spring and winter which commonly occurs in this type of farming system, and also provides the opportunity to exchange nutrient elements between different components of the farming system.

10.2.3 Farming systems based on crop/pasture rotation

This type of mixed system exists mainly in the transition zone between the nomadic and cropping areas and between the nomadic and forest areas in Asia. They are part of the Eurasian steppe and have relatively sufficient rainfall and heat, and have therefore been cultivated for crop production for a long time. Potato (*Solanum tuberosum* L.), maize, some small grains such as oat (*Avena chinensis* (Fisch. ex Roem. et Schult.) Metzg.), foxtail millet (*Tetaria italic* L.), broom millet (*Panicum miliaceum* L.) and legume crops such as soybean, pea and bean are the main crops in the region. The area planted to potato accounts for 73.4% of China's total potato crop (Hou et al., 2008). The main livestock are goats, sheep, beef cattle and donkeys (mule). Rainfed farming is dominant in a gulley area where the annual average rainfall is more than 250 mm and over 60% of the annual rainfall falls during the crop growing season. Frequent droughts are a key risk, especially in spring, because of large year-to-year variation in rainfall (Hou and Nan, 2006). A large number of farmers plant small grain crops in late summer or early autumn in order to utilize the rainfall and warmth for hay production (Hou and Nan, 2006). In most cases, this kind of farming operation takes place because of crop failure as a result of drought during spring or early summer. Legume crops are planted as part of the crop rotation in order to maintain or improve the fertility of cropland and to supply protein-rich fodder to livestock.

Crop production and livestock production are integrated into these systems in four ways (Hou et al., 2008): (i) forage crops (including some legume crops) and residues of other crops are fed to livestock in pens; (ii) livestock supply manure and draft power for crop production; (iii) livestock graze stubble cropland, fallow cropland and sparse rangeland; and (iv) livestock graze crops after failed harvests because of economic reasons as the result of serious disease or drought.

10.2.4 Agrosilvopastoral systems

Based on forest, this system is operated mainly in temperate forest areas, forest zones in the high mountains and some of the subtropical forest areas of Asia (Fig. 10.1). Dominant crops are wheat, soybean and maize in the temperate zone and rice and maize in the subtropical zone. The main livestock are cattle, goats, buffalo and deer (reindeer, wapiti, sika, river deer, etc.).

There are five ways in which livestock, crops and forestry enterprises mutually interact: (i) livestock graze in the forests; (ii) livestock graze the harvested cropland, forage cropland and fallow cropland; (iii) grain and crop residues are supplemented to livestock in pens; (iv) livestock supply draft power and manure both for crop production and timber production; and (v) forests provide shade and windbreaks for both crops and grazed livestock. Large areas of forest have been converted to cropland over a long period in these regions. Forests and cropland exchange nutrient elements through livestock movement, but there is a net nutrient flow from forestland to cropland because farmers collect manure from the pens where livestock sleep overnight, after grazing in the forest areas, and apply this manure to cropland. Both deer and goats browse trees, so they play a key role in the timber production of farming systems and forest conservation in some areas.

10.2.5 Farming systems based on ponds

Integrated systems based on ponds are located in the tropics and subtropics with good rainfall and relatively flat land (Fig. 10.1). This type of system has a relatively short history which can be traced back only about 600 years in inshore regions and gradually spreads to inland areas with abundant water resources in big river basins (Nie et al., 2003). This type of farming contributes over half of the rice, pork and chicken and most of the buffalo in the world, and the other main ruminant livestock are goats and cattle, which play a relatively minor role. The main crops are rice, tropical fruits and vegetables, among which the planted area of rice occupies nearly 60% of the total cropland in this region.

Interactions between livestock production and crop production in this type of system include: (i) crop residues are fed to livestock; (ii) livestock excrement together with some forage crops and crop residues are used as a resource for pond production; (iii) pond sludge together with livestock excrement are applied to cropland as fertilizers;

(iv) buffalo or cattle supply draft power for crop production; and (v) livestock graze the sparse rangeland and the cropland after being harvested. Obviously, the mixed farming systems originate from the pond production, which plays a key role in recycling nutrient elements and the economic allocation and energy exchange of the whole system (Pittaway et al., 1996).

10.3 Mixed Farming Systems, Climate Change and Adaptation

10.3.1 Mixed farming systems under global climate change

Global climate change threatens the sustainable productivity of farming systems at all scales, especially at the scale of species (crop cultivars or animal breeds) and ecosystems.

Impacts at species scale

Climatic factors play an important role in the productivity and distribution of crops and livestock. If the climate becomes warmer and drier, which has been identified as the main trend of global climate change in most areas of Asia (Ren et al., 2011), livestock with high adaptation to drought, such as goat, donkey, camel, deer, will extend their distributive areas, while other livestock with high susceptibility to climate change (such as horse, cattle, buffalo, sheep, and so on) will have their area of distribution reduced (Fan and Zhang, 1993). If the climate becomes warmer and wetter, the changes in distribution of both the above types of livestock will be reversed.

Global climate change will potentially affect the quality of animal products, although this topic has been largely ignored in much previous research. In cold regions of eastern and central Asia, livestock usually have higher meat production per capita, with higher fat content in animal products (Cheng, 1993). Global warming might result both in smaller livestock body weight and a decrease of meat production, but result in higher lean meat percentage (Cheng, 1993). In wetter regions including eastern Asia, South-east Asia and some of southern Asia, the quality of fur, wool and cashmere is usually poor, but the quality may improve if the climate becomes warmer and drier (Zhao and Qiu, 1999). In China, most of the fine-wool sheep have been bred in the cold regions, so a warmer climate could result in a negative influence on the yield and quality of fine wool. However, if precipitation increases more than evaporation, global climate change may promote animal production.

Impacts at ecosystem scale

Global climate change not only results in transforming the distribution, productivity and interaction of crop, rangeland and livestock but also affects the whole farming production system. Global warming with increased rainfall will raise the productivity of all types of farming systems, including both plant and animal production. In Asia, the area of rangeland has been forecast to expand and that of woodland to shrink under conditions of global warming (Schellnhuber et al., 2013). Furthermore, increased rainfall will boost the effects of global warming. However, other models have indicated that global warming will decrease the productivity of grassland in the farming–pastoral areas of northern China and exacerbate the drought in arid regions of central and western Asia (Qiu et al., 2001; IPCC, 2007).

Normally, farming systems are relatively stable on an environmental gradient because the existing farm management measures could minimize the drift of farming systems under conditions of limited climatic fluctuation. All types of integrated farming systems can be characterized as part of a successional framework responding to the interactions among biotic factors (crops, livestock, etc.), abiotic (environmental) factors (precipitation, heat, etc.) and social factors (economics, management, etc.; see Fig. 10.2). Global climate change is another factor exerting selection pressure on the succession of farming systems. If the climate becomes warmer, management of forage crops and of the interactions between herbivore and forage will determine the stable level of the integrated farming systems (Fig. 10.2). However, increased frequency of dry and hot periods associated with global warming could be disastrous for farming systems.

Global climate change is a slow and gradual process at a large timescale, so livestock and crops could adapt themselves slowly and simultaneously (Hou and Yang, 2006; FAO, 2007; Yadav et al., 2011). Humans have time enough to breed new cultivars or breeds and to develop innovative management practices. However, global climate change will also induce a natural selection on the new breeds of livestock and new cultivars of crops; the influence of this is little known.

10.3.2 Adaptation of mixed farming systems to global climate change

Varieties of crop and breeds of livestock with high stress resistance

It is generally recognized that both varieties of crops and breeds of livestock with high stress tolerance have more stable and higher productivity under global climate change; this provides more options to improving farm management. A number of studies have looked at the effects of climate change on forage and animal species, and on their potential to enhance adaptation both by traditional and genetic improvement (FAO, 2007; Yadav et al., 2011; Redden, 2013). Furthermore, forage and livestock breeding can also contribute to climate change mitigation through reducing emissions of greenhouse gases (GHG) and raising carbon

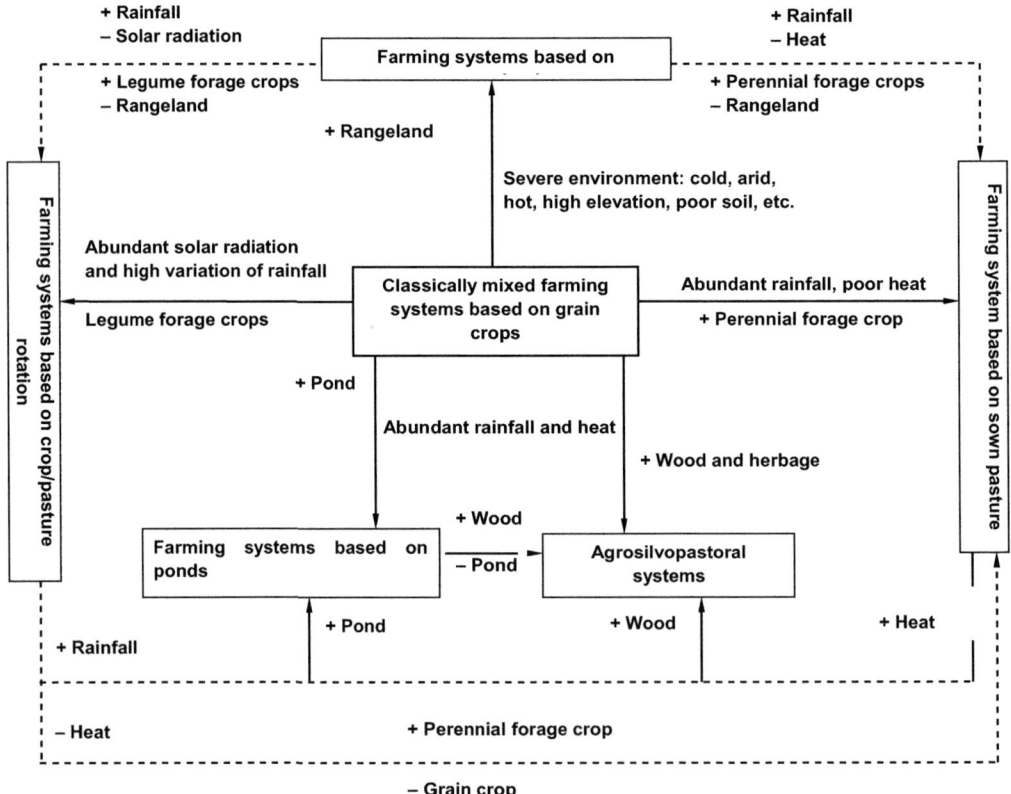

Fig. 10.2. Succession of the integrated farming systems. (Adapted from Hou *et al.*, 2009.)

(C) sequestration in both grassland and livestock production. Asia has one of the most abundant germplasm resources of forage and domestic animals in the world, which can serve as the basis of new breeds.

Improvements of forage and animal breeds will decrease GHG emissions and resource use per unit of animal product (Hou et al., 2009). High sugar ryegrass leads to a 7.5–21.0% increase in milk yield and a 7.1–25.7% decrease in excrement nitrogen (N) (Cheng et al., 2011). Re-seeding native grass species with those with higher productivity or C allocation to deeper roots, or introducing legumes into grazing lands, can all promote soil C in rangeland soils and reduce N emissions (Kell, 2011; Waha et al., 2013). Biological N fixation of the latter displaces the need for fertilizer N, which was often used to rehabilitate the seriously degraded alpine meadow in the Tibetan-Qinghai Plateau, Mongolian Plateau and mountainous rangeland of inland arid regions (Hou et al., 2009, 2014, unpublished results).

The adaptive farming system

In the face of global climate change, an adaptive farming system supplies opportunities, not only for new crop varieties and livestock breeds to manifest more sustainable productivity but also for more innovative management practices to be implemented. In most Asian regions, especially in developed regions of eastern Asia and South-east Asia, integrated crop–livestock farming systems possess higher productivity and stability under conditions of global climate change through the coupling of plant production and animal production, promoting efficient use of biotic and abiotic resources, prolonging the economic chain and strengthening the interaction of all components (Hou et al., 2009; Burney et al., 2013).

The inevitable evolution of agricultural systems in Asia towards enhanced productivity due to structural optimization or better application of existing breeds and technologies is generally associated with the integration of crop production and livestock production. However, with the largest and fastest growing population in the world, the increased demand in this region for animal products must be associated with decreasing emissions per unit of product, and by controlling the increase in emissions through establishing and improving mixed farming systems. Otherwise, a vicious circle inevitably emerges between mitigation and adaptation of global climate change.

10.4 Approaches to Mitigating Greenhouse Gases through Managing Integrated Farming Systems

Asian food systems, from rangeland utilization to fertilizer manufacturing to food storage and packaging, are responsible for nearly one-third of all human-caused GHG emissions (Vermeulen et al., 2012). However, in terms of the components of integrated farming system, GHG emissions could be mitigated through the use of management practices on a farm scale, including rangeland management, switching to no-till, reducing fallow, managing species composition on grazing lands, adjusting management of nitrogen fertilizer and improved manure management.

10.4.1 Rangeland management

Rangeland is the dominant component of mixed farming systems and also plays a key role in the livestock production of agro-silvopastoral systems. One of the main contributors to the emission of GHGs from rangeland is the severe degradation owing to overgrazing and cultivation for crop production (Fig. 10.3a). The latter operation accounted for 40% of the loss of world total soil organic carbon (SOC) from 1850 to 1980 (Houghton, 1995). Degradation of rangeland has caused 39% loss of biomass C and 25.4% loss of SOC, equal to 0.8–1.5 times the total cropland SOC in China.

Exclusion plays an important role in rehabilitating the carbon of vegetation and the soils of rangeland (Fig. 10.3b), but long-

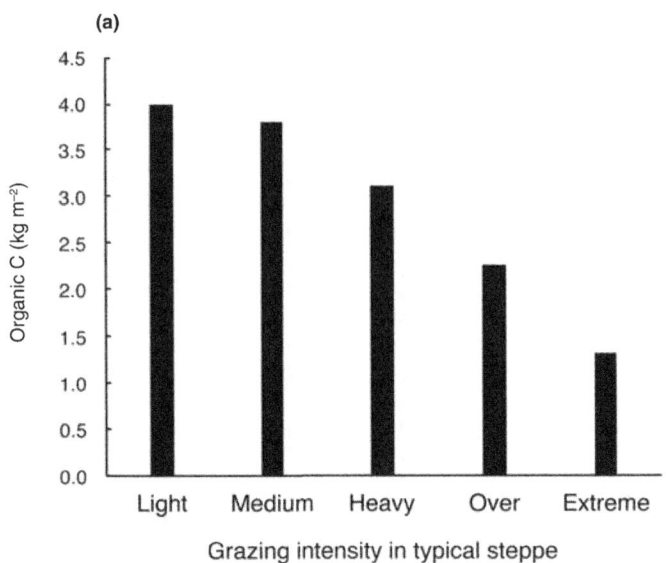

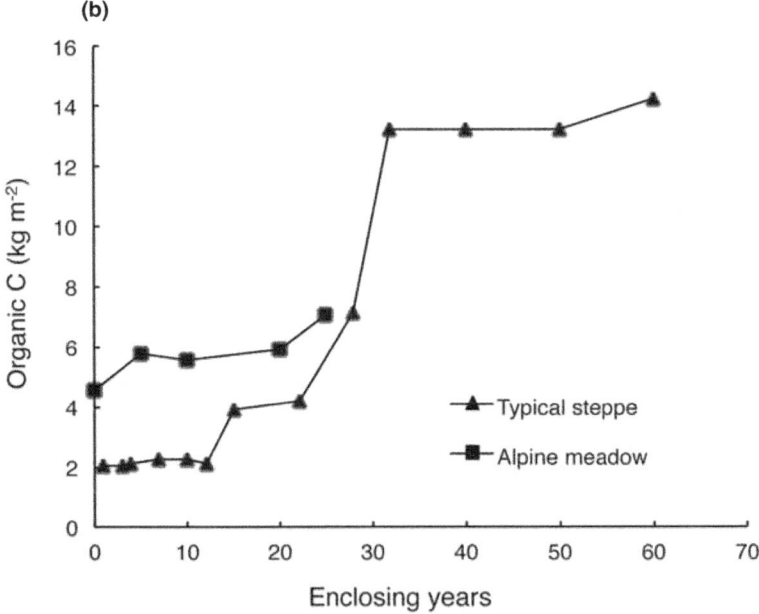

Fig. 10.3. Organic C content density (a) of typical steppe under different grazing intensities (adapted from Wang and Li, 1995) and (b) following grazing exclusion of typical steppe and alpine meadow (adapted from Jia *et al.*, 2009.)

term exclusion increases grazing pressure in other areas of the rangeland and destroys the continuity of nomadic culture and coupled human–rangeland systems (Hou and Yang, 2006; Ren *et al.*, 2011). Systemic integration of livestock production and forage crop production is necessary, both to balance the livestock demand and feed supply on the range and to reduce the grazing pressure of rangeland while improving the livelihood of ranchers. Forage crops could be planted in farming regions and then transported to pastoral regions after harvest and made into hay, and could be sown in pastoral regions without destroying the fragile environment through controlling the cultivated area of rangeland.

10.4.2 Nitrogen fertilizer application in crop production

The application of fertilizer is a common approach for enhancing the productivity and quality of sown pasture, which is important to livestock production in all mixed farming systems. Because the applied N is not always used efficiently by forage crops (Galloway et al., 2003), improving N-use efficiency can significantly reduce emissions of nitrous oxide (N_2O) generated by soil microbes largely from surplus N, and can indirectly reduce emissions of CO_2 from industrial N fertilizer production (Schlesinger, 1999). Operations in mixed farming systems that can improve N-use efficiency include the following: (i) precisely estimating application rates based on the need of the forage crop in crop production systems and the need of livestock in grazed sown pasture, together with a further need of economic profit; (ii) using slow-release N fertilizer forms; (iii) using nitrification inhibitors, which could slow the microbial processes effectively, leading to N_2O formation; (iv) avoiding time delays between N application and plant N uptake, mostly through improving the integration of grazing and N application with irrigation or rainfall; (v) placing N fertilizers into the soil more precisely, to make it more accessible to the roots of forage crops on the premise of not reducing the profit of the whole farming system; and (vi) avoiding excess N applications, or eliminating N applications under conditions of economic benefit (Smith et al., 2008).

10.4.3 Manure management

Livestock is responsible for 18% of GHG emissions in the world, and a significant portion of livestock emissions results from poor manure management (Steinfeld et al., 2006). The dramatically increased livestock production, which has been caused by the sharp rise both in population and living standards, is leading to increasing volumes of manure to be managed, which are a source of methane (CH_4) and N_2O (Hou et al., 2008). Net emissions of CH_4 and N_2O depend not only on manure composition and local management practices with respect to preliminary treatment, storage and field application but also on ambient climatic conditions. The diversity of livestock production systems and their associated manure management has resulted in various patterns of nutrient management and environmental regulation (Jungbluth et al., 2001; Heitschmidt et al., 2004; Garnett, 2009). Growth in livestock populations is projected to occur mainly in intensive production systems where the largest potential for GHG mitigation may be found (Jarvis and Pain, 1994; Hao et al., 2001; Jungbluth et al., 2001). In extensive systems, there is almost no excessive emission of CH_4 from manure because it is promptly involved in the N cycle of grazing systems. There is no conflict between efforts to improve food and feed production and those to reduce GHG emissions from manure management. However, emissions from manure might be curtailed, both by altering feeding practices and by composting the manure in livestock pen-feeding systems (VanderZaag et al., 2013).

10.4.4 Livestock management

Livestock are important sources of CH_4 because most CH_4 is produced primarily by enteric fermentation. Adjusting feeding rations can reduce GHG emissions from livestock through feeding more concentrates, which may increase daily CH_4 emissions per capita, but almost invariably reduce the CH_4 emissions per kilogram of feed intake and per kilogram of product (Smith et al., 2008). High sugar ryegrass has been fed to ruminant livestock because it could increase N-use efficiency in the intestine and reduce

N excretion. Feed additives such as coconut oil and garlic in the ration can also decrease the GHG emissions of ruminant livestock. However, the effect of chemical additives in livestock rations on the food safety of humans is widely feared in the developed countries of Asia. Uncertainties also remain as to the balance of benefits resulting from reduced animal numbers or younger age at slaughter for meat production, against how the practice affects emissions when producing and transporting concentrates and other fodders, and the cost of adjusting the livestock production system from one to another.

10.4.5 Management of sown pasture

Improved agronomic practices that increase yield and generate higher inputs of residue C can result in increased soil C storage (Follett et al., 2001). The practices that could be used are as follows: (i) growing improved crop species or varieties such as high-sugar ryegrass; (ii) expanding crop/forage rotations which mitigate GHG emissions by multiple pathways, including reducing chemicals for the control of weeds, diseases and pests, limiting grain crop production, most of which is for livestock production in developed countries, and promoting water-use efficiency in arid and semi-arid regions; (iii) planting perennial forage crops which allocate more C below-ground and reduce GHG emissions both from annual sowing and annually trampled soil; (iv) avoiding or reducing the re-cultivation of fallow cropland and the cultivation of rangeland; and (v) reducing the intensity of cropping systems can also reduce GHG emissions because of less inputs of chemicals and fertilizers (Smith et al., 2008).

10.5 Conclusion

In most of the developing countries of Asia, extensively mixed farming systems are currently predominant. Compared with the intensively mixed farming systems mainly operated in the developed countries of the world, extensive systems are characterized by low input, low output and low risk (Hou et al., 2009). Extensive systems manage carbon more positively than intensive systems, because the low input of carbon is associated with low GHG emissions (Table 10.1).

However, there is an increasing shift from extensive mixed crop–livestock systems to intensive systems, which has resulted from the increased demands for both quantity and quality of animal products, and has resulted in serious environmental problems, such as pollution of both underground and surface water. Currently, environmental problems arising from agricultural operations are one of the great challenges facing the human race, both in Asia and in other continents. This threatens the sustainability of farming systems, long-term food security and carbon

Table 10.1. Comparison between carbon inputs and outputs in extensive and intensive mixed crop–livestock systems.

Carbon	Extensive systems	Intensive systems
Input	Very low	High
Component of input	Human labour	Chemicals, machinery, fuel energy, human labour
Output	Low	High
Component of output	Animal products	Plant products, animal products
Output/input	High	Low
Balance	Positive	Negative
Risk of management	Low	High
Response to climate change	Resilient	Susceptible

balance. An equilibrium point between the 'traditional' operation of mixed farming systems and 'modern' ones must be found to resolve the above problems.

References

Blackburn, H. (1998) Mixed farming systems and the environment: livestock production, the environment and mixed farming systems. In: Nell, A.J. (ed.) *Proceedings of the International Conference on Livestock and the Environment*. International Agricultural Centre, Wageningen, the Netherlands, pp.113–123.

Burney, J.A., Davis, S.J. and Lobell, D. (2013) Greenhouse gas mitigation by agricultural intensification. *Proceedings of the National Academy of Sciences* 107, 12052–12057.

Cheng, C.S. (ed.) (1993) *Climate and Agriculture in China*. China Meteorological Press, Beijing.

Cheng, H., Kim, E.J. and Hou, F.J. (2011) Introduction of high soluble sugar ryegrass in Longzhong Loess plateau and Hexi Oasis. *Pratacultural Science* 28, 978–982.

Fan, J.Z. and Zhang, C.D. (1993) Effects of global climate change on animal production in pastoral region. In: Deng, G.Y. (ed.) *Effect of Climate Change on Agriculture in China*. Beijing Science and Technology Press, Beijing, pp. 406–416.

FAO (2007) *Adaptation to Climate Change in Agriculture, Forestry and Fisheries: Perspective, Framework and Priorities*. Rome, Italy (ftp.fao.org/docrep/fao/009/j9271e/j9271e.pdf, accessed 29 November 2013).

Follett, R.F., Kimble, J.M. and Lal, R. (2001) *The Potential of US Grazing Lands to Sequester Carbon and Mitigate the Greenhouse Effect*. Lewis Publishers, Boca Raton, Florida.

Galloway, J.N., Aber, J.D., Erisman, J.W., Seitzinger, S.P., Howarth, R.W., Cowling, E.B., et al. (2003) The nitrogen cascade. *BioScience* 54, 341–356.

Garnett, T. (2009) Livestock-related greenhouse gas emissions: impacts and options for policy makers. *Environmental Science and Policy* 12, 491–503.

Hao, X., Chang, C., Larney, F.J. and Travis, G.R. (2001) Greenhouse gas emissions during cattle feedlot manure composting. *Journal of Environmental Quality* 30, 376–386.

Heitschmidt, R.K., Vermeire, L.T. and Grings, E.E. (2004) Is rangeland agriculture sustainable? *Journal of Animal Science* 82 (Suppl.), 138–146.

Herrero, M., Thornton, P.K., Notenbaert, A.M., Wood, S., Msangi, S., Freeman, H.A., et al. (2010) Smart investments in sustainable food production: revisiting mixed crop–livestock systems. *Science* 327, 822–825.

Hou, F.J. (2000) Landscape classification of Hexi region. *Journal of Soil and Water Conservation* 14, 18–22.

Hou, F.J. and Li, G. (2001) Change of ecological landscape in Hexi region. *Journal of Soil and Water Conservation* 15, 53–57.

Hou, F.J. and Nan, Z.B. (2006) Improvements to rangeland livestock production on the Loess Plateau: a case study of Daliangwa village, Huanxian county. Invited keynote presentations by the 2nd China-Japan-Korea Grassland Conference. *Acta Prataculturae Sinica* 15 (Suppl.), 104–110.

Hou, F.J. and Yang, Z.Y. (2006) Effects of grazing on grassland. *Acta Ecologica Sinica* 26, 244–264.

Hou, F.J., Nan, Z.B., Xie, Y.Z., Li, X.L., Lin, H.L. and Ren, J.Z. (2008) Integrated crop–livestock production systems in China. *Rangeland Journal* 30, 221–231.

Hou, F.J., Nan, Z.B. and Ren, J.Z. (2009) Integrated crop–livestock production system. *Acta Prataculturae Sinica* 18, 211–234.

Hou, F.Q. (2007) Characteristics of matter, energy, economic and matter economic of oasis agricultural system in Shihezi Oasis. MSc Thesis, Lanzhou University, Lanzhou, China.

Houghton, R.A. (1995) Changes in the storage of terrestrial carbon since 1850. In: Lal, R., Kimble, J., Levine, E. and Stewart, B.A. (eds) *Soils and Global Change*. CRC Lewis Publishers, Boca Raton, Florida, pp. 45–65.

IPCC (2007) *Climate Change 2007 – The Physical Science Basis*. Contribution of Working Group I to the Fourth Assessment Report of the Intergovernmental Panel on Climate Change. Solomon, S., Qin, D., Manning, M., Chen, Z., Marquis, M., Averyt, K.B., Tignor, M. and Miller, H.L. (eds). Cambridge University Press, Cambridge, UK, and New York, USA.

Jarvis, S.C. and Pain, B.F. (1994) Greenhouse gas emissions from intensive livestock systems: their estimation and technologies for reduction. *Climatic Change* 27, 27–38.

Jia, H.T., Jiang, P.A., Zhao, C.Y., Hu, Y.K. and Li, Y. (2009) The influence of enclosing life on carbon distribution of grassland ecosystem. *Agricultural Research in the Arid Areas* 27, 33–36.

Jungbluth, T., Hartung, E. and Brose, G. (2001) Greenhouse gas emissions from animal houses and manure stores. *Nutrient Cycling in Agroecosystems* 60, 133–145.

Kell, D.B. (2011) Breeding crop plants with deep roots: their role in sustainable carbon, nutrient and water sequestration. *Annals of Botany* 108, 407–418.

Nan, Z.B. (2005) The grassland farming system and sustainable agricultural development in China. *Grassland Science* 51, 15–19.

Nie, C.R., Luo, S.M., Zhang, J.N., Li, H.S. and Zhao, Y.H. (2003) The dike-pond system in the Pearl River Delta: degradation following recent land use alternations and measures for their ecological restoration. *Acta Ecologica Sinica* 23, 1851–1860.

Pittaway, P.A., Wildin, J.H. and McDonald, C.K. (eds) (1996) *Beef Production from Ponded Pasture*. Tropical Grasslands Occasional Publication No 7. Tropical Grassland Society of Australia Inc, St Lucia, Australia.

Qiu, G.W., Zhao, Y.X. and Wang, S.L. (2001) The impacts of climate change on the interlock area of farming–pastoral region and its climatic potential productivity in Northern China. *Arid Zone Research* 18, 23–28.

Redden, R. (2013) New approaches for crop genetic adaptation to the abiotic stresses predicted with climate change. *Agronomy* 3, 419–432.

Ren, J.Z. (1985) *Survey and Planning of Grassland Resources*. Chinese Agriculture Press, Beijing.

Ren, J.Z., He, D.H., Zhu, X.Y. and Li, Z.Q. (1995) Models of coupling agro-grassland systems in desert-oasis region. *Acta Praticulturae Sinica* 4, 11–19.

Ren, J.Z., Hu, Z.Z., Zhang, Z.H., Hou, F.J. and Chen, Q.G. (1999) A preliminary discussion on grassland ecological-economic regions in China. *Acta Praticulturae Sinica* 8 (Suppl.), 12–22.

Ren, J.Z., Hou, F.J. and Xu, G. (2011) Discussion on inheritance of grassland culture. *Agricultural History of China* 30, 15–19.

Schellnhuber, H.J., Hare, B., Serdeczny, O., Schaeffer, M., Adams, S., Baarsch, F., et al. (2013) *Turn Down the Heat: Climate Extremes, Regional Impacts, and the Case for Resilience – Full Report*. World Bank, Washington, DC.

Schlesinger, W.H. (1999) Carbon sequestration in soils. *Science* 284, 2095.

Seré, C. and Steinfeld, H. (1996) *World Livestock Production Systems: Current Status, Issues and Trends*. FAO, Rome (ftp.fao.org/docrep/fao/005/w0027e/w0027e00.pdf, accessed 29 November 20013).

Smith, P., Martino, D., Cai, Z., Gwary, D., Janzen, H., Kumar, P., et al. (2008) Greenhouse gas mitigation in agriculture. *Philosophical Transactions of the Royal Society B* 363, 789–813.

Steinfeld, H., Gerber, P., Wassenaar, T., Castel, V., Rosales, M. and de Haan, C. (2006) *Livestock's Long Shadow*. FAO, Rome.

VanderZaag, A.C., MacDonald, J.D., Evans, L., Vergé, X.P.C. and Desjardins, R.L. (2013) Towards an inventory of methane emissions from manure management that is responsive to changes on Canadian farms. *Environmental Research Letters* 8, 035008.

Vermeulen, S.J., Campbell, B.M. and Ingram, J.S.I. (2012) Climate change and food systems. *Annual Review of Environment and Resources* 37, 195–222.

Waha, K., Müller, C., Bondeau, A., Dietrich, J.P., Kurukulasuriya, J., Heinke, H., et al. (2013) Adaptation to climate change through the choice of cropping system and sowing date in sub-Saharan Africa. *Global Environmental Change* 23, 130–143.

Wang, J. and Zhou, H. (2007) Improving the utilization ratio of straw, developing save-grain animal husbandry. *Grass-feeding Livestock* 34, 46–49.

Wang, R.Z. and Li, J.D. (1995) Grazing succession pattern of alkalized *Aneurolepidium chinense* grassland on Songnen plain. *Chinese Journal of Applied Ecology* 6(3), 277–281.

Yadav, S.S., Redden, R., Hatfield, J.L., Lotze-Campen, H. and Hall, A.J.W. (2011) *Crop Adaptation to Climate Change*. Wiley Blackwell, Hoboken, New Jersey.

Zhao, Y.X. and Qiu, G.W. (1999) A study of climate change impact on northern farming–pastoral region. *Meteorological Monthly* 27, 3–7.

11 Enhancing Climate Resilience of Cropping Systems

Heidi Webber,[1] Helena Kahiluoto,[2] Reimund Rötter[2] and Frank Ewert[1]

[1]*University of Bonn, Institute of Crop Science and Resource Conservation (INRES), Crop Science Group, Bonn, Germany;*
[2]*MTT Agrifood Research Finland, Plant Production Research, Mikkeli, Finland*

11.1 Introduction

How cropping systems will be impacted by the combination of rising temperatures, changing rainfall, more frequent extreme events and elevated CO_2 is highly uncertain (Tubiello *et al.*, 2007; Osborne *et al.*, 2013). The attribution of changes in crop productivity to climate change is difficult due to concurrent developments in technology and management (Howden *et al.*, 2007; IPCC, 2010) which occur in response to many sociocultural, environmental and market factors (Smit and Skinner, 2002; Mertz *et al.*, 2009). However, growing evidence suggests that at a regional scale, crop phenology (e.g. Siebert and Ewert, 2012) and yields (e.g. Lobell *et al.*, 2011) have already been impacted by increasing temperatures and days with extreme high temperatures (e.g. Reidsma *et al.*, 2009; Schlenker and Roberts, 2009; Asseng *et al.*, 2011; Lobell *et al.*, 2013). Furthermore, the majority of cereal production, particularly rice and maize, now occurs at mean temperatures above their optimal (Hatfield *et al.*, 2011), implying that increases in global mean temperatures would augment yield reductions (Lobell and Gourdji, 2012). These factors suggest that adaptations are inevitable if cropping systems are to continue providing acceptable yields and livelihoods within tolerable levels of risk of negative climate change impacts (Hall *et al.*, 2012; Dow *et al.*, 2013). However, if adaptation costs become too high or adaptations are no longer feasible to keep risks at tolerable levels, cropping will be unlikely to persist in its current form. If anticipated and planned for, transformational adaptation may lead to more sustainable social and environmental states, though representing a sharp break with present reality (Rickards and Howden, 2012).

Climate change reenforces inequity (Parry *et al.*, 2004) by posing the greatest risks to many of the world's poorest regions (Christensen *et al.*, 2007), though they have the least adaptive capacity and culpability (Mertz *et al.*, 2009; Godfray *et al.*, 2010; Thornton *et al.*, 2011; Beddington *et al.*, 2012). This reality makes the definition of a common reference frame in which to discuss climate change adaptations across regions challenging. In highly intensive and productive systems, cropping systems produce close to potential yield levels, with yields largely determined by weather and crop characteristics (van Ittersum and Rabbinge, 1997; Supit *et al.*, 2010; van Ittersum and Cassman, 2013; van Wart *et al.*, 2013). Improving yield potential (e.g. Reynolds *et al.*, 2011) and managing negative environmental impacts on water quality (Tilman *et al.*, 2002) or groundwater withdrawals (Gleeson *et al.*, 2012), for example, are priorities (Chen *et al.*, 2011). However, in many low-input, semi-subsistence cropping systems, farmers are mainly constrained by poverty (Gbetibouo et al., 2010) and improving crop productivity to close the yield gap primarily through better management of cropping systems (Lobell et al., 2009; Godfray et al., 2010; Tilman et al., 2011; Tittonell and Giller, 2013) is key to reducing poverty and improving food security (Hazell and Wood, 2008; Kolavalli et al., 2010). For these systems and farmers, the critical question is not to identify climate change adaptations, but rather how to mainstream climate change adaptation with other drivers of agricultural development and identify which current trends are the most robust under plausible future climatic conditions (Halsnæs and Trærup, 2009; Mertz et al., 2009; Nielsen and Reenberg, 2010; Tilman et al., 2011).

Nelson et al. (2007) and Ifejika Speranza (2010) have suggested that considering adaptations within a resilience framework (Holling, 1973; Folke, 2006) may allow going beyond the identification of interventions intended to keep specific climate-related risks at tolerable levels, but may also promote the system's development such that it will be robust to a range of possible and uncertain future changes. This chapter examines how a resilience approach may be used to learn about, and ultimately support, adaptations to increasing climate variability and uncertainty for cropping systems. It starts with an examination of several commonly mentioned crop and soil management adaptation options (Smit and Skinner, 2002; Howden et al., 2007; Olesen et al., 2011; Vermeulen et al., 2012) for their potential contributions to improving the climate resilience of cropping systems. Readers are referred to other chapters in this book for consideration of the interactions and trade-offs between cropping and the farming systems in which they are embedded. While this is a rather pragmatic approach, it is the essential first step to identifying which current approaches have the greatest potential to build climate resilience and also to identifying areas where new approaches and ways of conceptualizing adaptation are most needed.

11.2 Framing Adaptation for Improved Resilience

Globally, higher temperatures coupled with increased variability of precipitation imply that the frequency of days with extreme high temperatures and drought conditions will increase. Global-scale observational evidence suggests that the occurrence of extremes has already increased since the 1950s (Dai, 2011; Handmer et al., 2012; Hansen et al., 2012; IPCC, 2012; Seneviratne et al., 2012). At locations where temperatures are already at or beyond the optimum for crops (Burke et al., 2009; Thornton et al., 2011), even changes in mean temperatures are likely to produce marked negative impacts (Schlenker and Lobell, 2010; Seneviratne et al., 2012; Lobell et al., 2013). The implication is that for cropping systems to meet food security goals sustainably (Handmer et al., 2012), new approaches to management are required that can both anticipate and respond to extreme events, as well as foster flexibility and adaptability to face unexpected conditions.

The concept of resilience first emerged in ecology in relation to ecological stability theory (Holling, 1973), and many subsequent interpretations have evolved, as presented in Brand and Jax (2007). Resilience theory laid the foundation for adaptive resource management (Walters, 1986), and its use in natural resource management is now widespread (Carpenter et al., 1999; Walker et al., 2002; Tompkins and Adger, 2004; Drever et al., 2006; Marshall and Marshall, 2007). It has not yet found wide application in the agricultural sciences or management literature (for recent exceptions, see Ifejika Speranza, 2010; Lin, 2011; Hakala et al., 2012; Abson et al., 2013; Himanen et al., 2013a), although it shares many common elements with vulnerability analysis, which is more commonly applied in agriculture (Challinor et al., 2007; Reidsma and Ewert, 2008; Mertz et al., 2009; Rasmussen et al., 2009; Simelton et al., 2012), as discussed by Janssen and Ostrom (2006). Managing cropping systems for resilience may offer a new paradigm for adapting cropping systems to climate change (Nelson *et al.*, 2007) and perhaps a common reference framework to evaluate options across system levels, scales and regions (Ifejika Speranza, 2010). Two key aspects offer useful insights into how cropping systems could be managed in the context of climate change and diverse societal objectives for agriculture. First, the climate resilience of a cropping system reflects its capacity to experience high climatic variability and shocks (drought, flooding, high temperature) and still maintain its essential functions (Folke, 2006; Walker and Salt, 2006). The second aspect of resilience thinking characterizes cropping systems as existing in dynamic adaptive cycles with feedbacks and interactions with other systems at other scales (Holling, 1992; Folke, 2006). In this framework, at one extreme, systems exist in relatively disordered configurations, with many competitive actors and potential development pathways (Rötter and van Keulen, 1997; Rötter *et al.*, 1997; Carpenter *et al.*, 2001). At the other extreme, systems exist in highly specialized and productive configurations, though with greater vulnerability to shocks, as fewer response pathways exist. If a major perturbation causes the system to collapse, it can undergo a period of major and rapid change, followed by a period of renewal and rebuilding, thus starting a new iteration in the adaptive cycle (Carpenter *et al.*, 2001). This situation reflects what Rickards and Howden (2012) refer to as transformational adaptation. By contrast, incremental adaptation attempts to maintain a system in its current functioning in the face of climatic shocks by avoiding abrupt and major changes, and at the same time supporting the system's evolution to its desired state along the continuum between rapid growth and highly specialized extremes to satisfy its development goals (IPCC, 2012; Rickards and Howden, 2012).

While numerous theoretical frameworks are useful to conceptualize resilience (Holling, 1973; Walker *et al.*, 2002; Folke, 2006; Walker and Salt, 2006), its operationalization for real-world systems has proven challenging and received less attention (Carpenter *et al.*, 2001; Walker *et al.*, 2002; Bennett *et al.*, 2005). As a starting point, it is useful to envision both general and specific system resilience (Walker *et al.*, 2009). General resilience can be considered a system property and embracing adaptive capacity, representing the ability of a system to cope with a range of future and uncertain shocks (Walker *et al.*, 2002). Adaptive capacity is strengthened through learning, experimentation and the flexibility to try novel approaches (Walker *et al.*, 2002; Folke, 2006; Nelson, 2011). Diversity is emerging as key to building general resilience (Chapin *et al.*, 1997; Jarvis *et al.*, 2008), and has been demonstrated to reduce system sensitivity to shocks (Reidsma and Ewert, 2008; Himanen *et al.*, 2013b), increase flexibility and improve system resource-use efficiency (H. Kahiluoto and J. Kaseva, 2013, unpublished results).

Defining specific resilience provides an analytic lens to examine which aspects of a system are prone to moving the system into undesirable configurations when acted upon by external drivers (Walker *et al.*, 2009). With this understanding, management interventions can aim to steer the system in trajectories leading to desirable outcomes (Walker *et al.*, 2002). Defining specific resilience begins with specifying *what* valued aspect of the system is considered vulnerable to *which* type of external driver or change (Carpenter *et al.*, 2001). In the analysis that would follow, insights would emerge as to what could make the system more resilient to external changes, and also what might indicate or exemplify resilience for the system. For any given system, these aspects will depend on system boundaries, spatial and temporal scales considered, and the values and specific objectives of the stakeholders involved (Carpenter *et al.*, 2001). Implicit in the operationalization of resilience is that it is a context-specific and stakeholder-defined development, as the stakeholders have in-depth understanding of the context and decide the configuration they want for the system (Walker *et al.*, 2002; Ifejika Speranza, 2010). None the less, the analytical framework of Walker et al. (2002) has been used, building on an example in Bennett et al. (2005) and shown in Table 11.1, to derive some generic characteristics of cropping systems that can be used to evaluate whether an adaptation contributes to climate resilience. Two cases were considered: an intensive rainfed maize–wheat–soy system typical of the north-east of North America and a low-input, semi-subsistence system typical of the Sudan Savannah of West Africa. The generic characteristics of resilient cropping systems that emerge are:

- provision of stable yields
- provision of acceptable yields and the capacity to increase them
- conservation and maximization of productive use of soil water
- provision of stable, appropriate soil productivity/fertility/quality
- limitation of diseases, pests and weeds

- flexibility and adaptability, amenable to many possible system configurations (Smit and Skinner, 2002; Bennett et al., 2005; Anderies et al., 2006; Mertz et al., 2009)
- synergy with other interdependent systems and scales, contributing to their maintenance and resilience, including enhancing biodiversity (Bennett et al., 2005; Anderies et al., 2006; Lin, 2011; Nicholls and Altieri, 2013).

It is possible that other characteristics may be important for systems not considered here (e.g. rice cropping), and we propose this list to illustrate one possible approach to evaluating the contribution of an adaptation to climate resilience, rather than as a definite set of characteristics. An actual resilience analysis (which would involve the formalization of a system model and identification of controlling drivers, feedbacks, thresholds, shifts in and actual resilience) is case specific.

Table 11.1. Framing generic adaptation options within the context of resilience demonstrated for two case cropping systems. Based on an example in Bennett et al. (2005).

	Case cropping system	
	Intensive high-input North American maize, wheat and soy system	Low-input West African Sudan Savannah
Resilience of what?[a]	1. Yield stability. 2. Water quality.	Livelihoods from cropping (stable and adequate yield provision).
Resilience to what (external drivers)?	Climate.	Climate. Need to improve food security.[b]
What variables are changing (system variables)?	1. Grain yield, soil available water. 2. Algae populations in lakes.	Grain yield, stover yield, soil available water, soil organic carbon, soil fertility, disease, weed and pest levels.
What processes are causing the changes (internal system drivers)?	1. Drought stress, heat stress. 2. Erosion, leaching, high soil P and N status.	Water stress, heat stress, erosion, residue removal, net nutrient export, disease, pest and weed pressure.
Connection between the internal drivers and variables	1. Water conservation measures, irrigation, resistant varieties, diversity of varieties. 2. Soil cover, soil condition, fertilizer supply–plant demand balance, lake nutrient levels, intense rain events.	Water-conservation strategies, irrigation, resistant varieties, diversity of varieties, soil cover, soil condition, fertilizer availability, competing uses for residues, pest control, disease control, weed control.

Notes: [a]In practice, answering the question, 'resilience of what?', involves specifying the boundaries of the system considered, though for our aim of defining the generic characteristics of resilient cropping systems, we specify only the valued aspect considered vulnerable; [b]in investigating adaptations to build climate resilience, including increasing productivity as a second (likely primary) driver of change in low-input cropping systems, follows directly from a resilience approach to support the system's evolution to a desired configuration.

Further, farm management decisions are unlikely to be shaped only by climatic forcing, and adaptation to climate change must also be able to manage other sources of risk faced by farmers (Smit and Skinner, 2002; Mertz et al., 2009). Finally, the last two points do not emerge directly from the examples considered here to derive specified resilience, but are considered indicative of general resilience.

11.3 Soil and Crop Management Options Available to Increase Climate Resilience

Early reviews of adaptation options (Smit and Skinner, 2002; Howden et al., 2007) highlighted a range of management options for cropping systems, whereas more recent studies have focused on adaptation options for expected impacts in specific regions (e.g. Olesen et al., 2011, for Europe). A number of studies have reviewed actual cropping system adaptations in response to climate variability and change (Thomas et al., 2007; Reidsma and Ewert, 2008; Barbier et al., 2009), though none were found which offered a systematic critique of management options with respect to their potential contributions to resilience. The following reviews frequently discussed crop and soil management options that constitute incremental adaptations and consider their potential to improve the climate resilience of cropping systems.

11.3.1 Changing sowing dates

In the cooler climates of Europe, shifting sowing dates to earlier in the growing season (Olesen et al., 2011), enabled by higher spring temperatures, may stabilize yield levels by avoiding later summer drought and high-temperature stress, both of which are expected to be aggravated as evapotranspiration increases relative to summer precipitation and the likelihood of high-temperature events increases (Christensen and Christensen, 2007). However, van Oort et al. (2012) found that in the Netherlands, higher spring temperatures did not result in earlier planting of sugarbeet unless the opposite effect of fewer frost days (correlated with warmer temperatures) was also accounted for, as frost days have a positive influence on seedbed characteristics. In years with higher spring temperatures, there are likely to be fewer frost days, and farmers will either wait for them or perform extra cultivation operations, both of which result in delayed planting. Further, Siebert and Ewert (2012), who detected a shift of 2 weeks between 1959 and 2009 in phenological development between sowing and yellow ripeness in oats due to higher temperatures in Germany, estimated that sowing occurred only 1 day earlier over the same period. Caution should be used in concluding that this demonstrated there was no shift towards earlier planting as, concurrently, oat production shifted to more mountainous (cooler) regions during the same period. So, while accelerated phenological development with warming may offset the later negative impacts of drought, it may be that other factors (soil condition, spring rainfall, farmer preferences) limit a shift to earlier planting. Other examples exist, however, such as in maize cultivation in the USA (Sacks and Kucharik, 2011), where global warming has led to earlier sowing, which, in turn, has allowed the use of longer duration cultivars, with both factors contributing to significant yield gains. In rainfed systems in the tropics, cropping commences with the onset of the rainy season, and switching sowing dates may not be an option when the rains start very late, or may offer little benefit when the rains end early. However, Laux et al. (2010) suggested that when combined with the use of a number of varieties, shifting sowing dates could offset the negative impacts of drought and higher temperature, thus contributing to resilience by stabilizing yield in dry years.

11.3.2 Cultivar selection

Selecting new varieties of an existing crop may be one of the most important and easiest adaptation options to be implemented (Rötter et al., 2011; Semenov and Shewry, 2011). In the agricultural adaptation literature for cool climates, it is often suggested that the use of varieties with longer thermal time requirements represents a way to increase production potential (Olesen et al., 2011; Siebert and Ewert, 2012) in warming climates. In parts of sub-Saharan Africa (SSA), farmers have adopted drought-resistant maize varieties in response to drier conditions (e.g. Rötter and van Keulen, 1997). The use of shorter season cultivars, which mature quickly, has already emerged in oats in Germany, likely as an adaptation to the higher risk of summer drought (Siebert and Ewert, 2012), and in South Africa, where using shorter season varieties allows for the possibility of a short rainy season (Thomas et al., 2007). Within a resilience perspective, shifting to sets of cultivars with resistance to drought, high temperature, diseases or pests are good management options (e.g. Hakala et al., 2012; Himanen et al., 2013b). Within a farm or region, growing a diversity of cultivars can buffer against weather variation, extreme events and the risk of total crop failure (Himanen et al., 2013b). In addition, a set of cultivars with a diversity of responses to critical weather factors builds adaptive capacity to climate changes that are difficult to forecast (Hakala et al., 2012). To realize the potential of using different cultivars, farmers need access to, and knowledge of, different varietal options, such that they can develop their production system in response to multiple and changing stressors and opportunities (Zhu et al., 2000; Hakala et al., 2012). In the context of SSA smallholder systems, this means supporting seed production and storage initiatives, as well as improved functioning of input markets (McGuire and Sperling, 2013).

11.3.3 Crop selection

In the short to medium term, changing to new crop types is not expected to be an important adaptation for many parts of Europe, with the exception of the expansion of grain maize in Germany, Poland, the Czech Republic and parts of Eastern Europe (Olesen et al., 2011), as temperatures increases. Considering only climatic factors, switching to millet or sorghum from maize in parts of SSA would build climate resilience, as these crops are more drought tolerant, as well as better adapted to the low soil fertility common in many parts of SSA. Further, in most regions of the world, there are crop choices available that are tolerant to a range of conditions and pests. Many of these crops that were important in the past, though, have lost their presence in modern diets. However, to allow subsistence cropping to evolve to more productive states in order to reduce poverty, such switching would need to be accompanied by other management interventions. Reviewing current trends in switching to new crops in SSA, Thornton et al. (2011) point out that this is associated with considerable risk, and in many situations the required access to credit and/or markets is simply not available. Current trends in adopting new crops are likely influenced by expanded market opportunities and support (Thomas et al., 2007). A reasonable management for such cases may be to maintain a portfolio of crops with different stress tolerances to ensure a minimum level of yield in extreme years and to increase yield stability. Finally, switching from crops to extensive grassland cultivation can build soil resilience and adaptability of soil food webs, and would increase the climate resilience of cropping in highly degraded or erosion-prone areas (Jones and Thornton, 2009; de Vries et al., 2012).

11.3.4 Intercropping

Intercropping, in which two or more crops are grown together in various configurations in the same area, can increase climate resilience by reducing the risk of crop failure during droughts and stabilizing overall yield levels due to the different timing of crop phenology and the timing of peak water/nutrient uptake (Hauggaard-Nielsen and Jensen, 2001). While competition can result in yield penalties, overall land productivity is generally increased, potentially adding to soil organic carbon (SOC) and nitrogen levels, depending on subsequent residue use and the inclusion of legumes (Kaizzi et al., 2006; Rusinamhodzi et al., 2012). The impacts on nutrient-, water- or light-use efficiencies can be increased with the appropriate selection of crops, and can be tailored to climatic conditions (Malézieux et al., 2009; Lithourgidis et al., 2011; Malézieux, 2012). The increased ground cover of intercropped systems reduces soil evaporation, though it may lead to increased evapotranspiration in water-limited situations. However, the benefits of intercropping for managing the risk of crop failure is supported by evidence provided by Rusinamhodzi et al. (2012), who found that while maize–legume intercrops did not result consistently in improved yields of maize, they increased overall per unit of land productivity and reduced total crop failure in years with drought conditions. Likewise, intercropping with mixtures of cultivars (Kaut et al., 2008) is a good option to enhance resilience against pressure from pests and to stabilize yields, while being technologically less challenging than using crop mixtures (Tooker and Frank, 2012). Another case of intercropping with a high potential to increase climate resilience is agroforestry in which perennial woody plants are combined with annual crops, thus contributing to food security and a number of other ecosystem services (Akinnifesi et al., 2010; Nair and Garrity, 2012).

11.3.5 Reduced tillage

Reduced tillage techniques decrease soil loss from erosion associated with either wind or surface runoff, and when combined with soil residue retention (as is the case in most US systems), lead to increased SOC and reduced atmospheric emissions (Lal, 2004; Rosenzweig and Tubiello, 2007; Smith and Olesen, 2010). In intensive and mechanized systems, it can save fuel and reduce compaction caused by the use of machinery (Lal, 1985; Rosenzweig and Tubiello, 2007). Increased yields are explained by improvements in nutrient-use efficiency and soil physical properties associated with higher SOC levels (Smith and Olesen, 2010). Limiting erosion, accomplished by reduced tillage and other means, is critical to maintaining yield, as the depth of top soil is an important factor determining crop water availability and is related directly to maize yields (Cruse and Herndl, 2009). However, the success of reduced tillage in maintaining yield is largely a function of soil type, effective disease control and herbicide use (Lal, 1985; Giller et al., 2009). In systems typical of SSA, where herbicides are generally not available, it can lead to increased labour, particularly for women, as the weeding burden increases (Giller et al., 2009). Further, it is likely that improvements in SOC levels with no-till systems are due largely to the retention of large amounts of stover on the soil surface, which is currently not feasible in many parts of SSA, as discussed in Corbeels et al. (2006).

11.3.6 Soil residue retention

Maintaining soil cover by the retention of crop residues can produce many positive effects, ranging from improved soil structure (i.e. reducing erosion risk and increasing water-holding capacity), improved long-term fertility and nutrient-use efficiencies as SOC increases (Yamoah et al., 2002), higher crop water availability with increased infiltration and reduced non-consumptive evaporative losses, and reduced erosive soil loss as infiltration increases. All of these effects increase the climate resilience of cropping systems, especially to changes in rainfall intensity and drought (Lal, 2004; Cruse and Herndl, 2009; Smith and Olesen, 2010), and may moderate the negative impacts of higher temperatures on soils. However, in extremely dry or wet years, the water conservation potential of residue retention does not offset yield losses (Rusinamhodzi et al., 2011), suggesting that residue retention does not always buffer against extreme events. This point is critical for some smallholder producers in semi-arid areas, such as the Sudan Savannah of West Africa, as crop residues, produced in very low quantities due to various constraints, are a valuable source of fodder for livestock. Sale of livestock serves as an insurance mechanism in years when yields fail. A study comparing different farming systems in semi-arid Africa and Asia found that only sites with high population densities and significant production intensification produced sufficient stover biomass to satisfy the requirements of both livestock and soil conservation (Valbuena et al., 2012). In highland mixed farming systems of Ethiopia, net losses of carbon result via livestock grazing, biomass use for fuels and composting (K. Rimhanen and H. Kahiluoto, unpublished results). The use of residues for household energy needs represents a constant quantity on various farms, and would need to be substituted with another source to enable soil residue retention, Finally, residue retention may increase disease pressure, particularly in more humid or organic systems.

11.3.7 Contour stone bunds

The use of stone bunds along a field contour is a practice used by farmers in SSA to slow surface-water flow, leading to increased infiltration and reduced soil loss from cultivated slopes (Barbier et al., 2009). Farmers in South Africa report increased use of stone bunds in response to a perception of more intense and earlier rainfalls (Thomas et al., 2007). The technique is not practical when large tractors are used for cultivation, as it separates fields into shorter parts, but it is not expected to create problems for manual labour or animal traction.

11.3.8 Integrated weed, pest and disease management

Pest, weed and disease dynamics are likely to evolve in response to changing climatic conditions and other adapted crop management practices. Effective monitoring programmes, the use of resistant cultivars and targeted interventions are seen as being important to reduce the negative impacts of pests and diseases with climate change (Howden et al., 2007). Practices that have positive impacts on pest and/or disease suppression, as presented in the review of Lin (2011), include the diversification of landscapes (Thies and Tscharntke, 1999), crop types in rotation (Krupinsky et al., 2002), in-field crop varieties (Zhu et al., 2000) and the use of non-crop species in cropped fields (Rea et al., 2002). Finally, monitoring programmes foster learning and the development of farmers' adaptive capacity, as well as that of the institutions that support them, constituting a critical aspect of general resilience (Tschakert and Dietrich, 2010).

11.3.9 Use of landscape diversity

In some instances, diversified production can lead to increased overall productivity, as in the case of a combination of organic cropland and intensively managed grazing (Boody et al., 2009). Shifting cropping to higher elevations is an option to offset the negative impacts of increased temperatures while maintaining current crop varieties and cultivation timing (Thornton et al., 2011), and may represent a successful adaptation to increasing temperatures at a regional scale. At the farm scale, the resilience of production in a new area depends on farmers' access to land, the susceptibility of the newly cultivated land to erosion and the risk of the loss of biodiversity and/or other ecosystem services of the converted land. Farmers' current patterns of landscape utilization in the small farms of the West African Sudan Savannah already exemplify aspects of climate resilience, and intensification of these systems may also increase production. Legumes such as cowpea and groundnut, which are generally drought tolerant, are grown in the uplands with shallow soils, while sorghum, maize and root crops are common in the deeper, more fertile lowland soils, with rice sometimes cultivated at the valley bottoms (Windmeijer and Andriesse, 1993; Bationo et al., 1996). This strategy minimizes the risks of complete crop failure in the event of extreme weather conditions (Callo-Concha et al., 2012). Vegetated buffer zones around water bodies is a strategy that can prevent the transfer of nutrients, chemicals and soil from agricultural fields into water bodies, thus reducing non-point source pollution loads in areas where precipitation is expected to increase and/or intensify. Additionally, buffers can enhance biological diversity – providing habitats for natural predators and guarding against the dominance of invasive or undesirable native species with a competitive advantage in agricultural fields (Fischer et al., 2006). To build system resilience, Fischer et al. (2006) suggested maximizing landscape diversity in agricultural regions to protect biodiversity by ensuring that large areas of native vegetation remained, structural complexity was present throughout the area, with adequate connectivity, and cropping systems tried to mimic the patterns of the heterogeneity of the natural landscape. Enhancing biodiversity at the landscape scale is important for the climate resilience of cropping systems to maintain sources of genetic traits for crop adaptations, insects and birds for pollination, to facilitate groundwater recharge and to maintain healthy soils and habitats for the predators of crop pests (Fischer et al., 2006).

11.3.10 Irrigation

Improving the water productivity of crops is key to improving and stabilizing yields in rainfed environments where considerable opportunities exist in terms of soil-water conservation techniques to limit evaporation (Rockström and Barron, 2007). In irrigated systems, improvements in water productivity are likely to save water and can result from improved irrigation efficiencies and application techniques, crop breeding, manipulation of crop physiological responses (root or drought responses) to use more available soil water and from better management to match plant growth with times of maximum soil water availability (Morison et al., 2008). At the farm level, irrigation has great potential to increase climate resilience by buffering against drought and high-temperature events. By increasing production, irrigation can also improve SOC levels if residues are returned to the soil (Lal, 2004). Further, irrigation would enable subsistence farmers in many areas with high precipitation variability to stabilize yields, thus making other investments in intensification, such as new crops, fertilizer use or new machinery, less risky (Giordano et al., 2012). Deficit irrigation is an on-farm irrigation strategy that generally results in water savings, and aims to stabilize yields. When deficit irrigation leads to increases in water productivity, it can be an option to increase production in areas where water is the primary limiting resource (Fereres and Soriano, 2007; Geerts and Raes, 2009). In the Sahel region of northern Burkina Faso, farmers' preferred adaptation is to shift towards irrigated vegetable production in the dry season, when prices are higher, rather than relying on rainfed subsistence production of sorghum or millet (Barbier et al., 2009). The installation of any irrigation system must consider if the rate of water extraction is likely to have negative effects on water levels or downstream users and uses, and if drainage is sufficient to ensure salinization and waterlogging are avoided; both situations can lead to non-resilient states (Falloon and Betts, 2010). On a larger scale, irrigation is known to detract from system resilience. Large dams, particularly if shallow and flooded land is not cleared, have been shown to be large net greenhouse gas emitters as vegetation decays. The negative ecological and social consequences of large reservoirs are well documented, and large irrigation schemes can lead to highly non-resilient ecological systems due to problems with salinity and waterlogging (e.g. Walker et al., 2009).

11.3.11 Adjusting the intensity of cropping

The intensity of cropping at the farm level was determined by Reidsma and Ewert (2008) to be an important determinant of wheat yield stability with increasing temperatures across Europe, with medium intensive production systems being most resilient. Highly intensive systems achieve near potential yields and are very sensitive to climate, whereas in low-input systems, few technological or management options are available to respond to changing climatic conditions. This finding would imply that the intensification of subsistence cropping systems in SSA would have, in addition to indirect effects on soil fertility and structure (Lal, 2004; Vanlauwe and Giller, 2006) and meeting the region's larger development goals, direct positive impacts on climate resilience, as a greater number of management and technological options would be available to respond to climatic shocks (provided current options were still applicable under future conditions). For the highly intensive systems of North America and Europe, increased resilience to climatic shocks may necessitate less productive systems, especially if stress-tolerant cultivars are adopted that have a lower production potential. However, a high diversity of farm types in terms of size and intensity produced more stable yields at a regional scale in response to climatic variability, suggesting potentially antagonistic effects between scales (Reidsma et al., 2010). More deliberate integration of various farming activities, particularly animal husbandry and cropping, to close farm-level nutrient cycles and improve balances, can increase resource-use efficiency and minimize negative environmental impacts (e.g. Devendra and Thomas, 2002) and simultaneously enhance soil SOC levels, increasing soil water-holding capacity and buffering against moderate levels of drought stress (Rusinamhodzi et al., 2011), as well as reducing vulnerability to input-price volatility.

11.3.12 Climate forecasts

Agronomy and agroclimatic experts have listed seasonal weather forecasts as being critical adaptation tools for field crops in Europe (Olesen et al., 2011) and Africa (Rao, 2003). This is partly conflicting with findings from earlier studies investigating the use of climate forecast information with farmers in SSA and the USA. Bryan et al. (2009) found that, in South Africa, access to information such as climate forecasts was not an important factor leading to adaptation. Some suggest that if weather forecast information is developed without explicit consideration of how it will be used, and by whom, little real benefit of this work is likely to materialize (Patt and Gwata, 2002; Vogel and O'Brien, 2006; Patt et al., 2010; Crane et al., 2011). Farmers in the southeastern USA reported that they expected seasonal forecast information could influence their decisions regarding crop selection, cultivation timing and selection of inputs, and cultivation methods and varieties, but their attitudes and beliefs about the technologies and agencies delivering the information would influence whether they would accept the forecasts as legitimate and use the information (Patt and Gwata, 2002; Cash et al, 2003; Crane et al., 2010; Patt et al., 2010). In response to these insights, various initiatives are now under way to get climate information closer to those who need it most, i.e. farmers and agricultural extension services (see, for example, Rao, 2003).

11.3.13 Farmer experimentation

The literature on the resilience of social–ecological systems provides ample evidence that experimentation and learning are crucial to building the resilience of farms and farming communities (Osbahr et al., 2010; Tschakert and Dietrich, 2010; Kummer et al., 2012). The degree to which a system can build and increase the capacity for learning and adaptation has emerged as a key characteristic of resilient social–ecological systems (Folke, 2006; Nelson, 2011). It is believed that developing learning and experimentation skills is especially critical for farmers in large parts of SSA, where it is expected that by 2050 climatic conditions will have no present-day analogue to serve as examples of crop performance (Burke et al., 2009; Thornton et al., 2011). Additionally, encouraging and supporting increased experimentation by farmers may also directly improve the climate resilience of cropping systems as farmers develop first-hand experience and site-specific information about different management options in situations that integrate all the factors specific to their production system.

11.3.14 Is it possible to generalize the contributions of management options to climate resilience?

Our review of crop and soil management options highlights that many adaptations have the potential either to add to or to detract from climate resilience, as summarized in Table 11.2. Two important points emerge from examining cropping system resilience from this perspective. First, for many management options, the outcome depends on a host of biophysical parameters (soil type, climatic conditions, landscape position) and current practices (intensity of cropping, farm-type cropping embedded in, crop types), reinforcing the context-specific nature of resilience analysis and the need for participatory approaches in which affected stakeholders specify system attributes with the highest value. Second, there is often an antagonism between scales and system levels, as most clearly exemplified here for the case of irrigation. These feedbacks and interactions, considered together with the dynamic nature of cropping systems in which climatic (and other) drivers of change interact with the system's actual configuration, produce continuously shifting response landscapes. We propose that this necessitates (i) using modelling tools to facilitate understanding complex system behaviour and (ii) careful consideration of the effects of scale and the interactions between various system levels (e.g. interactions of cropping with soil, animal husbandry and food and earth systems) (Folke, 2006).

11.4 Scale Dependency?

The choice of scale at which to evaluate cropping systems' adaptations to increase climate resilience is critical. Desirable adaptations at one scale can produce negative effects at other scales, or on related systems (Smit and Skinner, 2002). There is often a mismatch in temporal scale between decisions taken by farmers (typically daily or seasonally), resource management and policy decisions (one or more years) and that over which climate signals emerge and can be discerned (decades) (Giorgi, 2005; Howden et al., 2007). Regarding spatial scales, differential responses to management interventions at the farm versus regional scale were illustrated by Reidsma and Ewert (2008) in the analysis of the correlation between wheat yield anomalies and rising temperatures in Europe. Regionally, high levels of farm diversity (size and intensity) stabilized yields to climate variability, thus increasing the resilience of cropping systems. Translation to the farm level necessarily revealed high year-to-year variability for some farm types and low climate resilience. Such a trade-off might be reduced through building higher resilience at farm level, and further through development of a diversity of resilient farming systems. A second example of possible antagonism between scales was found in South Africa, where more well-off households were better able to afford to take risks and try new cropping practices, giving them a competitive advantage over poorer households. Such a situation was evaluated as eroding general resilience at the community level (Osbahr et al., 2010). However, when groups were formed and collective initiatives and experimentation undertaken, a wider spectrum of the community was able to benefit from adaptations to increase climate resilience, though the benefits to individuals from more wealthy households were reduced.

11.5 Conclusions

Soil and crop management options commonly referred to in the climate change adaptation literature have been evaluated in terms of what they could contribute to the resilience of cropping systems in the face of

Table 11.2. Cropping systems' climate resilience with current trends in crop and soil management incremental adaptation options. Note that depending on the context and scale considered, some management options have the potential either to add to or to detract from climate resilience, as discussed for the various options in Section 11.3.

Characteristics of climate resilience	Crop and soil management options	
	Adding to climate resilience	Detracting from climate resilience
Provision of stable yield	Changing sowing dates; cultivar selection; crop selection; intercropping; soil residue retention; contour stone bunds; integrated weed, pest and disease management; use of landscape diversity; irrigation; medium-intensity cropping; climate forecasts	High-intensity cropping; cultivar selection
Provision of acceptable yields and the capacity to increase them	Changing sowing dates; cultivar selection; reduced tillage (soil type, long term); soil residue retention (long term); use of landscape diversity; irrigation; high-intensity cropping	Reduced tillage (soil type, time frame); soil residue retention (short term); low intensity of cropping
Conserve and maximize productive use of soil water	Changing sowing dates; cultivar selection; crop selection; intercropping; reduced tillage; soil residue retention; contour stone bunds; use of landscape diversity; irrigation; intensity of cropping	Irrigation; changing sowing dates; cultivar selection; intensity of cropping
Provision of stable appropriate soil conditions	Crop selection; intercropping; reduced tillage; soil residue retention; contour stone bunds; use of landscape diversity; irrigation; increasing intensity of cropping	Reduced tillage (lack of aeration); soil residue retention (tie up nitrogen on short term); irrigation (salinity)
Limit disease, pest and weed populations	Cultivar selection; crop selection; intercropping; soil residue retention (limit weeds); integrated weed, pest and disease management; use of landscape diversity; buffer zones; irrigation	Reduced tillage; soil residue retention; irrigation
Flexible and adaptable, amenable to many system configurations	Changing sowing dates; cultivar selection; intercropping; integrated weed, pest and disease management; use of landscape diversity; irrigation; climate forecasts; farmer experimentation	Crop selection; irrigation
Synergy with other interdependent systems and scales	Intercropping (agroforestry); reduced tillage (save fuel and energy – limit emissions); soil residue retention (reduce erosion and soil loss – pollution); contour stone bunds; use of landscape diversity; buffer zones; intensity of cropping (socio-economic status); climate forecasts; farmer experimentation	Reduced tillage (impact on labour division/timing); soil residue retention (compete with other uses); use of landscape diversity; irrigation; intensity of cropping (environment)

uncertain future climatic conditions. In an attempt to operationalize aspects of resilience theory, we suggest that adaptations that promote yield stability and yield improvement, promote, target and preserve crop and wild genetic diversity and maintain and build soils can lead to more resilient cropping systems. However, as future conditions are highly uncertain, building the general resilience of cropping systems by ensuring that management is flexible and adaptable to respond to a range of potential stressors (climatic and other) is critical. Implicit in the enhancement of resilience is that it is context specific, with the suitability of any adaptation depending on stakeholders' aims, the social, cultural and economic context and their interactions with genetic, management, climate and other environmental factors. Agricultural science has much to contribute to enhancing the resilience of cropping systems, but to date few examples exist in which cropping system resilience has been explicitly assessed.

References

Abson, D.J., Fraser, E.D. and Benton, T.G. (2013) Landscape diversity and the resilience of agricultural returns: a portfolio analysis of land-use patterns and economic returns from lowland agriculture. *Agriculture and Food Security* 2, 2.

Akinnifesi, F.K., Ajayi, O.C., Sileshi, G., Chirwa, P.W. and Chianu, J. (2010) Fertiliser trees for sustainable food security in the maize-based production systems of East and Southern Africa. A review. *Agronomy for Sustainable Development* 30, 615–629.

Anderies, J.M., Walker, B.H. and Kinzig, A.P. (2006) Fifteen weddings and a funeral: case studies and resilience-based management. *Ecology and Society* 11, 21.

Asseng, S., Foster, I. and Turner, N.C. (2011) The impact of temperature variability on wheat yields. *Global Change Biology* 17, 997–1012.

Barbier, B., Yacouba, H., Karambiri, H., Zoromé, M. and Somé, B. (2009) Human vulnerability to climate variability in the Sahel: farmer's adaptation strategies in Northern Burkino Faso. *Environmental Management* 43, 790–803.

Bationo, A., Rodes, E., Smaling, E.M.A. and Visker, C. (1996) Technologies for restoring soil fertility. In: Mukwunye, A.U., Jager, A. and Smaling, E.M.A. (eds) *Restoring and Maintaining Productivity in West Africa Soils: Key to Sustainable Development*. Miscellaneous Fertilizer Studies No 14. IFDC, Africa, pp. 61–72.

Beddington, J., Asaduzzaman, M., Clark, M., Bremauntz, A., Guillou, M., Jahn, M., et al. (2012) The role for scientists in tackling food insecurity and climate change. *Agriculture and Food Security* 1, 1–9.

Bennett, E.M., Cumming, G.S. and Peterson, G.D. (2005) A systems model approach to determining resilience surrogates for case studies. *Ecosystems* 8, 945–957.

Boody, G., Vondracek, B., Andow, D., Krinke, M., Westra, J., Zimmerman, J., et al. (2009) Multifunctional agriculture in the United States. *BioScience* 55, 27–38.

Brand, F.S. and Jax, K. (2007) Focusing the meaning (s) of resilience: resilience as a descriptive concept and a boundary object. *Ecology and Society* 12, 23.

Bryan, E., Deressa, T.T., Gbetibouo, G.A. and Ringler, C. (2009) Adaptation to climate change in Ethiopia and South Africa: options and constraints. *Environmental Science and Policy* 12, 413–426.

Burke, M.B., Lobell, D.B. and Guarino, L. (2009) Shifts in African crop climates by 2050, and the implications for crop improvement and genetic resources conservation. *Global Environmental Change* 19, 317–325.

Callo-Concha, D., Gaiser, T. and Ewert, F. (2012) Farming and cropping systems in the West African Sudanian Savanna. WASCAL research area: Northern Ghana, Southwest Burkina Faso and Northern Benin. *ZEF Working Paper 100*. ZEF, Bonn, Germany.

Carpenter, S., Walker, B., Anderies, J.M. and Abel, N. (2001) From metaphor to measurement: resilience of what to what? *Ecosystems* 4, 765–781.

Carpenter, S.R., Ludwig, D. and Brock, W.A. (1999) Management of eutrophication for lakes subject to potentially irreversible change. *Ecological Applications* 9, 751–771.

Cash, D.W., Clark, W.C., Alcock, F., Dickson, N.M., Eckley, N., Guston, D.H., et al. (2003) Knowledge systems for sustainable development. *Proceedings of the National Academy of Sciences* 100, 8086–8091.

Challinor, A., Wheeler, T., Garforth, C., Craufurd, P. and Kassam, A. (2007) Assessing the vulnerability of food crop systems in Africa to climate change. *Climatic Change* 83, 381–399.

Chapin, F.S., Walker, B.H., Hobbs, R.J., Hooper, D.U., Lawton, J.H., Sala, O.E., et al. (1997) Biotic control over the functioning of ecosystems. *Science* 277, 500–504.

Chen, X.P., Cui, Z.L., Vitousek, P.M., Cassman, K.G., Matson, P.A., Bai, J.S., et al. (2011) Integrated soil–crop system management for food security. *Proceedings of the National Academy of Sciences* 108, 6399–6404.

Christensen, J.H. and Christensen, O.B. (2007) A summary of the PRUDENCE model projections of changes in European climate by the end of this century. *Climatic Change* 81, 7–30.

Christensen, J.H., Hewitson, B., Busuioc, A., Chen, A., Gao, X., Held, I., et al. (eds) (2007) *Climate Change 2007: The Physical Science Basis. Contribution of Working Group I to the Fourth Assessment Report of the Intergovernmental Panel on Climate Change.* Cambridge University Press, Cambridge, UK, and New York, NY, pp. 847–940.

Corbeels, M., Scopel, E., Cardoso, A., Bernoux, M., Douzet, J.M. and Neto, M.S. (2006) Soil carbon storage potential of direct seeding mulch-based cropping systems in the Cerrados of Brazil. *Global Change Biology* 12, 1773–1787.

Crane, T.A., Roncoli, C., Paz, J., Breuer, N., Broad, K., Ingram, K.T., et al. (2010) Forecast skill and farmers' skills: seasonal climate forecasts and agricultural risk management in the southeastern United States. *Weather, Climate, and Society* 2, 44–59.

Crane, T.A., Roncoli, C. and Hoogenboom, G. (2011) Adaptation to climate change and climate variability: the importance of understanding agriculture as performance. *NJAS – Wageningen Journal of Life Sciences* 57, 179–185.

Cruse, R.M. and Herndl, C.G. (2009) Balancing corn stover harvest for biofuels with soil and water conservation. *Journal of Soil and Water Conservation* 64, 286–291.

Dai, A. (2011) Drought under global warming: a review. *Wiley Interdisciplinary Reviews: Climate Change* 2, 45–65.

Devendra, C. and Thomas, D. (2002) Crop–animal systems in Asia: importance of livestock and characterisation of agro-ecological zones. *Agricultural Systems* 71, 5–15.

Dow, K., Berkhout, F., Preston, B.L., Klein, R.J., Midgley, G. and Shaw, M.R. (2013) Limits to adaptation. *Nature Climate Change* 3, 305–307.

Drever, C.R., Peterson, G., Messier, C., Bergeron, Y. and Flannigan, M. (2006) Can forest management based on natural disturbances maintain ecological resilience? *Canadian Journal of Forest Research* 36, 2285–2299.

Falloon, P. and Betts, R. (2010) Climate impacts on European agriculture and water management in the context of adaptation and mitigation – the importance of an integrated approach. *Science of the Total Environment* 408, 5667–5687.

Fereres, E. and Soriano, M.A. (2007) Deficit irrigation for reducing agricultural water use. *Journal of Experimental Botany* 58, 147–159.

Fischer, J., Lindenmayer, D.B. and Manning, A.D. (2006) Biodiversity, ecosystem function, and resilience: ten guiding principles for commodity production landscapes. *Frontiers in Ecology and the Environment* 4, 80–86.

Folke, C. (2006) Resilience: the emergence of a perspective for social–ecological systems analyses. *Global Environmental Change* 16, 253–267.

Gbetibouo, G.A., Hassan, R.M. and Ringler, C. (2010) Modelling farmers' adaptation strategies for climate change and variability: the case of the Limpopo Basin, South Africa. *Agrekon* 49, 217–234.

Geerts, S. and Raes, D. (2009) Deficit irrigation as an on-farm strategy to maximize crop water productivity in dry areas. *Agricultural Water Management* 96, 1275–1284.

Giller, K.E., Witter, E., Corbeels, M. and Tittonell, P. (2009) Conservation agriculture and smallholder farming in Africa: the heretics' view. *Field Crops Research* 114, 23–34.

Giordano, M., de Fraiture, C., Weight, E. and van der Bliek, J. (eds) (2012) Water for wealth and food security: supporting farmer-driven investments in agricultural water management. Synthesis report of the AgWater Solutions Project. International Water Management Institute (IWMI), Colombo, Sri Lanka, 48 p. doi:10.5337/2012.207.

Giorgi, F. (2005) Interdecadal variability of regional climate change: implications for the development of regional climate change scenarios. *Meteorology and Atmospheric Physics* 89, 1–15.

Gleeson, T., Wada, Y., Bierkens, M.F. and van Beek, L.P. (2012) Water balance of global aquifers revealed by groundwater footprint. *Nature* 488, 197–200.

Godfray, H.C.J., Beddington, J.R., Crute, I.R., Haddad, L., Lawrence, D., Muir, J.F., et al. (2010) Food security: the challenge of feeding 9 billion people. *Science* 327, 812–818.

Hakala, K., Jauhiainen, L., Himanen, S.J., Rötter, R., Salo, T. and Kahiluoto, H. (2012) Sensitivity of barley varieties to weather in Finland. *Journal of Agricultural Science* 150, 145–160.

Hall, J.W., Brown, S., Nicholls, R.J., Pidgeon, N.F. and Watson, R.T. (2012) Proportionate adaptation. *Nature Climate Change* 2, 833–834.

Halsnæs, K. and Trærup, S. (2009) Development and climate change: a mainstreaming approach for assessing economic, social, and environmental impacts of adaptation measures. *Environmental Management* 43, 765–778.

Handmer, J., Honda, Y., Kundzewicz, Z.W., Arnell, N., Benito, G., Hatfield, J., et al. (2012) Changes in impacts of climate extremes: human systems and ecosystems. In: Field, C.B., Barros, V., Stocker, T.F., Qin, D., Dokken, D.J., Ebi, K.L., et al. (eds) *Managing the Risks of Extreme Events and Disasters to Advance Climate Change Adaptation. A Special Report of Working Groups I and II of the Intergovernmental Panel on Climate Change (IPCC).* Cambridge University Press, Cambridge, UK, and New York, NY, pp. 231–290.

Hansen, J., Sato, M. and Ruedy, R. (2012) Perception of climate change. *Proceedings of the National Academy of Sciences* 109, E2415–E2423.

Hatfield, J.L., Boote, K.J., Kimball, B., Ziska, L., Izaurralde, R.C., Ort, D., et al. (2011) Climate impacts on agriculture: implications for crop production. *Agronomy Journal* 103, 351–370.

Hauggaard-Nielsen, H. and Jensen, E.S. (2001) Evaluating pea and barley cultivars for complementarity in intercropping at different levels of soil N availability. *Field Crops Research* 72, 185–196.

Hazell, P. and Wood, S. (2008) Drivers of change in global agriculture. *Philosophical Transactions of the Royal Society B: Biological Sciences* 363, 495–515.

Himanen, S.J., Hakala, K. and Kahiluoto, H. (2013a) Crop responses to climate and socioeconomic change in northern regions. *Regional Environmental Change* 13, 17–32.

Himanen, S.J., Ketoja, E., Hakala, K., Rötter, R.P., Salo, T. and Kahiluoto, H. (2013b) Cultivar diversity has great potential to increase yield of feed barley. *Agronomy for Sustainable Development* 33, 519–530.

Holling, C.S. (1973) Resilience and stability of ecological systems. *Annual Review of Ecology and Systematics* 4, 1–23.

Holling, C.S. (1992) Cross-scale morphology, geometry and dynamics of ecosystems. *Ecological Monographs* 62, 447–502.

Howden, S.M., Soussana, J.-F., Tubiello, F.N., Chhetri, N., Dunlop, M. and Meinke, H. (2007) Adapting agriculture to climate change. *Proceedings of the National Academy of Sciences* 104, 19691–19696.

Ifejika Speranza, C. (2010) *Resilient Adaptation to Climate Change in African Agriculture.* Deutsches Institut für Entwicklungspolitik, Bonn, Germany, p. 336.

IPCC (2010) *Meeting Report of the Intergovernmental Panel on Climate Change Expert Meeting on Detection and Attribution Related to Anthropogenic Climate Change.* Stocker, T.F., Field, C.B., Qin, D., Barros, V., Plattner, G.-K., Tignor, M., et al. (eds) IPCC Working Group I Technical Support Unit, University of Bern, Bern.

IPCC (2012) *Managing the Risks of Extreme Events and Disasters to Advance Climate Change Adaptation.* A Special Report of Working Groups I and II of the Intergovernmental Panel on Climate Change (IPCC). Field, C.B., Barros, V., Stocker, T.F., Qin, Q., Dokken, D.J., Ebi, K.L., et al. (eds). Cambridge University Press, Cambridge, UK, and New York.

Janssen, M.A. and Ostrom, E. (2006) Resilience, vulnerability, and adaptation: a cross-cutting theme of the International Human Dimensions Programme on Global Environmental Change. *Global Environmental Change* 16, 237–239.

Jarvis, D.I., Brown, A.H.D., Cuong, P.H., Collado-Panduro, L., Latournerie-Moreno, L., Gyawali, S., et al. (2008) A global perspective of the richness and evenness of traditional crop-variety diversity maintained by farming communities. *Proceedings of the National Academy of Sciences* 105, 5326–5331.

Jones, P.G. and Thornton, P.K. (2009) Croppers to livestock keepers: livelihood transitions to 2050 in Africa due to climate change. *Environmental Science and Policy* 12, 427–437.

Kaizzi, C.K., Ssali, H. and Vlek, P.L.G. (2006) Differential use and benefits of Velvet bean (*Mucuna pruriens* var. *utilis*) and N fertilizers in maize production in contrasting agro-ecological zones of E. Uganda. *Agricultural Systems* 88, 44–60.

Kaut, A.H.E.E., Mason, H.E., Navabi, A., O'Donovan, J.T. and Spaner, D. (2008) Organic and conventional management of mixtures of wheat and spring cereals. *Agronomy for Sustainable Development* 28, 363–371.

Kolavalli, S., Flaherty, K., Al-Hassan, R. and Baah, K.O. (2010) *Do Comprehensive Africa Agriculture Development Program (CAADP) Processes Make a Difference to Country Commitments to Develop Agriculture? The Case of Ghana.* International Food Policy Research Institute (IFPRI), Washington, DC, Addis Ababa, and New Delhi.

Krupinsky, J.M., Bailey, K.L., McMullen, M.P., Gossen, B.D. and Turkington, T.K. (2002) Managing plant disease risk in diversified cropping systems. *Agronomy Journal* 94, 198–209.

Kummer, S., Milestad, R., Leitgeb, F. and Vogl, C.R. (2012) Building resilience through farmers' experiments in organic agriculture: examples from eastern Austria. *Sustainable Agriculture Research* 1, 308.

Lal, R. (1985) A soil suitability guide for different tillage systems in the tropics. *Soil and Tillage Research* 5, 179–196.

Lal, R. (2004) Soil carbon sequestration impacts on global climate change and food security. *Science* 304, 1623–1627.

Laux, P., Jäckel, G., Tingem, R.M. and Kunstmann, H. (2010) Impact of climate change on agricultural productivity under rainfed conditions in Cameroon – a method to improve attainable crop yields by planting date adaptations. *Agricultural and Forest Meteorology* 150, 1258–1271.

Lin, B.B. (2011) Resilience in agriculture through crop diversification: adaptive management for environmental change. *BioScience* 61, 183–193.

Lithourgidis, A.S., Dordas, C.A., Damalas, C.A. and Vlachostergios, D.N. (2011) Annual intercrops: an alternative pathway for sustainable agriculture. *Australian Journal of Crop Science* 5, 396–410.

Lobell, D.B. and Gourdji, S.M. (2012) The influence of climate change on global crop productivity. *Plant Physiology* 160, 1686–1697.

Lobell, D.B., Cassman, K.G. and Field, C.B. (2009) Crop yield gaps: their importance, magnitudes, and causes. *Annual Review of Environment and Resources* 34, 179–204.

Lobell, D.B., Schlenker, W. and Costa-Roberts, J. (2011) Climate trends and global crop production since 1980. *Science* 333, 616–620.

Lobell, D.B., Hammer, G.L., McLean, G., Messina, C., Roberts, M.J. and Schlenker, W. (2013) The critical role of extreme heat for maize production in the United States. *Nature Climate Change* 3, 497–501.

McGuire, S. and Sperling, L. (2013) Making seed systems more resilient to stress. *Global Environmental Change* 23, 644–653.

Malézieux, E. (2012) Designing cropping systems from nature. *Agronomy for Sustainable Development* 32, 15–29.

Malézieux, E., Crozat, Y., Dupraz, C., Laurans, M., Makowski, D., Ozier-Lafontaine, H., *et al.* (2009) Mixing plant species in cropping systems: concepts, tools and models. A review. *Agronomy for Sustainable Development* 29, 43–62.

Marshall, N.A. and Marshall, P.A. (2007) Conceptualizing and operationalizing social resilience within commercial fisheries in northern Australia. *Ecology and Society* 12, 1.

Mertz, O., Halsnæs, K., Olesen, J.E. and Rasmussen, K. (2009) Adaptation to climate change in developing countries. *Environmental Management* 43, 743–752.

Morison, J.I., Baker, N.R., Mullineaux, P.M. and Davies, W.J. (2008) Improving water use in crop production. *Philosophical Transactions of the Royal Society B* 12, 639–658.

Nair, P.K.R. and Garrity, D. (2012) Agroforestry – the future of global land use. In: Nair, P.K.R. (ed.) *Advances in Agroforestry* 9. Springer Science+Business Media, Dordrecht, the Netherlands, pp. 428.

Nelson, D.R. (2011) Adaptation and resilience: responding to a changing climate. *Wiley Interdisciplinary Reviews: Climate Change* 2, 113–120.

Nelson, D.R., Adger, W.N. and Brown, K. (2007) Adaptation to environmental change: contributions of a resilience framework. *Annual Review Environmental Resources* 32, 395–419.

Nicholls, C.I. and Altieri, M.A. (2013) Plant biodiversity enhances bees and other insect pollinators in agroecosystems. A review. *Agronomy for Sustainable Development* 33, 257–274.

Nielsen, J.Ø. and Reenberg, A. (2010) Cultural barriers to climate change adaptation: a case study from northern Burkina Faso. *Global Environmental Change* 20, 142–152.

Olesen, J.E., Trnka, M., Kersebaum, K., Skjelvåg, A., Seguin, B., Peltonen-Sainio, P., *et al.* (2011) Impacts and adaptation of European crop production systems to climate change. *European Journal of Agronomy* 34, 96–112.

Osbahr, H., Twyman, C., Adger, W.N. and Thomas, D.S.G. (2010) Evaluating successful livelihood adaptation to climate variability and change in Southern Africa. *Ecology and Society* 15, 27.

Osborne, T., Rose, G. and Wheeler, T. (2013) Variation in the global-scale impacts of climate change on crop productivity due to climate model uncertainty and adaptation. *Agricultural and Forest Meteorology* 170, 183–194.

Parry, M.L., Rosenzweig, C., Iglesias, A., Livermore, M. and Fischer, G. (2004) Effects of climate change on global food production under SRES emissions and socio-economic scenarios. *Global Environmental Change* 14, 53–67.

Patt, A. and Gwata, C. (2002) Effective seasonal climate forecast applications: examining constraints for subsistence farmers in Zimbabwe. *Global Environmental Change* 12, 185–195.

Patt, A.G., van Vuuren, D.P., Berkhout, F., Aaheim, A., Hof, A.F., Isaac, M., *et al.* (2010) Adaptation in integrated assessment modeling: where do we stand? *Climatic Change* 99, 383–402.

Rao, K.P.C. (2003) Better management of climate variability using simulation models and seasonal climate forecasts. SWMnet Discussion Paper. Soil and Water Management Network, Nairobi, pp. 44–46.

Rasmussen, K., May, W., Birk, T.L., Mataki, M., Mertz, O. and Yee, D. (2009) Climate change on three Polynesian outliers in the Solomon Islands: impacts, vulnerability and adaptation. *Geografisk Tidsskrift* 109, 1–13.

Rea, J.H., Wratten, S.D., Sedcole, R., Cameron, P.J., Davis, S.I. and Chapman, R.B. (2002) Trap cropping to manage green vegetable bug *Nezara viridula* (L.) (Heteroptera: Pentatomidae) in sweet corn in New Zealand. *Agricultural and Forest Entomology* 4, 101–107.

Reidsma, P. and Ewert, F. (2008) Regional farm diversity can reduce vulnerability of food production to climate change. *Ecology and Society* 13, 38.

Reidsma, P., Ewert, F., Boogaard, H. and van Diepen, K. (2009) Regional crop modelling in Europe: the impact of climatic conditions and farm characteristics on maize yields. *Agricultural Systems* 100, 51–60.

Reidsma, P., Ewert, F., Lansink, A.O. and Leemans, R. (2010) Adaptation to climate change and climate variability in European agriculture: the importance of farm level responses. *European Journal of Agronomy* 32, 91–102.

Reynolds, M., Bonnett, D., Chapman, S.C., Furbank, R.T., Manès, Y., Mather, D.E. and Parry, M.A. (2011) Raising yield potential of wheat. I. Overview of a consortium approach and breeding strategies. *Journal of Experimental Botany* 62, 439–452.

Rickards, L. and Howden, S. (2012) Transformational adaptation: agriculture and climate change. *Crop and Pasture Science* 63, 240–250.

Rötter, R. and van Keulen, H. (1997) Variations in yield response to fertilizer application in the tropics: II. Risks and opportunities for smallholders cultivating maize on Kenya's arable land. *Agricultural Systems* 53, 69–95.

Rötter, R., van Keulen, H. and Jansen, M. (1997) Variations in yield response to fertilizer application in the tropics: I. Quantifying risks and opportunities for smallholders based on crop growth simulation. *Agricultural Systems* 53, 41–68.

Rötter, R.P., Palosuo, T., Pirttioja, N.K., Dubrovsky, M., Salo, T., Fronzek, S., *et al.* (2011) What would happen to barley production in Finland if global warming exceeded 4 C? A model-based assessment. *European Journal of Agronomy* 35, 205–214.

Rockström, J. and Barron, J. (2007) Water productivity in rainfed systems: overview of challenges and analysis of opportunities in water scarcity prone savannahs. *Irrigation Science* 25, 299–311.

Rosenzweig, C. and Tubiello, F.N. (2007) Adaptation and mitigation strategies in agriculture: an analysis of potential synergies. *Mitigation and Adaptation Strategies for Global Change* 12, 855–873.

Rusinamhodzi, L., Corbeels, M., van Wijk, M., Rufino, M.C., Nyamangara, J. and Giller, K.E. (2011) Long-term effects of conservation agriculture practices on maize yields under rainfed conditions: lessons for southern Africa. *Agronomy for Sustainable Development* 31, 657–673.

Rusinamhodzi, L., Corbeels, M., Nyamangara, J. and Giller, K.E. (2012) Maize–grain legume intercropping is an attractive option for ecological intensification that reduces climatic risk for smallholder farmers in central Mozambique. *Field Crops Research* 136, 12–22.

Sacks, W.J. and Kucharik, C.J. (2011) Crop management and phenology trends in the US Corn Belt: impacts on yields, evapotranspiration and energy balance. *Agricultural and Forest Meteorology* 151, 882–894.

Schlenker, W. and Lobell, D.B. (2010) Robust negative impacts of climate change on African agriculture. *Environmental Research Letters*, 014010.

Schlenker, W. and Roberts, M.J. (2009) Nonlinear temperature effects indicate severe damages to US crop yields under climate change. *Proceedings of the National Academy of Sciences* 106, 15594–15598.

Semenov, M.A. and Shewry, P.R. (2011) Modelling predicts that heat stress, not drought, will increase vulnerability of wheat in Europe. *Science Report* 1, 66. doi:10.1038/srep00066.

Seneviratne, S.I., Nicholls, N., Easterling, D., Goodess, C.M., Kanae, S., Kossin, J., *et al.* (2012) Changes in climate extremes and their impacts on the natural physical environment. In: Field, C.B., Barros, V., Stocker, T.F., Qin, D., Dokken, D.J., Ebi, K.L., *et al.* (eds) *Managing the Risks of Extreme Events and Disasters to Advance Climate Change Adaptation. A Special Report of Working Groups I and II of the Intergovernmental Panel on Climate Change (IPCC)*. Cambridge University Press, Cambridge, UK, and New York, pp. 109–230.

Siebert, S. and Ewert, F. (2012) Spatio-temporal patterns of phenological development in Germany in relation to temperature and day length. *Agricultural and Forest Meteorology* 152, 44–57.

Simelton, E., Fraser, E.D., Termansen, M., Benton, T.G., Gosling, S.N., South, A., *et al.* (2012) The socioeconomics of food crop production and climate change vulnerability: a global scale quantitative analysis of how grain crops are sensitive to drought. *Food Security* 4, 163–179.

Smit, B. and Skinner, M.W. (2002) Adaptation options in agriculture to climate change: a typology. *Mitigation and Adaptation Strategies for Global Change* 7, 85–114.

Smith, P. and Olesen, J.E. (2010) Synergies between the mitigation of, and adaptation to, climate change in agriculture. *Journal of Agricultural Science* 148, 543–552.

Supit, I., Van Diepen, C., De Wit, A., Kabat, P., Baruth, B. and Ludwig, F. (2010) Recent changes in the climatic yield potential of various crops in Europe. *Agricultural Systems* 103, 683–694.

Thies, C. and Tscharntke, T. (1999) Landscape structure and biological control in agro-ecosystems. *Science* 285, 893–895.

Thomas, D.S., Twyman, C., Osbahr, H. and Hewitson, B. (2007) Adaptation to climate change and variability: farmer responses to intra-seasonal precipitation trends in South Africa. *Climatic Change* 83, 301–322.

Thornton, P.K., Jones, P.G., Ericksen, P.J. and Challinor, A.J. (2011) Agriculture and food systems in sub-Saharan Africa in a 4 C+ world. *Philosophical Transactions of the Royal Society A: Mathematical, Physical and Engineering Sciences* 369, 117–136.

Tilman, D., Cassman, K.G., Matson, P.A., Naylor, R. and Polasky, S. (2002) Agricultural sustainability and intensive production practices. *Nature* 418, 671–677.

Tilman, D., Balzer, C., Hill, J. and Befort, B.L. (2011) Global food demand and the sustainable intensification of agriculture. *Proceedings of the National Academy of Sciences* 108, 20260–20264.

Tittonell, P. and Giller, K.E. (2013) When yield gaps are poverty traps: the paradigm of ecological intensification in African smallholder agriculture. *Field Crops Research* 143, 76–90.

Tompkins, E.L. and Adger, W. (2004) Does adaptive management of natural resources enhance resilience to climate change? *Ecology and Society* 9, 10.

Tooker, J.F. and Frank, S.D. (2012) Genotypically diverse cultivar mixtures for insect pest management and increased crop yields. *Journal of Applied Ecology* 49, 974–985.

Tschakert, P. and Dietrich, K.A. (2010) Anticipatory learning for climate change adaptation and resilience. *Ecology and Society* 15, 11.

Tubiello, F.N., Soussana, J.-F. and Howden, S.M. (2007) Crop and pasture response to climate change. *Proceedings of the National Academy of Sciences* 104, 19686–19690.

Valbuena, D., Erenstein, O., Homann-Kee Tui, S., Abdoulaye, T., Claessens, L., Duncan, A.J., et al. (2012) Conservation agriculture in mixed crop–livestock systems: scoping crop residue trade-offs in Sub-Saharan Africa and South Asia. *Field Crops Research* 132, 175–184.

van Ittersum, M. and Cassman, K. (2013) Yield gap analysis – rationale, methods and applications – Introduction to the Special Issue. *Field Crops Research* 143, 1–3.

van Ittersum, M. and Rabbinge, R. (1997) Concepts in production ecology for analysis and quantification of agricultural input–output combinations. *Field Crops Research* 52, 197–208.

van Oort, P., Timmermans, B. and van Swaaij, A. (2012) Why farmers' sowing dates hardly change when temperature rises. *European Journal of Agronomy* 40, 102–111.

van Wart, J., van Bussel, L.G., Wolf, J., Licker, R., Grassini, P., Nelson, A., et al. (2013) Use of agro-climatic zones to upscale simulated crop yield potential. *Field Crops Research* 143, 44–55.

Vanlauwe, B. and Giller, K.E. (2006) Popular myths around soil fertility management in sub-Saharan Africa. *Agriculture, Ecosystems and Environment* 116, 34–46.

Vermeulen, S.J., Aggarwal, P., Ainslie, A., Angelone, C., Campbell, B.M., Challinor, A., et al. (2012) Options for support to agriculture and food security under climate change. *Environmental Science and Policy* 15, 136–144.

Vogel, C. and O'Brien, K. (2006) Who can eat information? Examining the effectiveness of seasonal climate forecasts and regional climate-risk management strategies. *Climate Research* 33, 111–122.

Vries, F. de, Liiri, M.E., Bjørnlund, L., Bowker, M.A., Christensen, S., Setälä, H.M., et al. (2012) Land use alters the resistance and resilience of soil food webs to drought. *Nature Climate Change* 2, 276–280.

Walker, B. and Salt, D. (2006) *Resilience Thinking: Sustaining Ecosystems and People in a Changing World*. Island Press.

Walker, B., Carpenter, S., Anderies, J., Abel, N., Cumming, G., Janssen, M., et al. (2002) Resilience management in social-ecological systems: a working hypothesis for a participatory approach. *Conservation Ecology* 6, 14.

Walker, B.H., Abel, N., Anderies, J.M. and Ryan, P. (2009) Resilience, adaptability, and transformability in the Goulburn-Broken Catchment, Australia. *Ecology and Society* 14, 12.

Walters, C. (1986) *Adaptive Management of Renewable Resource.* Macmillan Publishing, New York.

Windmeijer, P.N. and Andriesse, W. (1993) *Inland Valleys in West Africa: An Agro-ecological Characterization of Rice Growing Environments.* International Institute for Land Reclamation and Improvement. Wageningen, the Netherlands.

Yamoah, C.F., Bationo, A., Shapiro, B. and Koala, S. (2002) Trend and stability analyses of millet yields treated with fertilizer and crop residues in the Sahel. *Field Crops Research* 75, 53–62.

Zhu, Y., Chen, H., Fan, J., Wang, Y., Li, Y., Chen, J., et al. (2000) Genetic diversity and disease control in rice. *Nature* 406, 718–722.

12 Shaping Sustainable Intensive Production Systems: Improved Crops and Cropping Systems in the Developing World

Clare Stirling,[1] Jon Hellin,[2] Jill Cairns,[3] Elan Silverblatt-Buser,[2] Tadele Tefera,[4] Henry Ngugi,[2] Sika Gbegbelegbe,[4] Kindie Tesfaye,[5] Uran Chung,[2] Kai Sonder,[2] Rachael A. Cox,[2] Nele Verhulst,[2] Bram Govaerts,[2] Phillip Alderman[2] and Matthew Reynolds[2]

[1] *International Maize and Wheat Improvement Center (CIMMYT), Wales, UK;* [2] *International Maize and Wheat Improvement Center (CIMMYT), Mexico, DF, Mexico;* [3] *International Maize and Wheat Improvement Center (CIMMYT), Harare, Zimbabwe;* [4] *International Maize and Wheat Improvement Center (CIMMYT), Nairobi, Kenya;* [5] *International Maize and Wheat Improvement Centre (CIMMYT), Addis Ababa, Ethiopia*

12.1 Introduction

Population growth will reach nearly 9 billion by 2050. Crop production will need to increase by approximately 100% from 2005 to 2050 levels (Tilman *et al.*, 2011) because of this growth in population, as well as global changes in diet and the increasing use of bioenergy (Godfray *et al.*, 2010). Maize and wheat are critical for global food security and poverty reduction. Maize and wheat, together with rice, provide at least 30% of food calories to about 4.5 billion people in 100 developing countries. In Africa, maize is the most widely grown staple crop, and its importance is increasing in Asia. Wheat is the most widely grown of any crop, with around 220 million hectares (Mha) cultivated annually in environments ranging from very favourable in Western Europe to severely stressed in parts of Asia, Africa and Australia (Braun *et al.*, 2010).

Many small-scale maize farmers in Africa, Asia and Latin America grow maize under rainfed conditions, and the crop is, therefore, very vulnerable to climatic variability and change (Bänziger and Araus, 2007). Fluctuations in maize production often give rise to price fluctuations that can affect both poor producers and consumers adversely. Wheat is also very susceptible to climate change, and is especially sensitive to heat. Low growth in productivity or stagnation, for example in the Green Revolution areas of South Asia, will make it more challenging to meet the growing demand for wheat (Rosegrant *et al.*, 2009).

Agricultural practices also contribute to upwards of 25% of all global greenhouse gas emissions and are linked to losses in biodiversity and carbon storage, and to soil degradation (Foley *et al.*, 2005; Burney *et al.*, 2010). Climate change, therefore, poses huge challenges to food security and the livelihood security of millions of people. Furthermore, production increase will need to take place with less land available per capita than ever before. In 1950, there were 13.5 ha person^{-1}, in 2005 the figure had fallen to 3.2 ha person^{-1}, and by 2050 it is projected that there will only be 1.5 ha person^{-1} (Beddington et al., 2012).

Climate change and extreme weather events will worsen food production systems that are already under threat from soil and land degradation. In addition to the billions of hectares of arable land that are currently suboptimal due to land degradation, 12 Mha of farmable land are annually becoming unusable due to land degradation (Bai et al., 2008).

Tilman et al. (2011) suggested that if crop production were to meet the demands of the future by continuing current trends – greater agricultural intensification in richer nations and greater land clearing in poorer nations – this would result in the clearance of approximately 1 billion additional hectares of land by 2050. Hence, there is growing interest in the concept of sustainable intensification. Sustainable agricultural intensification consists of producing more output from the same area of land while reducing the negative environmental impacts and, simultaneously, increasing contributions to natural capital and the flow of environmental services (Godfray et al., 2010). If crop production systems were to change by closing the yield gaps through intensification on existing croplands – including breeding for maximum yield and more resilient and resource-efficient crops, improved agronomy and better postharvest storage technologies – only 0.2 billion ha of land would be cleared by 2050, significantly reducing potential negative impacts (Tilman et al., 2011).

However, to secure future food security, focus should be not only on the production side of the equation but also on the consumption side, including wastage. Increasing the efficiency of food supply chains (FSC) will expand the amount of food available in the world without having to increase inputs such as fertilizers, water and pesticides. Cutting losses at the harvest and postharvest level as well as addressing the growing gap between food production and consumption will have substantial impacts on food security (Parfitt et al., 2010). Some estimates suggest that up to one-third of food is never consumed. Infrastructural gaps, adverse climatic conditions and pest infestations all contribute to this phenomenon (Lundqvist et al., 2008). Food waste can also have immense environmental impacts. In the USA, for example, food waste is the single largest component of municipal solid waste and, as a result, accounts for almost 25% of US methane emissions (Gunders, 2012). Additionally, food waste in the USA accounts for approximately 300 million barrels of oil per year and upwards of one-quarter of freshwater consumption (Hall et al., 2009).

The gap between the amount of food produced and food consumed continues to increase in more developed countries. For example, in 2000, the entire continent of Europe had more than 3300 kcal of food available per day per capita when daily food requirements are around 2000 kcal day^{-1} (FAO, 2012). Smil (2004) suggested that this difference in available over consumed food in Europe could supply 350 million people with diets normally found in the developed world and easily twice as many people on a vegetarian diet. Today, the overconsumption of food has led to alarming increases in obesity. While agriculture needs to play its role, the fact is that sustainable intensification and food security cannot be achieved without addressing the significant problem of food wastage (losses and overconsumption). Reducing food waste would reduce the pressure to increase food production, thereby saving land, water and fertilizer use and carbon emissions.

12.2 Predicted Impacts of Climate Change

12.2.1 High temperatures

There is broad agreement among climate scientists that climate change will result in increased temperatures (see Rummukainen,

Chapter 1, this volume). During the last century, sub-Saharan Africa (SSA) has warmed by an average of 0.5°C, with maximum warming occurring over the interior of southern Africa, which is a major maize producing region (Hulme et al., 2001). Similarly, since 1980, average maximum temperatures during the growing season have increased between 0.6 and 3°C for major wheat production areas in the Indo-Gangetic Plains (IGP) and other parts of Asia (Lobell and Gourdji, 2012).

Regarding future trends, the Intergovernmental Panel on Climate Change (IPCC) Fourth Assessment Report showed a general trend of warming over SSA, IGP and Latin America based on results from 21 climate models (IPCC, 2007). Further, one prominent study by Battisti and Naylor (2009) suggests a high probability (>90%) that by the end of this century growing season temperatures will exceed the most extreme seasonal temperatures recorded in the past century. In SSA, maximum temperatures are predicted to increase by an average of 2.6°C across maize mega-environments, and in IGP, current climate projections show increases in temperature between 0.6 and 3.1°C by 2050 (Cairns et al., 2012; New et al., 2012). Likewise, maize growing season temperatures in Central America are projected to increase by 1°C by 2020 and by 2°C by 2050 (Schmidt et al., 2012).

12.2.2 Rainfall variability

Climate change projections for rainfall are less consistent. Overall, increasing variability in rainfall patterns is likely to lead to increased water scarcity in the coming decades (Lobell et al., 2008). High interannual rainfall variability is already a major feature in a large proportion of croplands in the tropics, and there is general agreement that this variability will increase with climate change, further exacerbating yield instability and losses (Ericksen et al., 2011). Cairns et al. (2013) report potential changes in monthly precipitation by 2050 in 16 key maize production areas across mega-environments in SSA. The direction and magnitude of change in precipitation varies with location. In East Africa, a general increase in rainfall during the maize growing season is predicted across mega-environments; however, in the dry lowlands, rainfall is projected to decrease during the maize reproductive stage, with the onset of the short rainy season also delayed. Cairns et al. (2013) predict a notable delay in the onset of the rainy season in southern Africa, with an early cessation of the rainy season in many parts, in agreement with Shongwe et al. (2009).

Wheat production areas show a similar divergence, with increases in precipitation of between 0.1 and 0.5 mm day^{-1} for IGP and decreases in precipitation between 0.1 and 0.4 mm day^{-1} in wheat growing regions of Mexico and Central Asia (IPCC, 2007). In some regions, changes in rainfall distribution will result in temporary excessive soil moisture or waterlogging in maize production areas. Currently, waterlogging regularly affects over 18% of the total maize production area in South and South-east Asia (Hellin et al., 2012).

12.2.3 Crop production and abiotic stress

The impact of climate change on agricultural production will be greatest in the tropics and subtropics. Compared to the situation without climate change, changes in climate are likely to reduce maize production globally by 3–10% by 2050 (Rosegrant et al., 2009). Many parts of Africa will be particularly vulnerable. Climate change projections suggest that by 2030 maize yields in southern Africa will be 50% of the average yields achieved at the beginning of this century (Lobell et al., 2008). An analysis of more than 20,000 historical maize trial yields in Africa over an 8-year period showed maize yields were reduced by 1% and 1.7% for every degree day above 30°C under optimal and drought conditions, respectively (Lobell et al., 2011).

CIMMYT has undertaken a modelling study using the Decision Support System for Agrotechnology Transfer (DSSAT, 2012) to assess the impact of climate change on maize in Africa. The results indicate that maize yields will be affected negatively by climate change, with potential reductions as high as 14% in 2050 in the lowland maize mega-environments. The results also indicated that the impacts of climate change in Africa would differ across regions (Plate 8). Both the production and yield of maize would be greatly reduced by climate change in western and southern Africa compared to eastern and central Africa. Results also indicated that Africa would face a reduction in maize yield of 5–11%, with a corresponding fall in maize production of 5–11% in 2050.

Climate change is projected to reduce wheat production in developing countries by 29–34% (Rosegrant et al., 2009). However, the effects of climate change on wheat production will vary. While future climate scenarios may be beneficial for the wheat crop in high latitudes, global warming will reduce productivity in zones where favourable temperatures already exist, for example in the IGP of South Asia. The IGP accounts for 15% of global wheat production. By 2050, and due to climate change, 51% of the region might suffer from a significant reduction in wheat yields unless farmers adopt appropriate cultivars and crop management practices (Ortiz et al., 2008). Global warming is likely to increase productivity and open up new cropping opportunities at high latitudes in areas of Canada and Russia.

CIMMYT has used bioeconomic modelling to quantify the impact of promising wheat technology on regional and global food security. Preliminary results from the biophysical analysis imply that wheat yields in most parts of the developing world are expected to decrease under all of the climate change models considered in the study, assuming that farmers continue to use traditional wheat varieties. However, drought-tolerant wheat cultivars would enhance wheat yields under climate change as compared to the benchmark cultivars (Plates 9 and 10).

The analysis suggests that seed adoption would begin 15 years after the investment of about US$10 million into the development, production and dissemination of drought-tolerant wheat. The maximum adoption rate of the improved wheat varieties is 80% of the total wheat area in each country. The targeted region for the adoption study is Central West Asia and North Africa (CWANA), as it is one of the major wheat producing regions in the developing world (FAO, 2012); it is also the region where wheat production is mostly affected by water shortage and drought. Within this area, the targeted countries are Turkey, Pakistan and Iran, the largest wheat producers in CWANA (FAO, 2012).

Preliminary analyses have been performed for two scenarios; the CSIRO scenario based on a model produced by Australia's Commonwealth Scientific and Industrial Research Organisation (CSIRO) and the Model for Interdisciplinary Research on Climate (MIROC) produced by the University of Tokyo's Center for Climate System Research. The CSIRO scenario has smaller but more evenly distributed increases in precipitation. The MIROC General Circulation Model (GCM) has greater increases in precipitation on average, but important agricultural regions see decreased rainfall by 2050. The results indicate that the drought-tolerant wheat varieties would increase wheat production and hence improve national food security in Turkey, Pakistan and Iran. Similarly, global food and nutrition would be enhanced (Table 12.1). The number of people at risk of hunger would decrease worldwide by about 4.6 million if drought-tolerant wheat cultivars were delivered to farmers in Turkey, Pakistan and Iran under the 'CSIRO-A1B' climate change scenario. This number rises to 8.4 million people under the 'MIROC-A1B' scenario.

12.2.4 Crop production and biotic stress

Climate change is also likely to increase outbreaks of pests and diseases (Legrève and Duveiller, 2010) (see Collier and Else, Chapter 6, this volume). There have been many projections made of the effects of climate change on the biotic stresses of

Table 12.1. Simulated global impact from the adoption of drought-tolerant (DT) wheat cultivars in Turkey, Pakistan and Iran under climate change – preliminary results from the IMPACT[a] model.

Variable	Change (CSI-A1; DT cultivars)	Change (MIR-A1; DT cultivars)
World wheat price (US$ t^{-1})	−2.60	−4.97
Global producer surplus (US$ billion)	−6.4411	−11.9370
Global consumer surplus (US$ billion)	14.0145	26.3502
Global social welfare (US$ billion)	7.5734	14.4132
Project costs (US$ billion)	0.0051	0.0051
Global food and nutrition security		
	DT versus conventional cultivars under CSI-A1	DT versus conventional cultivars under MIR-A1
Net benefit (US$ billion)	7.5683	14.4081
Change in kilocalories per capita in 2050	3.4245	5.9076
Change in population at risk of hunger in 2050 (people)	−4,600,531	−8,423,055

Note: [a]Where IMPACT is the IFPRI International Model for Policy Analysis of Agricultural Commodities and Trade (Rosegrant, 2012).

wheat and maize, and these have been based either on the speculation of expert opinion or on simulation modelling (Juroszek and von Tiedemann, 2012). These predictions can be grouped into two broad categories, as illustrated with rust diseases in wheat. First, climate change may impact host–pathogen interactions directly by modifying host resistance and/or pathogen virulence or fitness (Coakley et al., 1999; Chakraborty and Newton, 2011; Chakraborty et al., 2011). For example, climate change may modify the effectiveness of many resistance genes currently deployed for management of rust diseases, which constitute some of the most important diseases limiting the productivity of wheat. This prediction is premised on the fact that the expression of many genes for resistance to leaf rust (*Puccinia triticina*) (Kolmer, 1996), stripe rust (*P. striiformis*) (Datta et al., 2009) and stem rust (*P. graminis*) (Leonard and Szabo, 2005) can vary depending on temperature and/or plant development stage. Some genes become less effective at elevated temperature (e.g. *Sr15* against stem rust; Roelfs, 1988) and others more effective (e.g. *Yr18* against stripe rust; Park et al., 1992).

Data from annual rust surveys in the USA have also shown that races of the stem rust fungus *P. striiformis* f. sp. *tritici* collected in recent times have different temperature optima profiles from those collected prior to 2000, suggesting a capacity for this pathogen to adapt to elevated temperatures (Markell and Milus, 2008). In this case, the outcome of the projected increases in global temperature will likely depend on the specific host–pathogen interaction.

Evidence for the impact of climate change on biotic stresses includes increased frequency of aflatoxicosis outbreaks in eastern Africa (Lewis et al., 2005), increased risks of invasions by new pathogens or migrant pests (e.g. Ug99; Pretorius et al., 2000; Wanyera et al., 2006), new races of rust in the eastern USA (Markell and Milus, 2008) and maize lethal necrosis disease complex (MLN) in eastern Africa (Wangai et al., 2012). In the USA, annual disease surveys provide evidence that the rust fungi have become adapted to elevated temperatures (Milus et al., 2006).

Recent reviews have summarized the current understanding of the potential effects of climate change on wheat diseases, and this knowledge is broadly applicable to maize diseases (Charkraborty et al., 2011; Juroszek and von Tiedemann, 2012). For predictions relating to diseases of wheat in Europe, Juroszek and von Tiedemann, (2012) further classified whether the predictions were speculations based on expert knowledge or the results of simulation modelling. Conclusions across the studies are that: (i) predictions of the impacts of climate change on diseases of wheat and maize are uncertain; (ii) predictions also differ depending on geographic and time scales; and (iii) the impacts of climate change will be positive, negative or neutral, depending on the host–pathogen interaction. Similar general predictions can be made regarding insect pests of wheat and maize, although most projections point to increased negative impacts of insect pests (Diffenbaugh et al., 2008; Singh et al., 2013). That is, there is so much uncertainty around the current predictions about the impacts of climate change on the insect pests and diseases of maize and wheat that any general conclusions must be treated with caution.

12.3 Technologies to Close the Yield Gap to Achieve Sustainable Intensification

12.3.1 Breeding

Per capita access to prime land resources is becoming increasingly limited in many countries and so yield gap reductions, technological improvements and efficiency gains will be needed to increase food production. Extensive breeding efforts over the past 50 years have tripled the production of cereal crops in the developing world, with only a 30% increase in cultivated land area (Wik et al., 2008). The impact of improved varieties on increasing yields and reducing food insecurity are well documented (Evenson and Gollin, 2003; Pingali, 2012). Scientists need to direct crop breeding efforts at heat and drought resistance. Furthermore, maintaining foliar and root health through genetic resistance to pests and diseases is a prerequisite when providing public-good germplasm to resource-poor farmers, and is already a major focus of the CIMMYT breeding effort in conjunction with global disease monitoring (Braun et al., 2010). The development of climate-adapted germplasm is possible through a combination of conventional, molecular and transgenic breeding approaches.

Maize

For maize, yields have increased steadily in over 70% of the world's maize growing regions (Ray et al., 2012). Under high-input conditions, gains in maize grain yields have been estimated at 94.7 kg ha^{-1} year^{-1} in China (Ci et al., 2011), 132 kg ha^{-1} year^{-1} in Argentina (Luque et al., 2006), 80 kg ha^{-1} year^{-1} in Canada (Bruulsema et al., 2000) and 65–75 kg ha^{-1} year^{-1} in the USA (Duvick, 2005). Further yield improvements are required to offset potential losses under climate change. Varieties with increased tolerance to abiotic stresses including heat and drought stress will play an important role in adaptation to climate change (Fedoroff et al., 2010). However, further increases in yield potential will also be required to ensure food security under future climates.

Maize yield is a function of assimilate supply to the kernel (source) and the potential of the kernel to accommodate this assimilate (sink potential) (Jones and Simmons, 1983). Yield improvements in modern temperate hybrids have been associated with more efficient resource capture and use of resources, particularly under stress (Tollenaar and Lee, 2006). Improved resource capture has been associated with increased interception of radiation through both early canopy closure and delayed senescence. During senescence, chlorophyll content declines and leaves lose their greenness. Leaf senescence in maize begins before the leaf area is fully developed and continues at an increased rate during grain filling (Borrás et al., 2003). A reduced rate of canopy senescence during grain filling has been associated with breeding progress for grain yield in temperate maize (Duvick and Cassman, 1999; Tollenaar and Lee, 2006).

Conventional breeding has increased maize yields steadily in drought-prone environments (Bänziger et al., 2006). This success in tropical maize has been attributed largely to the application of proven drought breeding methodologies in managed stress screening (Bänziger et al., 2006). Selection based on grain yield and increased flowering synchronization (reduced anthesis-silking interval, ASI) resulted in gains of up to 144 kg ha^{-1} year^{-1} under drought stress (Edmeades et al., 1999). More research is needed on the interaction of heat and drought stress in cereals (including maize) (Barnabás et al., 2008). Research is required into the identification of traits associated with combined heat and drought tolerance, and the development of improved germplasm for high-temperature, water-limited environments.

Wheat

While wheat is classed as a temperate species, with over 200 Mha grown worldwide, it is also the most widely grown of any crop, being cultivated in near optimal situations in Western Europe and northwest Mexico, and highly stressed regions of Asia, Africa and Australia (Braun et al., 2010). Global warming is likely to increase productivity as well as open up new cropping opportunities at high latitudes. However, wheat yields decline at above-optimal temperatures, and significant breeding effort would be required to maintain productivity in regions closer to the equator (Cossani and Reynolds, 2012). None the less, wheat productivity is well adapted to environments with water deficit, such as Australia and throughout the Mediterranean region.

Wheat breeding has had considerable impact in marginal environments as well as temperate ones, which is evidence of the potential for wheat to adapt to climate change. For example, analysis of data from CIMMYT's international nursery network shows clear and steady progress in the performance of both bread and durum wheat under drought (Manes et al., 2012). In addition, analysis of germplasm released by CIMMYT for hot, irrigated environments shows significant progress, with many of the lines that perform well at the hottest sites also expressing good yield potential under more temperate conditions (Lillemo et al., 2005; Lopes et al., 2012), an important consideration given typical year-to-year variation in temperature. Recent effort has focused on breeding for earlier maturing cultivars that escape terminal heat stress and encompass resistance to diseases associated with warm, humid environments (Joshi et al., 2007), as well as the highly virulent Ug99 stem rust strain.

The adaptability of wheat is backed by both economic and physiological analysis. Between 1964 and 1979, approximately 25% of global wheat production increase came from improved production in marginal environments, while between 1979 and 1998, impacts on drought- and heat-affected environments showed annual yield gains of 2–3% year^{-1} (Lantican et al., 2003). Impacts were achieved initially by combining genes of major effect associated with agronomic type, phenology and disease resistance into good-yielding backgrounds, and more recently through targeting specific heat- and drought-adaptive traits.

A key strategy has been to change the phenological pattern of the crop so that critical growth stages do not coincide with stressful conditions, or simply to finish the life cycle early before severe stress conditions occur. Another is to minimize the occurrence of stress through the development of a good root system, which is crucial under both drought and heat stress. In the case of drought, this permits water to be accessed deeper in the soil. In the case of heat, it permits transpiration rates that match evaporative demand better, thereby permitting maximal carbon fixation with the added benefit of cooler plants (Reynolds et al., 2010). In environments where 'extra' water is not available to mitigate stress, other stress-adaptive strategies include a range of leaf canopy traits such as epicuticular wax, pigment composition, leaf angle and rolling, etc., which influence radiation load and photosynthetic response, while increased transpiration efficiency permits available water to be used more effectively (Richards, 2006). Strategies for increasing heat adaptation have been reviewed recently (Cossani and Reynolds, 2012) and in addition to those mentioned above, include longer-term goals such as genetic modification of Rubisco to reduce wasteful photorespiration at high temperature and screening genetic resources for more heat-stable metabolism, including lowering carbon losses associated with respiration.

12.3.2 Improved agronomy

Adaptation to climate change will require improved agronomic management and more resilient agronomic systems that can adapt and thrive under new conditions. In areas prone to drought or low rainfall, climate change models predict there will be more frequent water stress as a result of decreased precipitation and increased evaporation from higher temperatures (Cooper et al., 2008). Long-term experiments can help to identify agronomic practices that are more resilient to changing climatic conditions and result in high and stable yields. In a semi-arid, rainfed, maize-based system in the central Mexican highlands, conservation agriculture practices based on minimal soil movement, crop residue retention and rotation had yields that were higher and more stable over 15 years than practices involving tillage (Govaerts et al., 2005; Verhulst et al., 2011a). Averaged over 1997–2009, these practices had a maize yield advantage of approximately 1.5 Mg ha^{-1} over practices involving conventional tillage and zero tillage with residue removal (Verhulst et al., 2011a). In an arid, irrigated, wheat-based system in north-west Mexico, wheat yields were more stable under various climate scenarios in permanent beds with residue retention than in conventionally tilled beds (Verhulst et al., 2011b). Wallace and Batchelor (1997) gave an overview of techniques to improve agricultural water-use efficiency. Water-use efficiency can be increased (i) by using more of the water resource as transpiration or (ii) by fixing more carbon per unit of water transpired (Wallace, 2000). Agronomic practices will mainly have an effect on the first option.

In rainfed conditions, the water used as transpiration can be increased by optimizing the infiltration of rainfall and reducing runoff, evaporation and deep drainage. Optimizing the capture and storage of rainfall in the soil can help buffer dry periods during the growing season. Conservation agriculture contributes to improved rainfall infiltration in multiple ways. Leaving crop residues at the soil surface increases organic matter in the topsoil and improves the water stability of aggregates. Crop residue forms a physical barrier to protect the aggregates from raindrop impact and reduces the likelihood of the formation of a crust layer. Various studies have shown that zero-tillage systems with residue left on the surface have higher infiltration rates than conventional systems (McGarry et al., 2000; Govaerts et al., 2009). A mulch cover also reduces water loss through evaporation (Hatfield et al., 2001). The use of mixed cropping systems such as agroforestry and intercrops has some potential for reducing soil evaporation by shading. Moreover, the deeper root systems of perennial plants such as trees can reduce drainage (Wallace and Batchelor, 1997).

In areas where climate change will cause intense rains and excessive quantities of water, improved agronomic practices through conservation agriculture will also be useful for adapting systems. With improved water infiltration under heavy rains, more water infiltrates the soil instead of staying on the fields in ponds or causing increased erosion by water runoff. In irrigated conditions, capturing the available rainfall in the ways described above can help to reduce irrigation needs. Similarly, irrigation efficiency can be increased by practices that improve water infiltration, and irrigation water needs can be reduced by decreasing evaporation at the soil surface. Specifically for irrigated systems, evaporation losses can be further reduced by limiting the number of irrigation applications or using systems like drip irrigation that decrease the amount of exposed wet soil surface (Wallace, 2000).

Irrigation systems that reduce application losses, improve distribution uniformity, or both, can further increase water-use efficiency, especially when combined with demand-based irrigation scheduling (Howell, 2001). Grassini et al. (2011) evaluated options to increase water-use efficiency in irrigated maize systems of central Nebraska, USA. Applied irrigation was 41% and 20% less under pivot and conservation agriculture than under surface irrigation and conventional tillage, respectively. Additional water could be saved by improving irrigation schedules so that they are better synchronized with crop water requirements. Although sprinklers and drip irrigation offer improved water efficiency, they are not as highly accessible in developing countries because they are technology and knowledge intensive. The technology required for these systems is often cost prohibitive for farmers in developing countries.

In a review of zero-tillage studies for rice–wheat systems in the IGP, Erenstein and Laxmi (2008) reported irrigation water savings for zero tillage in the range of 20–35% in the wheat crop compared to conventional tillage, reducing water usage by about 10 cm ha^{-1}. The savings arise because, with zero tillage, it is possible to sow wheat just after the rice harvest, making use of residual moisture for wheat germination, potentially saving a pre-sowing irrigation. Moreover, irrigation efficiency increases for the first irrigation because water advances faster in untilled soil than in tilled soil, and irrigation can be stopped once the field is covered.

It is important to note that none of the techniques to improve water-use efficiency discussed above will be effective if the overall management is of a low standard (Wallace and Batchelor, 1997). Optimizing the use of other inputs like fertilizer, seeds and pesticides is needed to achieve optimal crop production and thus improve water-use efficiency. Precision agriculture aims to optimize input use by applying more precise amounts of fertilizer and other chemicals at the optimal time and only to the areas of the field that need the application, at the same time also contributing to the reduction of greenhouse gas emissions.

In addition to precipitation changes under climate change, most areas will also experience increases in temperature. Temperature changes may affect planting dates, seed germination, crop growth and pests and diseases. Conservation agriculture improves flexibility in planting dates because it is unnecessary to spend time completing tillage operations in between crop cycles. This allows more rapid turnover between crop cycles. This becomes especially important when rising temperatures at the end of the growing season reduce grain filling, as in northern Mexico or the IGP. Erenstein and Laxmi (2008) reported a 5–7% wheat yield increase, due mostly to timelier planting of wheat, for zero-tillage wheat after rice compared to conventional tillage in the IGP.

Additionally, residue left on the soil surface helps to regulate soil temperature by reflecting solar radiation and insulating the soil surface. The amplitude of soil temperature variation tends to be lower with zero tillage and residue retention than with conventional tillage, resulting in lower soil temperatures during the daytime and higher temperatures at night. In tropical hot soils, mulch cover reduces soil peak temperatures that are too high for optimum growth and development to an appropriate level, favouring biological activity, initial crop growth and root development during the growing season (Oliveira et al., 2001).

In addition to temperature changes, the management practices proposed to increase cropping system resilience to climate change, such as leaving crop residues at the soil surface, may change disease and pest pressure. Research is needed to determine how pest and disease pressure will change and to identify sustainable ways to manage them. Identifying viable crop rotations that disrupt the life cycle of pests or diseases can help to make the cropping system more resilient to the pests and diseases (Cook, 2006).

Improved agronomy also includes precision agriculture tools that lead to a more efficient use of nitrogen. The result is a reduction in the emission of nitrous oxide, a powerful greenhouse gas. The Yaqui Valley of Mexico is an area with intensive agriculture that has an agroecosystem typical of about 40% of all the wheat producing areas in the developing world. In the Yaqui Valley, the use of precision agriculture tools, together with improved timing of nitrogen application, has reduced emissions of nitric and nitrous oxide by 50% (Matson et al., 1998).

12.3.3 Postharvest storage technologies

Potential impact of climate change on postharvest losses

Sustainable intensification systems to improve food security and farm income diversification could contribute to food security and offer greater income opportunities to smallholder producers; however, this may not be fully realized unless farmers' postharvest storage facilities, processing and marketing are not improved. In SSA, for instance, poor postharvest management leads to 20–30% loss of grains, with an estimated monetary value of more than US$4 billion annually (FAO, 2012). The lack of effective storage management technologies often forces smallholders to sell their produce immediately after harvest at low prices, leaving them in a poverty trap. Poverty reduction and greater livelihood and food security would therefore be achieved if secure grain storage could be provided. Postharvest losses also contribute to soaring food prices and the sporadic food shortages that have fuelled public unrest recently in many parts of the world. Postharvest losses also have an impact on environmental degradation and climate change, because land, water and non-renewable resources such as fertilizer and energy are used to produce, process, handle and transport food that no one consumes (Tefera, 2012).

Climate change, especially the rise in temperatures, is likely to intensify the problem of postharvest losses by facilitating a higher prevalence of postharvest insect pests and pathogens, due to favourable environmental conditions and favourable ecological habitats in which new pests and pathogens can proliferate (Paterson and Lima, 2010). Effective postharvest management of food grains is, therefore, an important component of sustainable food security and economic growth. Improving postharvest management technologies and practices and reducing losses of food grains would enhance food security and increase the adaptive capacity of resource-poor farmers to the changing climate. In Africa, most smallholder farmers rely on sun drying to ensure that grains are dry before storage. If unfavourable weather conditions prevent grains from drying sufficiently, losses will be high.

Factors causing postharvest losses

Postharvest insect pests, mainly the maize weevil, *Sitophilus zeamais*, the larger grain borer (LGB), *Prostephanus truncatus*, the angoumois grain moth, *Sitotroga cerealella*, and the lesser grain weevil, *S. oryzae*, cause an estimated 20–30% loss of maize, thus impacting food security and income generation (Markham et al., 1994). The incidence of pest attack on stored grains is also linked to mycotoxin contamination and poisoning. Mycotoxin contamination makes grain unsafe for food and animal feed, thus impacting food and feed safety adversely (Lewis et al., 2005). Socio-economic factors including access to technology, rural road infrastructure and market information also contribute to postharvest losses. These barriers are exacerbated by the lack of policy-based incentives for private sector involvement in postharvest technology development and dissemination, and adoption of postharvest management practices (Tefera, 2012)

Postharvest technologies

Moisture content plays a vital role in grain deterioration by creating favourable conditions for mycotoxin contamination; therefore, the first important attempt towards reducing postharvest grain losses is maintaining grain moisture content at

12–13%. CIMMYT has put considerable efforts into developing high-yielding maize varieties with resistance to the maize weevil and the LGB as options for minimizing postharvest losses of the maize. These varieties also reduce the risk associated with the consumption of maize treated with insecticides; they possess chemical compounds that deter insects from feeding or inhibit their growth, or they have tight husk cover and ear tips that make them inaccessible to insect infestation (Tefera et al., 2011).

Pest treatment and exclusion methods are also essential for protecting crops postharvest

Metal silos are an important technology for enhancing food security, particularly for small-scale farmers in developing countries. They have proven to be effective in protecting harvested grains from attack not only from storage insects but also from rodent pests (CIMMYT, 2009; Tefera et al., 2011). A silo is a simple structure that allows grains to be kept for long periods and prevents attack from pests, and they are promising to become a key technology for improving food security in Africa, Asia and Latin America. A metal silo generally holds between 100 and 3000 kg and can be built on-farm from locally available material. They are airtight and, if properly used, they eliminate postharvest losses, avoiding the use of insecticides and enabling smallholders to take advantage of fluctuating grain prices (CIMMYT, 2009).

Hermetic bags (super grain bags) produced by GrainPro Inc, Purdue University, USA (known as PICS bags: Purdue Improved Cowpea Storage), and MashAgrik, South Africa, are suitable for cereal storage (Kimenju et al., 2009; Tefera, 2012). Both super bags and metal silos work hermetically, whereby a lack of oxygen inside the container causes the death of insect pests. In addition, actellic super dust is a cocktail of 1.6% pirimiphos-methyl and 0.3% permethrin and has been promoted as a chemical that is effective against the LGB in combination with practices such as immediate shelling and treating (Farrell and Schulten, 2002). This technology has been adopted by small-scale farmers for grain storage in Africa.

12.3.4 Targeted deployment of technologies

Enhancing adaptation capacity

Socio-economic and spatial agroclimatic research is needed to understand and map climate hotspots, the vulnerability of livelihoods, current adaptation options and the institutional and policy mechanisms that promote the adoption of new technologies and enhance local adaptive capacity to climate change. The salutary reality is that farmers will not benefit from existing and future technology options if they are unable to access improved seed and other technological innovations (see Ingram, Chapter 16, this volume). The reasons behind farmer (non-)adoption of climate-smart technologies are complex, but we can learn much from previous research on farmers' reluctance to adopt soil and water technologies (Hudson, 1991).

Adoption studies related to smallholder production systems have shown that risk is an important component in farmers' decision making (Kaliba et al., 2000). Thus, farmers tend first to adopt simple technologies or components and then move progressively to more complex and more costly technologies (see Deressa, Chapter 17, this volume). An important factor is rural labour shortages (Zimmerer, 1993). Many farmers depend on both production from their land and off-farm income-generating activities. This has implications for labour availability at different times of year and can determine farmers' acceptance of practices such as conservation agriculture systems, especially if farmers are not able to purchase herbicides (a labour-saving technology) to control weeds (Giller et al., 2009).

While agricultural extension, education and training can help many farmers maximize the potential of their productive assets through the adoption of climate-smart technologies, the promotion of these technologies has coincided with deep cuts to publicly funded extension services in the developing world (Hellin, 2012). In the majority of cases, the private sector has proven incapable of replacing previous state services due to high transaction costs, dispersed clientele and low (or non-existent) profits (Muyanga and Jayne, 2008). There is a need for agricultural extension modalities that resource-poor farmers can access on a more sustainable basis. This requires novel, flexible research and extension approaches that differ from those more commonly used by policy makers, donors, researchers and extension agents (Ekboir et al., 2009). There is growing interest in the emergence of agricultural innovation systems.

An innovation system consists of a web of dynamic interactions among researchers, extension agents, equipment manufacturers, input suppliers, farmers, traders and processors (Hall et al., 2005). It is a network of organizations and individuals that focus on bringing new products, new processes and new forms of organization into social and economic use. The institutions and policies that affect their behaviour and performance are also part of the innovation system. Innovation systems depend on learning processes, feedback loops and iterative interactions that are decidedly non-linear (Davis et al., 2008; Spielman et al., 2008). Innovations systems have emerged around conservation agricultural practices across the developing world (Dixon et al., 2008; Erenstein et al., 2008) as well as in market access (Devaux et al., 2009; Hellin, 2012).

There is also the issue of enhancing adaptive capacity. Eakin and Lemos (2006) posit that the high uncertainties in climate change scenarios mean that there is growing interest in improving adaptive capacity as an alternative focus of policy efforts, rather than the promotion of specific adaptation options per se. There is an expectation that nation states will improve their capacity and that of their citizens to adapt to climatic change (Eakin and Lemos, 2006). Hence, while specific adaptation technologies and practices are critical, there is a need to direct more attention to the institutional changes that empower states to design and implement policy to increase adaptive capacity. Furthermore, the adaptive capacity of nation states is linked to the complex relationships that exist between the state and the private sector and civil society.

Multidisciplinary approaches for better targeting of technologies

Multidisciplinary approaches can be very effective in improving the targeting of technologies and enhancing farmers' adaptive capacity. A recent example comes from Central America and a study led by the Catholic Relief Services (CRS) involving researchers from CIAT (the International Center for Tropical Agriculture) and CIMMYT (Schmidt et al., 2012). Maize and beans are essential to the subsistence of over 1 million farms in El Salvador, Guatemala, Honduras and Nicaragua. Farmers grow the two crops on over 2.4 Mha. Central American small-scale maize and bean farmers are particularly vulnerable to the effects of climate change, due to their geographical location and limited adaptive capacities.

Most of the maize–bean production in Central America, however, occurs on shallow degraded soils and erosion-prone sloping terrain. Any changes to this fragile system make stakeholders especially vulnerable to the major determinates of poverty, such as: geographic isolation; lack of access to services, extension and infrastructure, credit and input and output markets; low education levels; and dependency on family labour. Climate change – in the form of higher temperatures and less precipitation – is likely to have a negative impact on crop production in Central America. Climate predictions, however, are too broad and are not specific enough to inform decision makers and smallholder farmers on local adaptation strategies.

Researchers used a cross-disciplinary approach – which combined biophysical crop modelling using future climate data based on predictions, together with detailed socio-economic analysis – to provide a comprehensive framework for enhancing climate change adaptive capacity in Central America and to inform policy makers about the necessary changes to achieve this. Researchers used existing global climate models to create outputs that were specific to local scales (5 km^2 resolution) across Central America. They then used the downscaled outputs of a number of climate models for the 2020s and 2050s in these areas. They mapped how climate change would affect maize and bean production, and in turn stakeholders, from both a production and economic level. Through a series of qualitative and quantitative research tools (including focus groups and a survey), researchers explored with farming communities suitable adaptation strategies via focus group meetings and a questionnaire.

Results suggest that soil quality will be one of the most important variables determining future yields. Researchers identified three types of hotspots in Central America. First, they identified areas where it would be extremely difficult for farmers to grow maize or beans in the future. Second, adaptation areas were determined where farmers would be able to continue to produce these crops if they pursued adaptation strategies now. Finally, pressure areas were recognized where uncultivated land or land currently cultivated with other crops would become more attractive for maize and bean farming. Based on farmers responses and the type of hotspot, the project recommended different climate change adaptation and mitigation strategies such as sustainable agricultural intensification and crop diversification.

12.4 Conclusion

Maize and wheat are among the three most important crops for global food security. Climate change will have variable impacts on supply and demand patterns for these crops. While wheat production may expand in high-latitude temperate regions, global warming will reduce production in low-rainfall tropical growing regions. Maize production in developing countries will suffer significantly from climate change.

Climate change will therefore undermine food and livelihood security and complicate efforts to fight poverty, hunger and environmental degradation.

The relationship between climate change, agriculture and food security is a complex one, shaped by economic policies and political decisions. Appropriate climate change research, therefore, involves researchers from a broad spectrum of disciplines, along with other stakeholders. Participatory and interdisciplinary research is needed in order to provide farmers, policy makers, donors and other stakeholders with the knowledge, tools and approaches required to meet the challenge of ensuring future food security.

Despite some uncertainties on the spatially differentiated impact of climate change on agricultural production, there is little doubt that new germplasm, more suited to future climates, is critical, along with improved agronomic and crop management practices. There is an urgent need to develop climate-adaptable crop varieties with improved tolerance to heat stress, and combined heat and drought stress. In some cases, climate change may create new biotic stresses brought by new conditions conducive to pest and disease infestations. Decision support systems (crop modelling) may help project any likely effects of climate change on the outbreak and spread of disease and pest epidemics. This may require proper forecasting and early warning systems.

The development and dissemination of climate-responsive germplasm may take several years, because the process consists of several steps including breeding, on-farm testing, release of varieties and germplasm dissemination. It is very important to facilitate farmers' adoption of these technologies. Such an effort has often been the missing link and has prevented farmers fully benefiting from investment in agricultural research. In addition to enhancing adaptation and reducing vulnerabilities, improved agricultural innovations such as conservation agriculture may also contribute towards mitigating global warming and climate change.

The performance of agricultural systems coupled with the introduction of climate change-adaptable varieties is determined by the complex interaction of agroecosystems and human activity. Due to variation in biophysical and socio-economic environments, it is important to model the vulnerabilities as well as the impact of proposed technological and policy interventions at spatially disaggregated scales. Such analysis, coupled with economic analysis, would enhance the ability to measure the impact of climate change on human well-being, as well as the potential of alternative response options for climate change adaptation and mitigation.

Acknowledgements

We wish to acknowledge funding from the CGIAR Research Programs on Climate Change, Agriculture and Food Security (CCAFS), MAIZE and WHEAT, with additional funding for the modelling work from Global Futures for Agriculture.

References

Bai, Z., Dent, D.L., Olsson, L. and Schaepman, M.E. (2008) Global assessment of land degradation and improvement. 1. Identification by remote sensing. Report 2008/01, ISRIC – World Soil Information, Wageningen (http://www.isric.org/sites/default/files/Report%202008_01_GLADA%20international_REV_Nov%202008.pdf, accessed 16 December 2013).

Bänziger, M. and Araus, J.L. (2007) Recent advances in breeding maize for drought and salinity stress tolerance. In: Jenks, M.A., Hasegawa, P.M. and Jain S.M. (eds) *Advances in Molecular Breeding Toward Drought and Salt Tolerant Crops*. Springerlink, Berlin, pp. 587–601.

Bänziger, M., Setimela, P.S., Hodson, D. and Vivek, B. (2006) Breeding for improved abiotic stress tolerance in Africa in maize adapted to southern Africa. *Agriculture and Water Management* 80, 212–214.

Barnabás, B., Jäger, K. and Fehér, A. (2008) The effect of drought and heat stress on reproductive processes in cereals. *Plant, Cell and Environment* 31, 11–38.

Battisti, D.S. and Naylor, R.L. (2009) Historical warnings of future food insecurity with unprecedented seasonal heat. *Science* 323, 240–244.

Beddington, J.R., Asaduzzaman, M., Clark, M.E., Bremauntz, A.F., Guillou, M.D., Jahn, M.M., et al. (2012) The role for scientists in tackling food insecurity and climate change. *Agriculture and Food Security* 1, 10 (doi:10.1186/2048-7010-1-10).

Borrás, L., Westgate, M.E. and Otegui, M.E. (2003) Control of kernel weight and kernel water relations by post-flowering source–sink ratio in maize. *Annals of Botany* 91, 857–867.

Braun, H.J., Atlin, G. and Payne, T. (2010) Multi-location testing as a tool to identify plant response to global climate change. In: Reynolds, M.P. (ed.) *Climate Change and Crop Production*. CAB International, Wallingford, UK, pp. 115–138.

Bruulsema, T.W., Tollenaar, M. and Heckman, J.R. (2000) Boosting crop yields in the next century. *Better Crops* 84, 9–13.

Burney, J.A., Davis, S.J. and Lobell, D.B. (2010) Greenhouse gas mitigation by agricultural intensification. *Proceedings of the National Academy of Sciences* 107, 12052–12057.

Cairns, J.E., Sonder, K., Zaidi, P.H., Verhulst, N., Mahuku, G., Babu, R., et al. (2012) Maize production in a changing climate: impacts, adaptation, and mitigation strategies. *Advances in Agronomy* 144, 1–58.

Cairns, J.E., Hellin, J., Sonder, K., Araus, J.L., Macrobert, J.F., Thierfelder, C., et al. (2013) Adapting maize to climate change in sub-Saharan Africa. *Food Security* 5, 345–360.

Chakraborty, S. and Newton, A.C. (2011) Climate change, plant diseases and food security: an overview. *Plant Pathology* 60, 2–14.

Chakraborty, S., Luck, J., Hollaway, G., Fitzgerald, G. and White, N. (2011) Rust-proofing wheat for a changing climate. *Euphytica* 179, 19–32.

Ci, X., Li, M., Liang, X., Xie, Z., Zhang, D., Li, X., et al. (2011) Genetic contribution to advanced yield for maize hybrids released from 1970 to 2000 in China. *Crop Science* 51, 13–20.

CIMMYT (International Maize and Wheat Improvement Center) (2009) *Annual Report. Effective Grain Storage for Better Livelihoods of African Farmers Project*. CIMMYT, Nairobi.

Coakley, S.M., Scherm, H. and Chakraborty, S. (1999) Climate change and plant disease management. *Annual Review of Phytopathology* 37, 399–426.

Cook, R.J. (2006) Toward cropping systems that enhance productivity and sustainability. *Proceedings of the National Academy of Sciences* 103, 18389–18394.

Cooper, P.J.M., Dimes, J., Rao, K.P.C., Shapiro, B., Shiferaw, B. and Twomlow, S. (2008) Coping better with current climatic variability in the rainfed farming systems of sub-Saharan Africa: an essential first step in adapting to future climate change? *Agriculture, Ecosystems and Environment* 126, 24–35.

Cossani, C.M. and Reynolds, M.P. (2012) Physiological traits for improving heat tolerance in wheat. *Plant Physiology* 160, 1710–1718.

Datta, D., Nayar, S.K., Prashar, M. and Bhardwaj, S.C. (2009) Inheritance of temperature-sensitive leaf rust resistance and adult plant stripe rust resistance in common wheat cultivar PBW343. *Euphytica* 166, 277–282.

Davis, K.E., Ekboir, J. and Spielman, D.J. (2008) Strengthening agricultural education and training in sub-Saharan Africa from an innovation systems perspective: a case study of Mozambique. *Journal of Agricultural Education and Extension* 14(1), 35–51.

Devaux, A., Horton, D., Velasco, C., Thiele, G., López, G., Bernet, T., et al. (2009) Collective action for market chain innovation in the Andes. *Food Policy* 34, 31–38.

Diffenbaugh, N.S., Krupke, C.H., White, M.A. and Alexander, C.E. (2008) Global warming presents new challenges for maize pest management. *Environmental Research Letters* 3, 044007. doi:10.1088/1748-9326/3/4/044007.

Dixon, J., Hellin, J., Erenstein, O., Kosina, P. and Nalley, L.L. (2008) Innovation systems and impact pathways for wheat. In: Reynolds, M.P., Pietragalla, J. and Braun, H.-J. (eds) *International Symposium on Wheat Yield Potential: Challenges to International Wheat Breeding*. CIMMYT, Mexico, pp. 175–180.

DSSAT (2012) Decision Support System for Agrotechnology Transfer (DSSAT) (http://dssat.net/, accessed 4 December 2013).

Duvick, D.N. (2005) Genetic progress in yield of United States maize (*Zea mays* L.). *Maydica* 50, 193–200.

Duvick, D.N. and Cassman, K.G. (1999) Post-green revolution trends in yield potential of temperate maize in the North-Central United States. *Crop Science* 39, 1622–1630.

Eakin, H. and Lemos, M.C. (2006) Adaptation and the state: Latin America and the challenge of capacity-building under globalization. *Global Environmental Change* 16, 7–18.

Edmeades, G.O., Bolaños, J., Chapman, S.C., Lafitte, H.R. and Bänziger, M. (1999) Selection improves drought tolerance in tropical maize populations. 1. Gains in biomass, grain yield and harvest index. *Crop Science* 39, 1306–1315.

Ekboir, J.M., Dutrénit, G., Martínez, V.G., Vargas, A.T. and Vera-Cruz, A.O. (2009) Successful organizational learning in the management of agricultural research and innovation: the Mexican Produce Foundations. IFPRI Research Report No 162. International Food Policy Research Institute, Washington, DC.

Erenstein, O. and Laxmi, V. (2008) Zero tillage impacts in India's rice–wheat systems: a review. *Soil and Tillage Research* 100, 1–14.

Ericksen, P., Thornton, P., Notenbaert, A., Cramer, L., Jones, P. and Herrero, M. (2011) Mapping hotspots of climate change and food insecurity in the global tropics. CCAFS Report No 5. CGIAR Research Program on Climate Change, Agriculture and Food Security (CCAFS). Copenhagen (www.ccafs.cgiar.org, accessed 4 December 2013).

Evenson, R.E. and Gollin, D. (2003) Assessing the impact of the green revolution, 1960 to 2000. *Science* 300, 758–762.

FAO (2012) FAOSTAT. Food and Agriculture Organization of the United Nations, Rome (http://faostat3.fao.org/faostat-gateway/go/to/home/E, accessed 4 December 2013).

Farrell, G. and Schulten, G.G.M. (2002) Larger grain borer in Africa; a history of efforts to limit its impact. *Integrated Pest Management Reviews* 7, 67–84.

Fedoroff, N.V., Battisti, D.S., Beachy, R.N., Cooper, P.J.M., Fischhoff, D.A., Hodges, C.N., et al. (2010) Radically rethinking agriculture for the 21st Century. *Science* 327, 833–834.

Foley, J.A., DeFries, R., Asner, G.P., Barford, C., Bonan, G., Carpenter, S.R., et al. (2005) Global consequences of land use. *Science* 309, 570–574.

Giller, K.E., Witter, E., Corbeels, M. and Tittonell, P. (2009) Conservation agriculture and smallholder farming in Africa: the heretics' view. *Field Crops Research* 114, 23–34.

Godfray, H.C.J., Beddington, J.R., Crute, I.R., Haddad, L., Lawrence, D., Muir, J.F., et al. (2010) Food security: the challenge of feeding 9 billion people. *Science* 327, 812–818.

Govaerts, B., Sayre, K.D. and Deckers, J. (2005) Stable high yields with zero tillage and permanent bed planting? *Field Crops Research* 94, 33–42.

Govaerts, B., Sayre, K.D., Goudeseune, B., De Corte, P., Lichter, K., Dendooven, L., et al. (2009) Conservation agriculture as a sustainable option for the central Mexican highlands. *Soil and Tillage Research* 103, 222–230.

Grassini, P., Yang, H., Irmak, S., Thorburn, J., Burr, C. and Cassman, K.G. (2011) High-yield irrigated maize in the Western US Corn Belt: II.

Irrigation management and crop water productivity. *Field Crops Research* 120, 133–141.

Gunders, D. (2012) *Wasted: How America Is Losing Up to 40 Percent of Its Food from Farm to Fork to Landfill.* Natural Resources Defense Council, New York.

Hall, A., Mytelka, L. and Oyeyinka, B. (2005) *Innovation Systems: Implications for Agricultural Policy and Practice. Institutional Learning and Change Initiative Brief 2.* CGIAR, Rome.

Hall, K.D., Guo, J., Dore, M. and Chow, C.C. (2009) The progressive increase of food waste in America and its environmental impact. *PloS One* 4(11), e7940.

Hatfield, J.L., Sauer, T.J. and Prueger, J.H. (2001) Managing soil to achieve greater water use efficiency: a review. *Agronomy Journal* 93, 271–280.

Hellin, J. (2012) Agricultural extension, collective action and innovation systems: lessons on network brokering from Peru and Mexico. *Journal of Agricultural Education and Extension* 18, 141–159.

Hellin, J., Shiferaw, B., Cairns, J.E., Reynolds, M., Ortiz-Monasterio, I., Banziger, M., et al. (2012) Climate change and food security in the developing world: potential of maize and wheat research to expand options for adaptation and mitigation. *Journal of Development and Agricultural Economics* 4, 311–321.

Howell, T.A. (2001) Enhancing water use efficiency in irrigated agriculture. *Agronomy Journal* 93, 281–289.

Hudson, N.W. (1991) *A Study of the Reasons for Success and Failure of Soil Conservation Projects. Soils Bulletin 64.* Food and Agriculture Organization of the United Nations, Rome.

Hulme, M., Doherty, R.M., Ngara, T., New, M. and Lister, D. (2001) African climate change: 1900–2100. *Climate Research* 17, 145–168.

IPCC Fourth Assessment Report (2007) *Climate Change 2007: Synthesis Repor* (http://www.ipcc.ch/pdf/assessment-report/ar4/syr/ar4_syr.pdf, accessed 27 March 2013).

Jones, P.G., Thornton, P.K. and Heinke, J. (2010) Characteristically generated monthly climate data using downscaled climate model data from the fourth assessment report of the IPCC (www.ccafs-climate.org/data/, accessed 16 December 2013).

Jones, R.J. and Simmons, S.R. (1983) Effect of altered source–sink ratio on growth of maize kernels. *Crop Science* 23, 129–134.

Joshi, A.K., Ortiz-Ferrara, G., Crossa, J., Singh, G., Sharma, R.C., Chand, R., et al. (2007) Combining superior agronomic performance and terminal heat tolerance with resistance to spot blotch (*Bipolaris sorokiniana*) in the warm humid Gangetic plains of south Asia. *Field Crop Research* 103, 53–61.

Juroszek, P. and von Tiedemann, A. (2012) Climate change and potential future risks through wheat diseases: a review. *European Journal of Plant Pathology* doi:10.1007/s10658-012-0144-9.

Kaliba, A.R.M., Verkuijl, H. and Mwangi, W. (2000) Factors affecting adoption of improved maize seeds and use of inorganic fertilizer for maize production in the intermediate and lowland zones of Tanzania. *Journal of Agricultural and Applied Economics* 32, 35–47.

Kimenju, S.C., De Groote, H. and Hellin, H. (2009) *Preliminary Economic Analysis: Cost Effectiveness of the Use of Improved Storage Methods by Small Scale Farmers in East and Southern Africa Countries.* International Maize and Wheat Improvement Center (CIMMYT), Nairobi, p. 17.

Kolmer J.A. (1996) Genetics of resistance to wheat leaf rust. *Annual Review of Phytopathology* 34, 435–455.

Lantican, M.A., Pingali, P.L. and Rajaram, S. (2003) Is research on marginal lands catching up? The case of unfavourable wheat growing environments. *Agricultural Economics* 29, 353–361.

Legrève, A. and Duveiller, E. (2010) Preventing potential disease and pest epidemics under a changing climate. In: Reynolds, M.P. (ed) *Climate Change and Crop Production.* CAB International, Wallingford, UK, pp. 50–70.

Leonard, K.J. and Szabo, L.J. (2005) Stem rust of small grains and grasses caused by *Puccinia graminis*. *Molecular Plant Pathology* 6, 99–111.

Lewis, L., Onsongo, M., Njapau, H., Schurz-Rogers, H., Luber, G., Kieszak, S., et al. and the Kenya Aflatoxicosis Investigation Group (2005) Aflatoxin contamination of commercial maize products during an outbreak of acute aflatoxicosis in eastern and central Kenya. *Environmental Health Perspectives* 113, 1763–1767.

Lillemo, M., van Ginkel, M., Trethowan, R.M., Hernandez, E. and Crossa, J. (2005) Differential adaptation of CIMMYT bread wheat to global high temperature environments. *Crop Science* 45, 2443–2453.

Lobell, D.B. and Gourdji, S.M. (2012) The influence of climate change on global crop productivity. *Plant Physiology* 160, 1686–1697.

Lobell, D.B., Burke, M.B., Tebaldi, C., Mastrandrea, M.D., Falcon, W.P. and Naylor, R.L. (2008) Prioritizing climate change adaptation needs for food security in 2030. *Science* 319, 607–610.

Lobell, D.B., Bänziger, M., Magorokosho, C. and Vivek, B. (2011) Nonlinear heat effects on African maize as evidenced by historical yield trials. *Nature Climate Change* 1, 42–45.

Lopes, M.S., Reynolds, M.P., Manes, Y., Singh, R.P., Crossa, J. and Braun, H.J. (2012) Genetic yield gains and changes in associated traits of CIMMYT spring bread wheat in a 'Historic' set representing 30 years of breeding. *Crop Science* 52, 1123–1131.

Lundqvist, J., de Fraiture, C. and Molden, D. (2008) *Saving Water: from Field to Fork: Curbing Losses and Wastage in the Food Chain. SIWI Policy Brief.* Stockholm International Water Institute, Stockholm.

Luque, S.F., Cirilo, A.G. and Otegui, M.E. (2006) Genetic gains in grain yield and related physiological attributes in Argentine maize hybrids. *Field Crops Research* 95, 383–397.

McGarry, D., Bridge, B.J. and Radford, B.J. (2000) Contrasting soil physical properties after zero and traditional tillage of an alluvial soil in the semi-arid subtropics. *Soil and Tillage Research* 53, 105–115.

Manes, Y., Gomez, H.F., Puhl, L., Reynolds, M., Braun, H.J. and Trethowan, R. (2012) Genetic yield gains of the CIMMYT international semi-arid wheat yield trials from 1994 to 2010. *Crop Science* 52, 1543–1552.

Markell, S.G. and Milus, E.A. (2008) Emergence of a novel population of *Puccinia striiformis* f. sp. *tritici* in eastern United States. *Phytopathology* 98, 632–639.

Markham, R.H., Bosque-Perez, N.A., Borgemeister, C. and Meikle, W.G. (1994) Developing pest management strategies for the maize weevil, *Sitophilus zeamais*, and the larger grain borer, *Prostephanus truncatus*, in the humid and sub-humid tropics. *FAO Plant Protection Bulletin* 42, 97–116.

Matson, P.A., Naylor, R. and Ortiz-Monasterio, I. (1998) Integration of environmental, agronomic, and economic aspects of fertilizer management. *Science* 280, 112–115.

Milus, E.A., Seyran, E. and McNew, R. (2006) Aggressiveness of *Puccinia stiiformis* f. sp. *tritici* isolates in the south-central United States. *Plant Disease* 90, 847–852.

Muyanga, M. and Jayne, T.S. (2008) Private agricultural extension system in Kenya: practice and policy lessons. *Journal of Agricultural Education and Extension* 14, 111–124.

New, M., Rahiz, M. and Karmacharya J. (2012) *Climate Change in Indo-Gangetic Agriculture: Recent Trends, Current Projections, Crop–Climate Suitability, and Prospects for Improved Climate Model Information.* CGIAR Research Program on Climate Change, Agriculture and Food Security (CCAFS), Copenhagen (http://ccafs.cgiar.org/, accessed 4 December 2013).

Oliveira, J.C.M., Timm, L.C., Tominaga, T.T., Cassaro, F.A.M., Reichardt, K., Bacchi, O.O.S., et al. (2001) Soil temperature in a sugar-cane crop as a function of the management system. *Plant and Soil* 230, 61–66.

Ortiz, R., Sayre, K.D., Govaerts, B., Gupta, R., Subbarao, G.V., Ban, T., et al. (2008) Climate change: can wheat beat the heat? *Agriculture, Ecosystems and Environment* 126, 46–58.

Parfitt, J., Barthel, M. and Macnaughton, S. (2010) Food waste within food supply chains: quantification and potential for change to 2050. *Philosophical Transactions of the Royal Society London B: Biological Sciences* 365, 3065–3081.

Park, R.F., Ash, G.J. and Rees, R.G. (1992) Effects of temperature on the response of some Australian wheat cultivars to *Puccinia striiformis* f. sp. *tritici*. *Mycological Research* 96, 166–170.

Paterson, R.R.M. and Lima, N. (2010) How will climate change affect mycotoxins in food? *Food Research International* 43, 1902–1914.

Pingali, P.L. (2012) Green Revolution: impacts, limits, and the path ahead. *Proceedings of the National Academy of Sciences* 109, 12302–12308.

Pretorius, Z.A., Singh, R.P., Wagoire, W.W. and Payne, T.S. (2000) Detection of virulence to wheat stem rust resistance gene *Sr31* in *Puccinia graminis* f. sp. *tritici* in Uganda. *Plant Disease* 84, 203 (http://dx.doi.org/10.1094/PDIS.2000.84.2.203B, accessed 16 December 2013).

Ray, D.K., Ramankutty, N., Mueller, N.D., West, P.C. and Foley, J.A. (2012) Recent patterns of crop yield growth and stagnation. *Nature Communications* 3, 1293 (doi: 10.1038/ncomms2296).

Reynolds, M.P., Hays, D. and Chapman, S. (2010) Breeding for adaptation to heat and drought stress. In: Reynolds, M.P. (ed.) *Climate Change and Crop Production.* CAB International, Wallingford, UK, pp. 71–91.

Richards, R.A. (2006) Physiological traits used in the breeding of new cultivars for water-scarce environments. *Agricultural Water Management* 80, 197–211.

Roelfs, A.P. (1988) Genetic control of phenotypes in wheat stem rust. *Annual Review of Phytopathology* 26, 351–367.

Rosegrant, M.W. (2012) *International Model for Policy Analysis of Agricultural Commodities and Trade (IMPACT): Model Description.* International Food Policy Research Institute, Washington, DC.

Rosegrant, M.W., Ringler, C. and Zhu, T. (2009) Water for agriculture: maintaining food security under growing scarcity. *Annual Review of Environment and Resources* 34, 205–222.

Schmidt, A., Eitzinger, A., Sonder, K. and Sain, G.

(2012) Tortillas on the roaster: Central American maize–bean systems and the changing climate. Technical Report (http://newswire.crs.org/wp-content/uploads/2012/10/Tortillas-on-the-Roaster-full-technical-report-minimum-size.pdf, accessed 4 December 2013).

Shongwe, M.E., van Oldenborgh, G.J., van den Hurk, B.J.J.M., de Boer, B., Coelho, C.A.S. and van Aalst, M.K. (2009) Projected changes in mean and extreme precipitation in Africa under global warming. Part I: Southern Africa. *Journal of Climate* 22, 3819–3837.

Singh, R.P., Prasad, P.V.V. and Reddy, K.R. (2013) Impacts of changing climate and climate variability on seed production and seed industry. *Advances in Agronomy* 118, 49–110.

Smil, V. (2004) Improving efficiency and reducing waste in our food system. *Environmental Sciences* 1, 17–26.

Spielman, D.J., Ekboir, J., Davis, K. and Ochieng, C.M.O. (2008) An innovation systems perspective on strengthening agricultural education and training in sub-Saharan Africa. *Agricultural Systems* 98, 1–9.

Tefera, T. (2012) Post-harvest losses in African maize in the face of increasing food shortage. *Food Security* 4, 267–277.

Tefera, T., Kanampiu, F., De Groote, H., Hellin, J., Mugo, S., Kimenju, S., *et al.* (2011) The metal silo: an effective grain storage technology for reducing post-harvest insect and pathogen losses in maize while improving smallholder farmers' food security in developing countries. *Crop Protection* 30, 240–245.

Tilman, D., Balzer, C., Hill, J. and Befort, B.L. (2011) Global food demand and the sustainable intensification of agriculture. *Proceedings of the National Academy of Sciences* 108, 20260–20264.

Tollenaar, M. and Lee, E.A. (2006) Dissection of physiological processes underlying grain yield in maize by examining genetic improvement and heterosis. *Maydica* 51, 399–408.

Verhulst, N., Nelissen, V., Jespers, N., Haven, H., Sayre, K.D., Raes, D., *et al.* (2011a) Soil water content, maize yield and its stability as affected by tillage and crop residue management in rainfed semi-arid highlands. *Plant and Soil* 344, 73–85.

Verhulst, N., Sayre, K.D., Vargas, M., Crossa, J., Deckers, J., Raes, D., *et al.* (2011b) Wheat yield and tillage-straw management system × year interaction explained by climatic co-variables for an irrigated bed planting system in northwestern Mexico. *Field Crops Research* 124, 347–356.

Wallace, J.S. (2000) Increasing agricultural water use efficiency to meet future food production. *Agriculture, Ecosystems and Environment* 82, 105–119.

Wallace, J.S. and Batchelor, C.H. (1997) Managing water resources for crop production. *Philosophical Transactions of the Royal Society London B: Biological Sciences* 352, 937–947.

Wangai, A.W., Redinbaugh, M.G., Kinyua, Z.M., Miano, D.W., Leley, P.K., Kasina, M., *et al.* (2012) First report of *maize chlorotic mottle virus* and maize lethal necrosis in Kenya. *Plant Disease* 96, 1582.

Wanyera, R., Kinyua, M.G., Jin, Y. and Singh, R.P. (2006) The spread of stem rust caused by *Puccinia graminis* f. sp. *tritici*, with virulence on *Sr31* in wheat in eastern Africa. *Plant Disease* 90, 113.

Wik, M., Pingali, P. and Broca, S. (2008) *Background Paper for the World Development Report 2008: Global Agricultural Performance: Past Trends and Future Prospects.* World Bank, Washington, DC (http://wdronline.worldbank.org/worldbank/a/nonwdrdetail/100, accessed 4 December 2013).

Zimmerer, K.S. (1993) Soil erosion and labor shortages in the Andes with special reference to Bolivia, 1953–91: implications for 'conservation-with-development'. *World Development* 21, 1659–1675.

13 The Role of Modelling in Adapting and Building the Climate Resilience of Cropping Systems

Helena Kahiluoto,[1] Reimund Rötter,[1] Heidi Webber[2] and Frank Ewert[2]

[1]*MTT Agrifood Research Finland, Plant Production Research, Mikkeli, Finland;* [2]*University of Bonn, Institute of Crop Science and Resource Conservation (INRES), Crop Science Group, Bonn, Germany*

13.1 Introduction

Historical weather records show (Coumou and Rahmsdorf, 2012) and climate models predict (Rummukainen, 2012; see, also Rummukainen, Chapter 1, this volume) that global warming is driving changes in rainfall and increasing the frequency and severity of extreme events. Extreme events, together with extreme impacts resulting from warming in areas where crops are currently grown at or beyond their temperature thresholds (Schlenker and Lobell, 2010; Hatfield *et al.*, 2011), make cropping even more risky than it already is at present (Rötter *et al.*, 2013a). Given that reliable and affordable food is central to human well-being and the stability of societies, it will be paramount to adapt cropping systems effectively to climate change and simultaneously develop and utilize their considerable mitigation potential (Smith and Olesen, 2010). However, there are distinct uncertainties involved in projecting climate change for the future – especially at the regional and local scales where farmers act (Tebaldi and Knutti, 2007; Rummukainen, 2012). In addition to uncertainties in climate projections, there are several other sources of uncertainty, such as imperfect impact models, which constitute a considerable share of overall uncertainty in projections of climate change impacts on crop production (Asseng *et al.*, 2013; Rötter *et al.*, 2013a). Too high a degree of uncertainty in projecting climate change impacts has been considered a major reason for delaying or neglecting the development of adaptation strategies (Dow *et al.*, 2013). The high uncertainty in future conditions will require new ways of managing cropping systems that are flexible, adaptable and amenable to the occurrence of extreme events and higher volatility, if food security targets are to be achieved (see Webber *et al.*, Chapter 11, this volume). Managing cropping systems for resilience (Holling, 1973; Folke, 2006) is emerging as an appropriate approach under such highly uncertain conditions, to take into account the variety of drivers constantly shaping cropping systems, including adaptation to climate change (Nelson *et al.*, 2007; Nelson, 2011). In a recent attempt to operationalize approaches to build cropping systems' resilience (see Webber *et al.*, Chapter 11, this volume), the potential value of modelling tools, particularly process-based crop models, was highlighted. It was suggested that such tools could facilitate the understanding of complex system behaviour by quantifying the integrated impact of the many factors (i.e. climate, environment and management) at various scales and considering associated systems acting on cropping systems, to better inform management interventions.

The question arises of whether modelling tools are up to this task (Rötter et al., 2011a). Despite having been applied – individually or in combination – for several decades to assess the impacts of climate change and the adaptations of crop production systems (Fischer et al., 2005; Nelson et al., 2009), dynamic climate, agricultural crop (White et al., 2011; Asseng et al., 2013) and trade models all still suffer from distinct deficiencies (Tebaldi and Knutti, 2007; Asseng et al., 2013; Rosenzweig et al., 2013). Further, it is unlikely that current impact modelling studies capture, or are capable of capturing, the reality of farmers' approaches to cropping system adaptations, particularly in semi-subsistence systems (McCown, 2001; Patt et al., 2010; H. Webber et al., 2013, unpublished results), because the uncertainty in the management, social and economic aspects of future cropping systems likely outweigh by far the uncertainties of climate and associated crop physiology (Vermeulen et al., 2013). Hence, projections of climate change impacts on crop production are hardly reliable, as they represent the outcome of a 'cascade' of uncertainties (see, for example, Osborne et al., 2013; Rötter et al., 2013a). For example, assessments of climate change impact on yields of food crops at the global and continental scale by the middle of this century differ by several orders of magnitude, depending on whether only the effects of changed climate and elevated CO_2 are considered (−30 to +20%) or, in addition, technology trends under different assumptions of progress are taken into account (+10 to 90%) (Ewert et al., 2005; Müller et al., 2011; Rötter et al., 2013a).

While the lack of scientific certainty may delay adaptation actions (Dow et al., 2013), it has also placed new emphasis on more holistic 'capacity approaches' (Vermeulen et al., 2013), which acknowledge that adaptation decisions and actions must be made in spite of imperfect scientific understanding. This is in contrast to traditional 'impact approaches' in which adaptation measures appear as a 'postscript' to impact studies. Furthermore, 'capacity approaches' give explicit consideration to environmental sustainability and food security, in addition to productivity and economic performance, thus reflecting more fully the reality experienced by farmers and producers.

Both capacity and impact approaches are likely to be needed and should be employed to support adaptation decisions and enhance the resilience of cropping systems (Vermeulen et al., 2013). In fact, various crop modelling approaches could serve as tools to enhance learning, and thus build the capacity of farmers to manage the resilience of their cropping systems, and of administrators, politicians and scientists to enable it. Models could facilitate dialogues between the scientific and actor communities, with the potential benefit of improving the relevance of the models themselves. However, beyond the recognition of the value of capacity approaches, better integration of the soft and highly context-specific knowledge of how farmers are likely to adapt their cropping systems (McCown, 2001) with impact modelling studies is needed if they are to contribute critically to the development of adaptation science and building resilience into cropping systems (Meinke et al., 2009), though it represents a challenge that is outside any one disciplinary domain.

The following tries to answer the overarching question of this chapter: 'What can modelling contribute to improving cropping systems' resilience to climate change?' To do so, two different modelling studies are considered: one evaluating alternative crop rotations under climate change in the North American Midwest using several process-based models, and a second employing an empirical modelling approach in assessing and managing the resilience of barley cropping in Finland.

13.2 Current Applications of Modelling Tools for Cropping System Assessment under Climate Change

Experimental and observational studies provide evidence that a range of feasible possibilities exist to build resilience to specific climatic drivers (see Webber et al., Chapter 11, this volume). While interactions of climatic, other environmental, genetic and management factors in real production settings make it difficult to generalize their performance under future climatic conditions, various approaches can offer insights into probable outcomes of adaptations. These include process-based crop modelling studies (Asseng et al., 2013), statistical models based on large (spatial and/or temporal) observational data sets (Hakala et al., 2012; Himanen et al., 2013a) or the use of agroclimatic indices (Trnka et al., 2011; Rötter et al., 2013a), all of which can provide a basis for quantifying the impacts of climate change and variability on adapted management practices to allow more objective evaluation of the robustness of alternative options.

The analysis of time-series and cross-sectional panel data has provided a number of useful insights into changes in crop phenology (Siebert and Ewert, 2012; Palosuo et al., 2013), crop productivity (Schlenker and Roberts, 2009; Lobell et al., 2011) and the relative area share of crops (Elsgaard et al., 2012) in response to changing temperatures and precipitation. This has also generated new knowledge about the relative importance of climate, management and socio-economic factors in determining yields (Reidsma and Ewert, 2008; Sacks and Kucharik, 2011; Himanen et al., 2012b) and vulnerability to drought (Simelton et al., 2012) at regional and global scales. As statistical analyses are based on observations, they capture all the effects that influence yield, without the need to make assumptions about uncertain management variables, and they represent time-invariant processes as fixed effects, thus avoiding omitted variable bias (Schlenker and Roberts, 2009). Their limitations include being invalid beyond the conditions under which they were developed, confounding effects (e.g. Bakker et al., 2005) and not illuminating underlying processes or interactions driving cropping systems' responses. These deficits can, however, be tackled through validating the models under different conditions (Kahiluoto et al., 2014) and through careful planning and hypothesis-driven research.

Process-based crop simulation modelling overcomes the primary limitations of statistical models; they provide insights into the underlying mechanisms for high correlations between crop yield reduction and climate (see, for example, Lobell et al., 2013), thus potentially being able to unravel the interactions between key environment, genotype and management processes governing crop growth, while also being applicable beyond present conditions (Rötter et al., 2011a; Asseng et al., 2013), and suggesting areas where the underlying science should be strengthened (Boote et al., 2013; Lobell et al., 2013). However, the ability of crop models to do so correctly depends on their adequate representation of the key processes driving crop growth at various scales and under a range of biophysical and socio-economic conditions. Rötter et al. (2011a) note that most crop models currently do not yet adequately incorporate the state-of-the-art knowledge in agronomy, plant physiology or soil science; for example, with respect to how crops respond to extreme high temperatures, interactions between elevated CO_2 concentrations and water use, and insect and disease damage (Boote et al., 2013). Crop modelling is most fruitful when conducted in conjunction with experimentation, or when being confronted with empirical analysis (van Oort et al., 2012; van Ittersum et al., 2003). In recent years, crop modelling has also been applied successfully in separating and quantifying the effects of climate, specific incremental adaptation measures (e.g. earlier sowing and cultivar choice) and other shifts in management practices on crop phenology, yield and other agroecosystem variables (e.g. Asseng et al., 2011; Sacks and Kucharik, 2011).

When crop models are linked to hydrological, farm system or livestock models, it is possible to widen the scope of the potential impacts of adaptations to climatic changes, as well as to probe the interactions of cropping systems with other systems in which farmers operate (Janssen and van Ittersum et al., 2007; Ewert et al., 2009). Crop models have been used extensively to assess climate change impacts and selected adaptation options (Reilly et al., 2003; Fischer et al., 2005; Challinor et al., 2010). However, there are few, if any, examples of crop modelling studies that have analysed adaptations explicitly for their contributions to climate resilience. An early exception are the studies of Rötter et al. (1997) and Rötter and van Keulen (1997), who used combinations of long-term field trials and simulation modelling to evaluate the risk and opportunities for farmers associated with investing in site-specific nutrient management for food crops in different agroecological zones of Kenya (see Fig. 13.1). Their results indicated that the application of manure together with mineral fertilizer was able to increase yield, build soil productivity and buffer against the variability associated with using only mineral fertilizers. However, the risk of crop failure in years with failed rainy seasons would render the investment in fertilization prohibitive (for conditions at the time of analysis) for the most risk-adverse farmers. (Source: Rötter and van Keulen, 1997, p. 79.)

13.3 How Can Modelling Contribute to Enhancing the Resilience of Cropping Systems?

Despite few examples of crop modelling studies explicitly investigating resilience, many crop models are able to simulate a number of biophysical variables, such as seed production and total above-ground dry

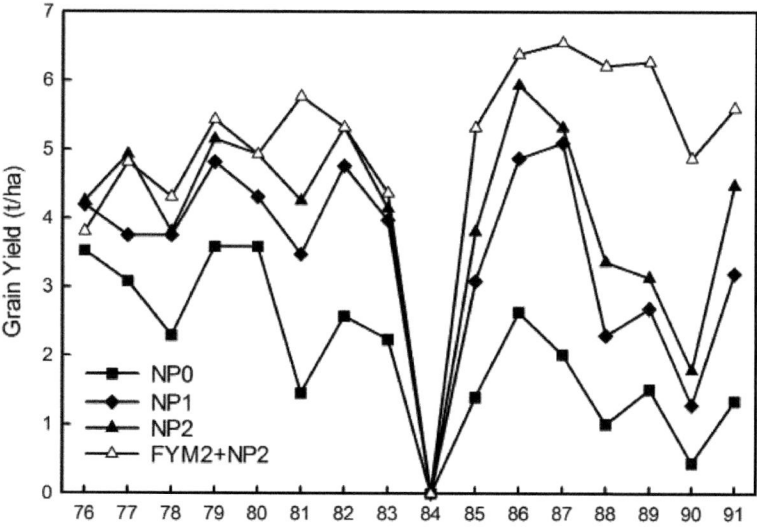

Fig. 13.1. Observed maize yields from Nairobi between 1976 and 1991 with NP0, NP1 and NP2 referring to fertilization levels of 0 kg ha^{-1} N and 0 kg ha^{-1} P; 60 kg ha^{-1} N and 26 kg ha^{-1} P; and 120 kg ha^{-1} N and 56 kg ha^{-1} P, respectively. FYM2 is 10 t ha^{-1} manure.

matter yield, growth duration, water use, nitrogen uptake, soil organic carbon levels, etc. (see, for example, Eckersten et al., 2001) and to quantify changes in cropping systems' climate resilience with various adaptations. To do so, emphasis would need to be placed on defining system boundaries, key variables, associated indicators and thresholds (e.g. Porter and Gawith, 1999). The reality of what models can simulate would need to be balanced with the reality of feedbacks and interactions with other systems at larger scales, including many socio-economic factors. We propose that a conceptual assessment framework to evaluate the resilience of cropping systems would need to consist of four key components: resilience theory, crop models, models of interdependent systems linked to crop models, and meaningful stakeholder input and involvement, as illustrated in Fig. 13.2. The key elements of each component are indicated in Fig. 13.2, as is the scientific basis and societal context that underlies each. The ability of integrated models to evaluate the climate resilience of the adapted management clearly depends on the models reflecting the state-of-the-art knowledge in a number of disciplines, a correct characterization of the key drivers, variables and processes governing the configuration of the cropping system, as well as the meaningful participation of farmers, decision makers and other key actors (McCown et al., 2009).

To illustrate aspects of such an assessment framework, two adaptation studies are reviewed in the following section. In the first, process-based models are used to

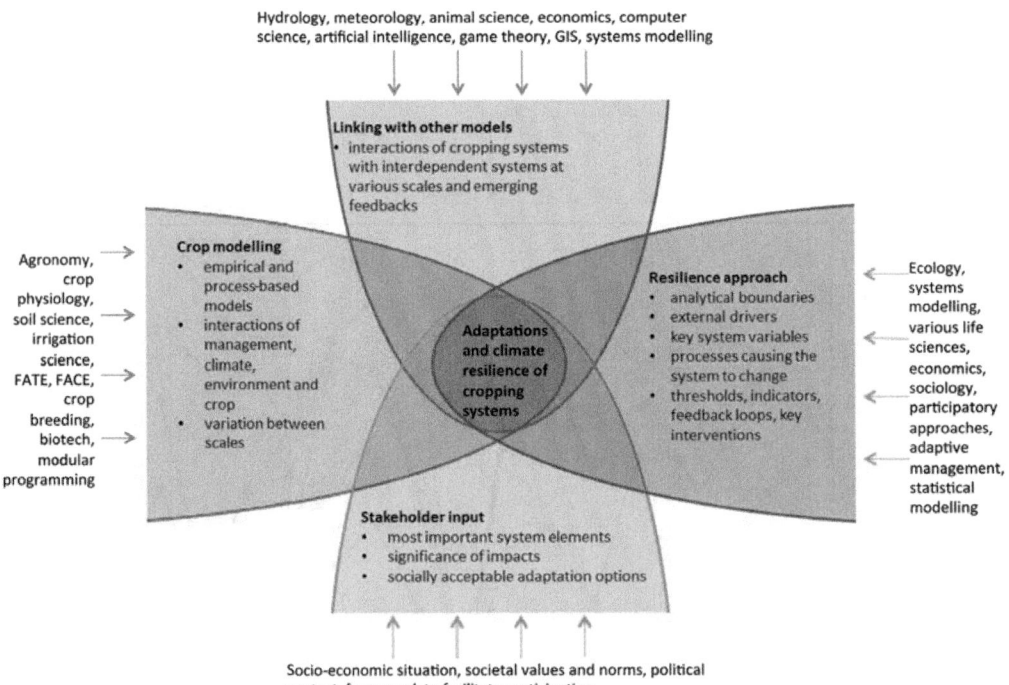

Fig. 13.2. Conceptual assessment framework to identify and evaluate the potential of management interventions to enhance the climate resilience of cropping systems. The four key components (resilience theory, crop models, models of interdependent systems linked to crop models and meaningful stakeholder input and involvement) and their primary element constitute the assessment framework, and are shown together with their underlying scientific basis and societal context for stakeholder input.

enable a preliminary analysis of specific system resilience at the farm level, whereas in the second, an empirical modelling approach is applied to identify the most efficient use of diversity to build climate-specific resilience, as well as general resilience at subnational scale. We focus on the methodological aspects of these studies that can support adaptation decision making for improved climate resilience.

13.3.1 Example 1

The impact of shifting crop shares in common rotations in the Midwest USA and the resulting changes in crop productivity and sowing dates on erosion and soil loss under future climatic conditions was evaluated using a combination of climate, economic, crop and erosion models (O'Neal et al., 2005). This study is interesting on a number of counts and represents the direction modelling studies should increasingly pursue to provide the relevant scientific information that is required to evaluate options to enhance climate resilience. The crop rotations investigated in this study were selected based on estimates of their economic profitability adjusted for future climatic conditions. In 10 out of 11 locations analysed, erosion and soil loss increased from present-day levels, as expected from increased precipitation, which leads to lower soil water infiltration rates and, consequently, increases in overland flow and erosion. However, erosion estimates were also affected by expected changes in rotations and crop productivity: maize yields were found to decrease in response to periods of water and temperature stress near silking (see, for example, Lobell et al., 2013), leading to reduced ground cover and increased erosion with late summer rainfalls. Sowing dates were selected to give optimal yields, which led to later planting for both maize and soybean, whereas the later sowing date resulted in increased soil loss due to decreased ground cover during the intense April/May rains. A final note regarding their modelling approach is its heuristic value; their results suggested decreased erosion with a soybean monocrop (higher ground cover but lower canopy cover than with maize canopy), which was contrary to the results of field trials (Laflen and Colvin, 1981), thus leading the authors to suggest that the erosion model they used, i.e. the Water Erosion Prediction Project-Carbon Dioxide (WEPP-CO2) model, needed to be improved to account better for canopy architecture (O'Neil et al., 2005). The results of the study investigating yield response alone would have come to different conclusions about the success of adaptation, and it was only through the combined use of different models – and knowledge from different disciplines – that a more nuanced analysis accounting for feedbacks could emerge. Further improvements to their methodology to enable the evaluation of resilience would be to include feedbacks of the changed patterns of soil loss and water infiltration to the crop models under various climate scenarios to assess the impacts on crop productivity, and to consider the impacts of using resilient varieties (see, for example, Hakala et al., 2012). Of course, the major deficiency with the study as an example of identification of adaptation to enhance resilience is the absence of information about stakeholder input. Similar examples of studies using multiple agroecosystem and economic models in the analyses have been reported elsewhere (e.g. Eckersten et al., 2001; Mandryk et al., 2012).

13.3.2 Example 2

The second example deals with the need to reconsider cropping from the viewpoint of increased variability and more frequent anomalies with which a farmer must cope. Weather variability and extremes are related directly to climate change, but product and input price volatility and shifting production levels of bioenergy crops are also influenced by climate change. To facilitate addressing these challenges related to climate variability, an empirical modelling approach was developed (Kahiluoto et al., 2014). This generic approach can be applied to manage for resilience through a targeted diversification of responses to the most critical change factors. Such a targeted diversification can concern crop cultivars, crops, cropping systems and/or land use by farmers, even farm activities, retail suppliers and so forth, to manage system response to weather or price volatility, for example. The approach was exemplified by quantifying the response diversity of different barley cultivars to critical weather events in the 16 Finnish regions during the most recent decades. A positive relationship between barley cultivar diversity and regional yield was established (Fig. 13.3), with the strongest correlations emerging with precipitation anomalies (Himanen et al., 2013a), thus illustrating the potential benefit of cultivar diversity in stabilizing yields. Therefore, to test precisely the benefit of the hypothetically most relevant diversity, the sensitivity of yield of Finnish barley cultivars to various agroclimatic parameters was determined, and a number of bottlenecks for the adaptive capacity of the available set of cultivars were identified for the present and a range of future scenarios (Hakala et al., 2012). Furthermore, a model was developed for the response of cultivars to the agroclimatic parameters using multi-annual, multi-location trials and validated with farm data. In a subsequent analysis, regional cultivar diversity in response to weather was quantified and compared with mere cultivar diversity. It turned out that while cultivar diversity in all regions had increased rapidly, the response diversity showed a declining trend during the latter half of the period. The striking difference between mere cultivar diversity and the response diversity demonstrated the added value of the approach. A generic approach for quantification of response diversity as a tool to manage for resilience was proposed (Kahiluoto et al., 2014). In a follow-up study, the approach was applied to developing tools to facilitate cultivar choices by farmers and other actors in the food and fodder industry. Furthermore, it served as a basis for discussions between various actors, breeders and the Finnish Security Authority to contribute to social learning and building the resilience of Finland's cropping systems.

13.4 Challenges for Modelling

The two case studies presented here illustrate the potential of process-based and statistical models to identify opportunities to build system resilience, as well as building adaptive capacity and general resilience by promoting learning and building social capital (see also Ebi et al., 2011). None the less, several challenges exist. One type of challenge is for the more effective use of process-based crop models in evaluating management adaptations. Rötter et al. (2011a) have summarized some of the limitations of crop models to assess adequately the impacts of future climates on crop productivity, with the recommendation to utilize better the potential of the multi-model ensemble simulation techniques – as already practised for more than a decade by the climate modelling community – found to be promising in model intercomparisons (e.g. Palosuo et al., 2011; Rötter et al., 2012; Asseng et al., 2013; Rosenzweig et al., 2013). Improved awareness and quantitative information about risks and uncertainty and their sources has paved the way for concerted efforts to working on distinct options for impact model improvement (see, for example, Asseng et al., 2011, 2013; Boote et al., 2013; Lobell et al., 2013; Rötter et al., 2013b). Another type of challenge is to combine empirical statistical approaches better with process-based model simulations to examine and identify the underlying mechanisms for (strong) statistical relationships found by empirical models (e.g. van Oort et al., 2012; Lobell et al., 2013). In recognition of these challenges, concerted efforts have been emerging at the international level, exemplified most clearly by the Agricultural Model Intercomparison and Improvement Project (AgMIP) (Asseng et al., 2013; Rosenzweig et al., 2013) and the Modelling European Agriculture with Climate Change for Food Security project (MACSUR; www.macsur.eu) (Rötter et al., 2013b).

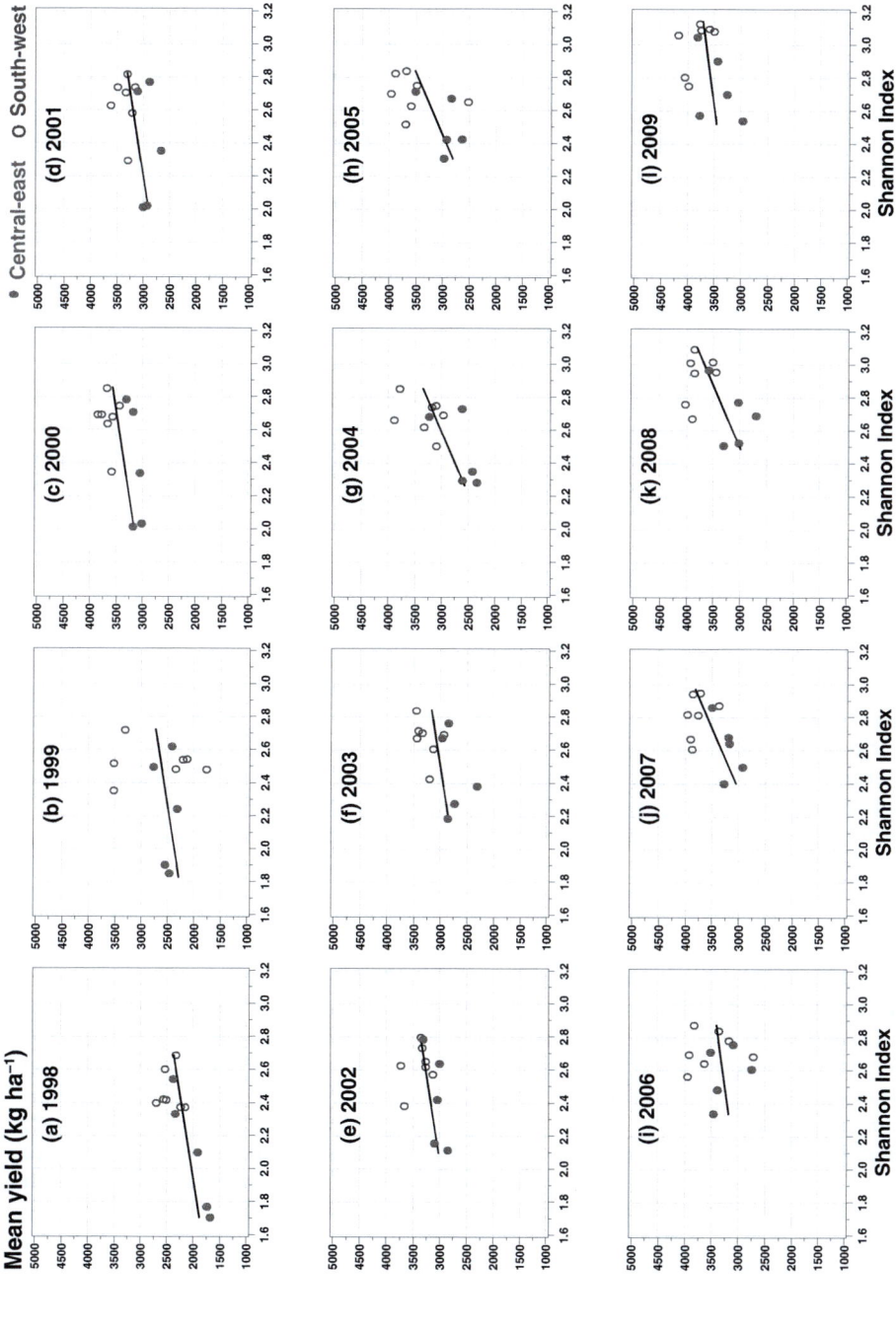

Fig. 13.3. Linear relationships between regional mean yield of feed barley and the Shannon Index describing cultivar diversity in the south-west and central-east regions of Finland for 1998–2009. The relation was strongest in years with a rainy growing season (2004, 2005, 2007 and 2008). The differences in the Shannon Index and in the cultivated area between the south-west and central-east regions accounted for the difference in mean yield between the two regional groups (source: Himanen et al., 2013a).

However, there are several other challenges to making crop modelling studies of adaptation in cropping systems relevant and able to contribute to increasing the climate resilience of cropping systems (Meinke et al., 2009). For example, what are the most important drivers of change in cropping systems and what are plausible alternative ways this situation is likely to evolve? Which potential adaptations should be evaluated and how can crop modelling studies capture the reality of current adaptations and constraints faced by farmers? How can differences in adaptation options evaluated at different scales be presented to policy makers and other decision makers so as to capture their impacts on food security, poverty reduction and the environment at other scales or related systems? What role can crop models and/or their combinations with (farm) economic models play in studying transformational change (e.g. Mandryk et al., 2012) as opposed to incremental and marginal changes that are more easily and, therefore, most frequently handled by models (Rötter et al., 2011b; White et al., 2011; Rickards and Howden, 2012)? The solution to these questions lies beyond the realm of crop modelling expertise and highlights the need for increased interaction between different disciplinary experts and the stakeholders affected (see, for example, Ewert et al., 2009, 2011). Failing to address these challenges will likely introduce bias in climate change adaptation studies as only easy to model options will be evaluated, while the reality of high uncertainty and increased volatility and extremes associated with future conditions requires as many options as possible to be available for farmers (Vermeulen et al., 2012; White et al., 2011). It has been suggested that the seeming impossibility to represent adequately the reality in which farmers operate may best be overcome by directly involving farmers themselves in modelling research (McCown, 2001; McCown et al., 2009; Patt et al., 2010).

13.5 Conclusions

Both statistical and empirically founded process-based crop climate models have the potential, and are needed, to enable science to support the enhancement of cropping system resilience. By integrating many factors, these models allow for *ex ante* evaluation of the performance of complex cropping systems under a broad range of plausible future conditions. While crop models are limited in their current capabilities to simulate yield under suboptimal management, widespread recognition of this has led to concerted efforts at the international scale. Integration of these improved crop models with farming systems, hydrology and market models will enable the analysis and assessment of cropping system resilience within the context of the larger systems of farms, watersheds, regions and the local to global markets in which they are embedded (Wolf et al., 2012), and potentially allow transformational adaptations to be addressed (Mandryk et al., 2012). Furthermore, both conceptual and empirically founded quantitative models could be tools to support learning and thus build the capacity of farmers to manage the resilience of their cropping systems, and of administrators, politicians and scientists to enable it. The use of models as management tools by farmers, politicians and other actors in agri-food systems has the potential to build social resilience by promoting learning (van Paassen et al., 2007) and experimentation (McCown, 2001; Meinke et al., 2009), and may ultimately improve science if their knowledge can be integrated with modelling studies and by considering the constraints and dynamics faced by farmers (McCown et al., 2009).

References

Asseng, S., Foster, I. and Turner, N.C. (2011) The impact of temperature variability on wheat yields. *Global Change Biology* 17, 997–1012.

Asseng, S., Ewert, F., Rosenzweig, C., Jones, J., Hatfield, J., Ruane, A., *et al.* (2013) Uncertainties in simulating wheat yields under climate change. *Nature Climate Change* 3, 827–832.

Bakker, M.M., Govers, G., Ewert, F., Rounsevell, M. and Jones, R. (2005) Variability in regional wheat yields as a function of climate, soil and economic variables: assessing the risk of confounding. *Agriculture, Ecosystems and Environment* 110, 195–209.

Boote, K., Jones, J., White, J.W., Asseng, S. and Lizaso, J.I. (2013) Putting mechanisms into crop production models. *Plant, Cell and Environment* 36, 1658–1672.

Challinor, A.J., Simelton, E.S., Fraser, E.D., Hemming, D. and Collins, M. (2010) Increased crop failure due to climate change: assessing adaptation options using models and socio-economic data for wheat in China. *Environmental Research Letters* 5, 034012.

Coumou, D. and Rahmsdorf, S. (2012) A decade of weather extremes. *Nature Climate Change* 2, 491–496.

Dow, K., Berkhout, F., Prestion, B.L., Klein, R.J.T., Midgeley, G. and Shaw, R.M. (2013) Limits to adaptation. *Nature Climate Change*, 3, 305–307.

Ebi, K.L., Padgham, J., Doumbia, M., Kergna, A., Smith, J., Butt, T., *et al.* (2011) Smallholders adaptation to climate change in Mali. *Climatic Change* 108, 423–436.

Eckertsen, H., Blombäck, K., Kätterer, T. and Nyman, P. (2001) Modeling C, N, water and heat dynamics in winter wheat under climate change in southern Sweden. *Agriculture Ecosystems and Environment* 142, 6–17.

Elsgaard, L., Børgesen, C.D., Olesen, J.E., Siebert, S., Ewert, F., Peltonen-Sainio, P., *et al.* (2012) Shifts in comparative advantages for maize, oat and wheat cropping under climate change in Europe. *Food Additives and Contaminants: Part A* 29, 1514–1526.

Ewert, F., Rounsevell, M.D.A., Reginster, I., Metzger, M.J. and Leemans, R. (2005) Future scenarios of European agricultural land use I. Estimating changes in crop productivity. *Agriculture, Ecosystems and Environment* 107, 101–116.

Ewert, F., van Ittersum, M.K., Bezlepkina, I., Therond, O., Andersen, E., Belhouchette, H., *et al.* (2009) A methodology for enhanced flexibility of integrated assessment in agriculture. *Environmental Science and Policy* 12, 546–561.

Ewert, F., van Ittersum, M.K., Heckelei, T., Therond, O., Bezlepkina, I. and Andersen, E. (2011) Scale changes and model linking methods for integrated assessment of agri-environmental systems. *Agriculture, Ecosystems and Environment* 142, 6–17.

Fischer, G., Shah, M., Tubiello, F.N. and van Velhuizen, H. (2005) Socio-economic and climate change impacts on agriculture: an integrated assessment, 1990–2080. *Philosophical Transactions of the Royal Society B: Biological Sciences* 360, 2067–2083.

Folke, C. (2006) Resilience: the emergence of a perspective for social–ecological systems analyses. *Global Environmental Change* 16, 253–267.

Hakala, K., Jauhiainen, L., Himanen, S.J., Rötter, R., Salo, T. and Kahiluoto, H. (2012) Sensitivity of barley varieties to weather in Finland. *Journal of Agricultural Science* 150, 145–160.

Hatfield, J.L., Boote, K.J., Kimball, B., Ziska, L., Izaurralde, R.C., Ort, D., *et al.* (2011) Climate impacts on agriculture: implications for crop production. *Agronomy Journal* 103, 351–370.

Himanen, S.J., Hakala, K. and Kahiluoto, H. (2013a) Crop responses to climate and socio-economic change in northern regions. *Regional Environmental Change* 13, 17–32.

Himanen, S.J., Ketoja, E., Hakala, K., Rötter, R.P., Salo, T. and Kahiluoto, H. (2013b) Cultivar diversity has great potential to increase yield of feed barley. *Agronomy for Sustainable Development* 33, 519–530.

Holling, C.S. (1973) Resilience and stability of ecological systems. *Annual Review of Ecology and Systematics* 4, 1–23.

Janssen, S. and van Ittersum, M.K. (2007) Assessing farm innovations and responses to policies: a review of bio-economic farm models. *Agricultural Systems* 94, 622–636.

Kahiluoto, H., Kaseva, J., Hakala, K., Himanen, S.J., Jauhiainen, L., Rötter, R.P., Salo, T. and Trnka, M. (2014). Cultivating resilience by empirically revealing response diversity. *Global Environmental Change* (in press), DOI 10.1016/j.gloenvcha.2014.02.002.

Laflen, J. and Colvin, T. (1981) Effect of crop residue on soil loss from continuous row cropping. *Transactions of the ASAE* 24, 604–609.

Lobell, D.B., Schlenker, W. and Costa-Roberts, J. (2011) Climate trends and global crop production since 1980. *Science* 333, 616–620.

Lobell, D.B., Hammer, G.L., McLean, G., Messina,

C., Roberts, M.J. and Schlenker, W. (2013) The critical role of extreme heat for maize production in the United States. *Nature Climate Change* 3, 497–501.

McCown, R. (2001) Learning to bridge the gap between science-based decision support and the practice of farming: evolution in paradigms of model-based research and intervention from design to dialogue. *Crop and Pasture Science* 52, 549–572.

McCown, R.L., Carberry, P.S., Hochman, Z., Dalgliesh, N.P. and Foale, M.A. (2009) Re-inventing model-based decision support with Australian dryland farmers. Changing intervention concepts during 17 years of action research. *Crop and Pasture Science* 60, 1017–1030.

Mandryk, M., Reidsma, P. and van Ittersum, M.K. (2012) Scenarios of long-term farm structural change for application in climate change impact assessment. *Landscape Ecology* 27, 509–527.

Meinke, H., Howden, S.M., Struik, P.C., Nelson, R., Rodriguez, D. and Chapman, S.C. (2009) Adaptation science for agriculture and natural resource management – urgency and theoretical basis. *Current Opinion in Environmental Sustainability* 1, 69–76.

Müller, C., Cramer, W., Hare, W.L. and Lotze-Campen, H. (2011) Climate change risks for African agriculture. *Proceedings of the National Academy of Sciences* 108, 4313–4315.

Nelson, D.R. (2011) Adaptation and resilience: responding to a changing climate. *Wiley Interdisciplinary Reviews: Climate Change* 2, 113–120.

Nelson, D.R., Adger, W.N. and Brown, K. (2007) Adaptation to environmental change: contributions of a resilience framework. *Annual Review of Environmental Resources* 32, 395–419.

Nelson, G.C., Rosegrant, M.W., Koo, J., Robertson, R., Sulser, T., Zhu, T., et al. (2009) Impact on agriculture and costs of adaptation. *IFPRI Food Policy Report*, Washington, DC.

O'Neal, M.R., Nearing, M., Vining, R.C., Southworth, J. and Pfeifer, R.A. (2005) Climate change impacts on soil erosion in Midwest United States with changes in crop management. *Catena* 61, 165–184.

Osborne, T., Rose, G. and Wheeler, T. (2013) Variation in the global-scale impacts of climate change on crop productivity due to climate model uncertainty and adaptation. *Agricultural and Forest Meteorology* 170, 183–194.

Palosuo, T., Kersebaum, K.C., Angulo, C., Hlavinka, P., Moriondo, M., Olesen, J.E., et al. (2011) Simulation of winter wheat yield and its variability in different climates of Europe: a comparison of eight crop growth models. *European Journal of Agronomy* 35, 103–114.

Palosuo, T., Rötter, R.P., Lehtonen, H., Virkajärvi, P. and Salo, T. (2013) How to assess climate change impacts on farmers' crop yields? Proceedings, Impacts World 2013: International Conference on Climate Change Effects, Potsdam, 27–30 May 2013 (http://www.climate-impacts-2013.org/files/hcaw_palosuo.pdf, accessed 5 December 2013).

Patt, A.G., van Vuuren, D.P., Berkhout, F., Aaheim, A., Hof, A.F., Isaac, M., et al. (2010) Adaptation in integrated assessment modeling: where do we stand? *Climatic Change* 99, 383–402.

Porter, J.R. and Gawith, M. (1999) Temperatures and the growth and development of wheat: a review. *European Journal of Agronomy* 10, 23–36.

Reidsma, P. and Ewert, F. (2008) Regional farm diversity can reduce vulnerability of food production to climate change. *Ecology and Society* 13, 38.

Reilly, J., Tubiello, F., McCarl, B., Abler, D., Darwin, R., Fuglie, K., et al. (2003) US agriculture and climate change: new results. *Climatic Change* 57, 43–67.

Rickards, L. and Howden, S.M. (2012) Transformational adaptation: agriculture and climate change. *Crop and Pasture Science* 63, 240–250.

Rosenzweig, C., Jones, J.W., Hatfield, J.L., Ruane, A.C., Boote, K.J., Thorburn, P., et al. (2013) The Agricultural Model Intercomparison and Improvement Project (AgMIP): protocols and pilot studies. *Agricultural and Forest Meteorology* 170, 166–182.

Rötter, R. and van Keulen, H. (1997) Variations in yield response to fertilizer application in the tropics: II. Risks and opportunities for smallholders cultivating maize on Kenya's arable land. *Agricultural Systems* 53, 69–95.

Rötter, R., van Keulen, H. and Jansen, M. (1997) Variations in yield response to fertilizer application in the tropics: I. Quantifying risks and opportunities for smallholders based on crop growth simulation. *Agricultural Systems* 53, 41–68.

Rötter, R.P., Carter, T.R., Olesen, J.E. and Porter, J.R. (2011a) Crop-climate models need an overhaul. *Nature Climate Change* 1, 175–177.

Rötter, R.P., Palosuo, T., Pirttioja, N.K., Dubrovsky, M., Salo, T., Fronzek, S., et al. (2011b) What would happen to barley production in Finland if global warming exceeded 4 C? A model-based assessment. *European Journal of Agronomy* 35, 205–214.

Rötter, R.P., Palosuo, T., Kersebaum, K.C., Angulo, C., Bindi, M., Ewert, F., *et al.* (2012) Simulation of spring barley yield in different climatic zones of Northern and Central Europe: a comparison of nine crop models. *Field Crops Research* 133, 23–36.

Rötter, R.P., Höhn, J.G., Trnka, M., Fronzek, S., Carter, T.R. and Kahiluoto, H. (2013a) Modelling shifts in agroclimate and crop cultivar response under climate change. *Ecology and Evolution* 3, 4197–4214.

Rötter, R., Ewert, F., Palosuo, Taru, Bindi, M., Kersebaum, K.C., Olesen, J.E., *et al.* (2013b) Challenges for agro-ecosystem modeling in climate change risk assessment for major European crops and farming systems. Proceedings, Impacts World 2013: International conference on climate change effects, Potsdam, 27–30 May 2013 (http://www.climate-impacts-2013.org/files/wism_roetter_1.pdf, accessed 5 December 2013).

Rummukainen, M. (2012) Changes in climate and weather extremes in the 21st century. *WIREs Climate Change* 3, 115–129.

Sacks, W.J. and Kucharik, C.J. (2011) Crop management and phenology trends in the US Corn Belt: impacts on yields, evapotranspiration and energy balance. *Agriculture and Forest Meteorology* 151, 882–894.

Schlenker, W. and Lobell, D.B. (2010) Robust negative impacts of climate change on African agriculture. *Environmental Research Letters* 5, 014010.

Schlenker, W. and Roberts, M.J. (2009) Nonlinear temperature effects indicate severe damages to US crop yields under climate change. *Proceedings of the National Academy of Sciences* 106, 15594–15598.

Siebert, S. and Ewert, F. (2012) Spatio-temporal patterns of phenological development in Germany in relation to temperature and day length. *Agricultural and Forest Meteorology* 152, 44–57.

Simelton, E., Fraser, E.D., Termansen, M., Benton, T.G., Gosling, S.N., South, A., *et al.* (2012) The socio-economics of food crop production and climate change vulnerability: a global scale quantitative analysis of how grain crops are sensitive to drought. *Food Security* 4, 163–179.

Smith, P. and Olesen, J.E. (2010) Synergies between the mitigation of, and adaptation to, climate change in agriculture. *Journal of Agricultural Sciences* 148, 543–552.

Tebaldi, C. and Knutti, R. (2007) The use of the multi-model ensemble in probabilistic climate projections. *Philosophical Transactions of the Royal Society A-Mathematical Physical and Engineering Sciences* 365, 2053–2075.

Trnka, M., Olesen, J.E., Kersebaum, K.C., Skjelvag, A.O., Eitzinger, J., Seguin, B., *et al.* (2011) Agroclimatic conditions in Europe under climate change. *Global Change Biology* 17, 2298–2318.

van Ittersum, M., Leffelaar, P., van Keulen, H., Kropff, M., Bastiaans, L. and Goudriaan, J. (2003) On approaches and applications of the Wageningen crop models. *European Journal of Agronomy* 18, 201–234.

van Oort, P., Timmermans, B., Meinke, H. and van Ittersum, M. (2012) Key weather extremes affecting potato production in the Netherlands. *European Journal of Agronomy* 37, 11–22.

van Paassen, A., Rötter, R.P., van Keulen, H. and Hoanh, C.T. (2007) Can computer models stimulate learning about sustainable land use? Experience with LUPAS in the humid (sub-) tropics of Asia. *Agricultural Systems* 94, 874–887.

Vermeulen, S.J., Aggarwal, P., Ainslie, A., Angelone, C., Campbell, B.M., Challinor, A., *et al.* (2012) Options for support to agriculture and food security under climate change. *Environmental Science and Policy* 15, 136–144.

Vermeulen, S.J., Challinor, A.J., Thornton, P.K., Campbell, B.M., Eriyagama, N., Vervoort, J.M., *et al.* (2013) Addressing uncertainty in adaptation planning for agriculture. *Proceedings of the National Academy of Sciences* 110, 8357–8362.

White, J.W., Hoogenboom, G., Kimball, B.A. and Wall, G.W. (2011) Methodologies for simulating impacts of climate change on crop production. *Field Crops Research* 124, 357–368.

Wolf, J., Reidsma, P., Schaap, B., Mandryk, M., Kanellopoulos, A., Ewert, F., *et al.* (2012) Assessing the adaptive capacity of agriculture in the Netherlands to the impacts of climate change under different market and policy scenarios (AgriAdapt project). Dutch National Research Programme Climate Changes Spatial Planning, Wageningen, the Netherlands.

14 Agroforestry Solutions for Buffering Climate Variability and Adapting to Change

Meine van Noordwijk,[1] Jules Bayala,[1] Kurniatun Hairiah,[2] Betha Lusiana,[1] Catherine Muthuri,[1] Ni'matul Khasanah[1] and Rachmat Mulia[1]

[1]World Agroforestry Centre (ICRAF), Bogor, Indonesia;
[2]Brawijaya University, Malang, Indonesia

14.1 Introduction

This chapter will focus on increasing the adaptive capacity of agricultural systems in tropical and subtropical regions through agroforestry. Agroforestry as a concept resists and tries to counteract the way agriculture has been segregated from forests and forestry. Understanding, using and improving agroforestry implies a focus on the interactions between trees, annual crops and domestic stock, given the local abiotic factors of climate, soils, water and nutrient balances, as well as the biotic context (pests, diseases, antagonists, predators, pollinators and dispersal agents), and the use of land, external inputs, labour and knowledge. We pose and review the hypothesis that the presence of trees increases the degree of buffering of climate variability from the perspective of an annual food crop, and that retention and the increase of trees in agricultural landscapes can be a relevant part of climate change adaptation strategies.

14.1.1 Intuitive appeal of linking trees to climate

People associate climate issues with trees. Tree planting as a ceremonial activity has intuitive appeal in the context of climate change and is popular among politicians who want to show that they are not just talking about climate, but are willing to act. At the micro-scale, this is a logical association, as we seek the shade of trees on a hot day, seek shelter under trees if surprised by a rainstorm (but some know that deep-rooted trees attract lightning), select tree-covered roads to cycle against the wind (if having grown up in a bicycle culture) and prefer trees around our houses to buffer both the heat of summer (or the day) and the cold of winter (or the night). Yet, trees have mostly been discussed in climate change in terms of their carbon storage and the contributions they make to the global carbon balance (Watson et al., 2000). Their more direct effect on micro- and mesoclimate is largely absent from the climate change debate, including that involving agriculture. This chapter will argue that important adaptive opportunities are missed if we continue to ignore trees in that context, as in a recent report by Beddington et al. (2011) and an overview by Vermeulen et al. (2012). The intuitive appeal of trees and tree planting may have to be channelled towards adaptation, however, rather than mitigation discussions of climate change (van Noordwijk et al., 2011a).

Path dependency of the international climate change discussion has led to a 'firewall' between the concepts and financing mechanisms of mitigation and adaptation.

When there was hope that mitigation efforts could contain global climate change to the level that adaptation would not be needed, this distinction made sense. However, the inadequacy of mitigation efforts implies that we are substantially beyond that option. Synergy of mitigation and adaptation actions across all sectors is needed, but in land use it has always been a very logical option (Verchot et al., 2007; van Noordwijk et al., 2011a; Matocha et al., 2012). Recent support for agroforestry as a part of agricultural adaptation strategies (Schoeneberger et al., 2012; Munang et al., 2013) is moving in this direction as well, as is the interest in landscape approaches that will drive towards the integration of sectors and issues.

14.1.2 Absence of trees in current climate science

Climate science has been cognizant of the substantial impacts of trees on wind speed, humidity, temperature and even rainfall, but has chosen to standardize its measurement effort on open-field situations (above a short grass cover), where such 'disturbing' factors are minimized. Thus, all climate maps and all climate models calibrated on synoptic weather station data refer implicitly to a tree-less landscape. There has been considerable discussion on the 'urban heat island' effect in less green environments (Arnfield, 2003), but less on its counterpart 'cool forest' effect that complements it (Bonan, 2008). Local topography influences microclimatic differences in interaction with vegetation, as has been recognized by some 'climate-smart' landscape designs (Bonan, 2002). However, there have to date been no attempts, to our knowledge, to show climate maps of the same area with different degrees of tree cover, obtained by adding the microclimatic variation caused by tree cover to the open-field climate data and their predicted patterns of change from global circulation models (GCMs). From the data reviewed in this chapter, it is suggested that the impacts of local tree cover on major weather variables at local scales is substantial – and that variations in tree cover in an agricultural landscape may, for the next 30 years or so, exceed the predicted patterns of climate change on key climatic variables for many locations (van Noordwijk et al., 2011a). A more directly empirical approach to the combination of local and global drivers of local climate change has been promoted by Pielke et al. (2007) and Bonan (2008). In the discussions of forest and climate, the net effects of changes in surface albedo and evaporative cooling imply that deforestation has a warming effect in the tropics, but can have a cooling effect at mid and high latitudes (Jackson et al., 2008; Swann et al., 2012).

14.1.3 Agroforestry as history and future of agricultural land use

Historically, agriculture in many parts of the world was compatible with the retention of valuable trees in cropped fields, and its impact on deforesting the world was gradual (Williams, 2006). It used only superficial soil tillage, usually in combination with a controlled fire that cleared the land but did not kill the larger trees (Cairns, 2007). In temperate zones with relatively mild climates, however, a different approach to growing crops emerged, 'non-conservation agriculture without trees', which was successful as it was readily scaled up, with horse-drawn ploughs replacing human tillage, and tractors with ever-more horse power, requiring the clearing of hedgerows and trees on field boundaries to make larger fields, drawing ever-deeper ploughs through a soil that responded by mineralizing a substantial part of its organic matter, feeding the crops. This yield benefit, however, was not sustainable, as it depleted the resource base – chemical fertilizer had to become the basis of plant nutrition. As tillage had killed many of the worms and other soil engineers, tillage became 'necessary' to create a structure compatible with crop roots. The trouble started when this tree-less, tillage-addicted form of agriculture became the norm, became known and taught worldwide as what agriculture was and should be, and was extended to parts of the world with less benign climates. Conservation agriculture in landscapes with trees, maintaining soil organic matter content, is now seen as 'climate-smart' agriculture (McCarthy et al., 2011). An important part of the solutions sought are part of a landscape agroforestry approach.

The term 'agroforestry' was coined in the mid-1970s when the 'green revolution' experience and debate had made clear that its perspective on intensifying crop production worked well in specific environments but not elsewhere (King, 1979). While a parallel approach to large-scale plantation forestry had success in some areas, it ran into major social conflicts and issues over land rights elsewhere. The idea that crops and trees were not necessarily incompatible was revolutionary for academically trained agronomists, while trained foresters had a hard time in seeing that the local people were not their major problem. In many parts of the tropics, these perspectives appeared to be self-evident, if only one took a good look around. Trees and crops, farmers and forest could somehow work together.

Yet advances in understanding the biophysical (Ong and Huxley, 1996), ecological, social and economic aspects of tree–soil–crop interactions were slow to become mainstreamed in the world of 'development' and 'modernization'. New forms of agroforestry, compatible with mechanization and focused on trees of high value, finally emerged in Europe, North America and Australia (Gordon and Newman, 1997; Eichhorn et al., 2006; Gold and Garrett, 2009) – challenging the rules and regulations that had been made on the concept of segregating trees and crops.

The climatic effects of trees vary with latitude, partly because the solar angle matters for the shade effect, while the windbreak effect operates at a near-zero angle of incidence at any latitude. Tree cover in agricultural landscapes, however, varies within the tropics as well. The scope of the additional adaptive benefits of enhancing agroforestry will be highest in areas that are currently low in tree cover, while in others retention is the first target. Analysis by Zomer et al. (2009) indicated that, globally, 46% of land classified as agricultural contained at least 10% tree cover, and in areas such as South-east Asia or Meso-America, half the agricultural land had at least 30% tree cover – sufficient to be classified as forest if it were not for the definitions of forestry that made agricultural use a disqualifying condition for land to be recorded as forest (de Foresta et al., 2013).

14.2 Supply and Demand of Buffering Functions in the Landscape

In the parklands of the Sahel (Africa), farmers have retained trees that provide edible fruits and grow their grain crops in between and underneath trees, despite the shortages of water that regularly occur during the growing season (Breman and Kessler, 1997; Boffa, 1999; Takimoto et al., 2008). Tree–soil–crop interactions in these systems, combining buffering and productivity, are a complex mixture of positive and negative effects, above and below ground (Kho, 2000; Kho et al., 2001).

The microclimatic effects of agroforestry have been studied through analysis of the energy and water balance for more than two decades (Monteith et al., 1991; Ong et al., 1991; Brenner, 1996). Analysis by Ong et al. (1991) suggested that atmospheric interactions in hedgerow cropping in the semi-arid tropics were positive but were of minor importance compared with below-ground, often competitive, interactions. Rao et al. (1997) concluded that the net positive effects of trees on crops were more likely in sequential rather than simultaneous agroforestry systems, as below-ground competition dominated tree–crop interactions for major food crops. When the focus of agroforestry research shifted from intensively mixed crop–hedgerow systems to field-boundary plantings of commercially important timber trees such as *Grevillea robusta*, a more positive evaluation of net above- plus below-ground interactions followed, even where maize was the dominant food crop (Ong et al., 2000). While the relative importance of below-ground relationships is not contested (van Noordwijk et al., 2004a), current interest in the climate change sensitivity of crops may make temperature shifts of the order of 2°C more relevant than they may have appeared in the past (Vermeulen et al., 2012): 'Agricultural production is highly vulnerable even to 2°C (low-end) predictions for global mean temperatures in 2100, with major implications for rural poverty and for both rural and urban food security'.

Active and flexible shade management has a long history in major commercial tree crops such as tea, coffee and cacao (Beer, 1987; Beer et al., 1997; Lin, 2007; Tscharntke et al., 2011). Appreciation of trees for their microclimatic effects in sylvopastoral agroforestry systems has been clear. Tree shade in relation to the daily cycle of livestock activity and movement can reduce direct heat exposure of the animals, and hence their sensitivity to climate variability (Thornton et al., 2009). Ongoing pasture intensification that removes trees, however, increases exposure and vulnerability (Harvey et al., 2011). Buffering functions are being compromised, just as they are likely to be needed more. Landscape multifunctionality, as contrasted with specialized monocultures, can be analysed in terms of the provision of and need for buffering across the five capitals of livelihood systems (Bebbington, 1999).

The buffer concept can be further analysed as assisting in persistence in the face of undesirable fluctuations, as well as innovations and sustainability (Jackson et al., 2010, 2012) in the face of directional change. A simple analytical frame for assessing vulnerability, supply and demand for buffer functions (van Noordwijk et al., 2011a) suggests that a true interdisciplinary effort is needed to build on current strength and address the interactive effects of climatic, demographic and socio-economic change in the ever-changing (inter)national policy environment. A positive appreciation of the multifunctional landscapes and agroforestry systems that have allowed smallholders to cope with external pressures is needed as a step to maintaining and supporting resilient socio-ecological systems for the (near) future.

Beyond a qualitative initial survey of buffer functions within multifunctional landscapes and livelihood systems, trends in the supply and demand for buffering can be assessed. Climate change and loss of buffer functions interact in increasing vulnerability, but buffer functions can be influenced locally and climate change requires global agreements that are extremely slow to emerge from complex negotiations. It is therefore logical that the 'adaptation' and maintenance or restoration of landscape buffer functions receives increased attention. A further quantification of buffering is based on the definition:

Buffering = 1 − Variation-with/Variation-without

For river flow, a more specific buffering factor has been based on the temporal autocorrelation of rivers (van Noordwijk et al., 2011b); for a buffering factor of zero, daily river flow is fully random; for buffering factor one, it is perfectly constant, regardless of rainfall. For nutrients, buffering relates to the temporal pattern of crop needs and soil supply, exposing nutrients to leaching and atmospheric losses under temporary excess conditions (van Noordwijk and Cadisch, 2002). A direct role by which trees can influence the buffering of soil water supply is by increasing the effective soil depth in which water can be stored during rainfall excess in the early part of the growing season for later use (Fig. 14.1) – as will be discussed in more detail below.

For economic portfolio effects in risk reduction, the degree of temporal correlation between commodity yields and/or prices determines combined variance, and thus buffer effects. Maintaining a diversified portfolio of activities is a safe and time-tested approach to reduce the risks (van Noordwijk et al., 1994). The inclusion of trees that provide annual harvests of fruits and/or long-term, high-value timber can reduce risk, even if the trade-off in resource capture is essentially neutral (Santos-Martin and van Noordwijk, 2011).

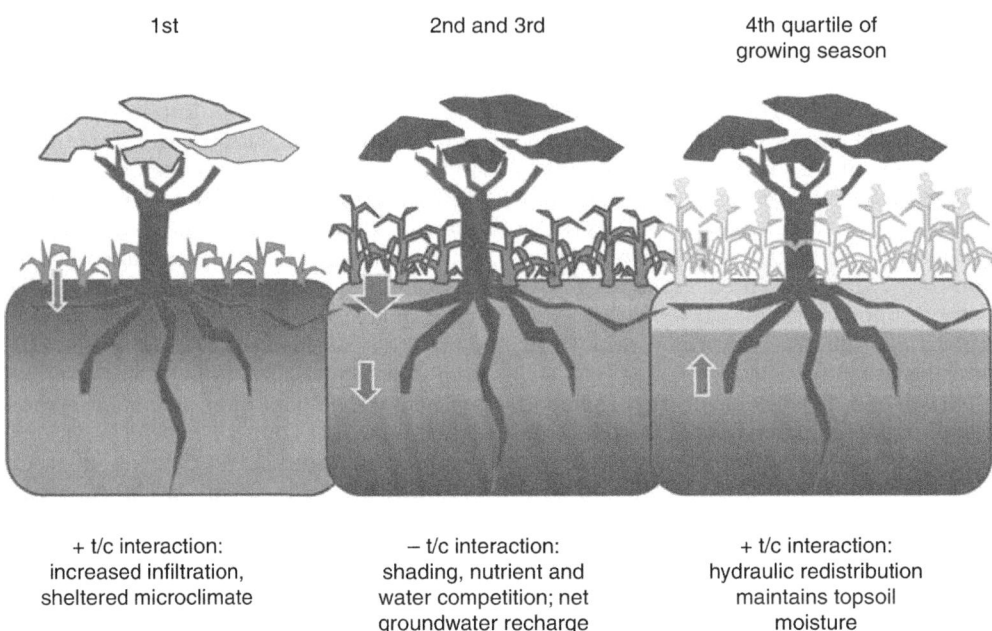

Fig. 14.1. Schematic representation of + and − tree–crop interactions (t/c) during a growing season based on the effects on infiltration, hydraulic redistribution, shading, crop phenology and harvest index.

14.3 Trees Modifying Wind Speed

The older literature (Caborn, 1965; Forman, 1990) has already established the direct utility of using shelterbelts and windbreaks to modify microclimates, with implications for landscape designs that minimize vulnerability. The effects of windbreaks on wind speed have been quantified in various environments, and the consequences of the ongoing removal of hedgerows have been documented (Sánchez et al., 2010).

A synthesis of older data by Nuberg and Bennell (2009) concluded that measurable reductions of wind speed could be expected from about 5 times the canopy height before and up to 20 times the tree canopy height after the trees in the direction of the wind, and that semi-transparent windbreaks were more effective than solid ones, as the latter might cause turbulence while the first could maintain laminar flow conditions (Cleugh, 1998). Models that include turbulence effects are now available (Yeh et al., 2010), and windbreak porosity can be easily quantified as a basis for optimizing management (Kenney, 1987). Connections are being made between the hedgerow fabric of landscapes and the social and policy dimensions (Larcher and Baudry, 2013).

Given the considerable benefits involved for orchard crops, the fine-tuning of recommendations to site-specific conditions is needed. Tamang et al. (2010), for example, installed automated weather stations on the leeside of single-row tree windbreaks in southern Florida, USA, and found the lowest wind speed (~5% of the open wind speed) at two to six times the distance of windbreak height, depending on tree species and porosity. They found statistically significant wind speed reduction up to 31 times the windbreak height.

14.4 Trees Buffering Temperature

While wind is a near-horizontal flux and tree effects can extend beyond 20 times the canopy height, the shading effect on temperature is restricted to a few multiples of tree height, depending on the solar elevation (Kohli and Saini, 2003). Compared to open-field agriculture, all land-use systems with trees have a reduced daily amplitude of air temperature, with a gradual dampening of the amplitude within the top layers of the soil. An example of a data set from East Java (Fig. 14.2.) shows a daily amplitude of air temperature in open-field agriculture of 10.7°C, in closed-canopy secondary (degraded) forest of 5.6°C and intermediate values for various types of agroforestry coffee-based systems. In the closed-canopy systems, the daily amplitude of soil temperature at 5 cm depth is less than 3°C, while it is up to 9°C in the open-field situation.

In related studies, Hairiah et al. (2006) compared the effects of shading on the litter layer soil temperature and its spatial variability in open- and closed-canopy coffee agroforestry systems in Lampung, with a natural forest comparison. Martius et al. (2004) measured the daily amplitude of temperature in the litter layer across various agroforestry systems and closed-canopy forest, and found that canopy closure marked a major shift in temperature regime with associated increase in the biomass of soil macrofauna.

Direct measurement with data loggers of temperature and air humidity in different positions with respect to trees in the parkland agroforestry systems of Sapone (Burkina Faso) gives evidence (Fig. 14.3) of the buffering effects of trees on maximum daily temperature (average 1°C up to 2.5°C on hot, cloudless days) and minimum air humidity (0–5%), with stronger differences on hotter and drier days.

Jonsson et al. (1999) reported that crops were less exposed to excessive temperature of above 40°C with 1–9 h week^{-1} under *karité* and *néré* trees, against 27 h week^{-1} in the open field. It is not quite clear which of the many ways of measuring temperature is the most relevant for predicting crop performance. In a *Grevillea*-based agroforestry system in semi-arid Kenya, the

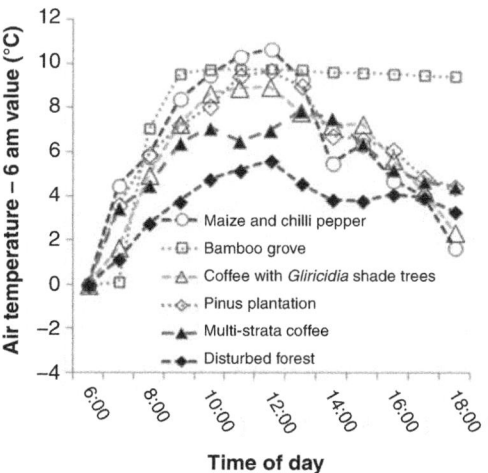
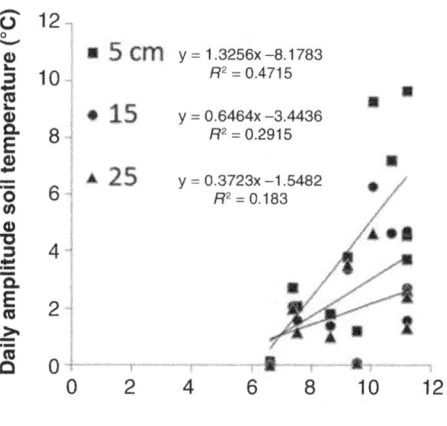

Fig. 14.2. (a) Temperature profile during a daily cycle in different land cover types in an East Java mountain location (Ngantang, East Java, Indonesia), including simple shade and multi-strata coffee agroforestry systems, compared to (degraded) forest and open-field agriculture (data were averaged for dry season and rainy season measurements); (b) relationship, across seasons and land-use systems, between the daily amplitude of air temperature and temperature at 5, 15 or 25 cm depth in the soil.

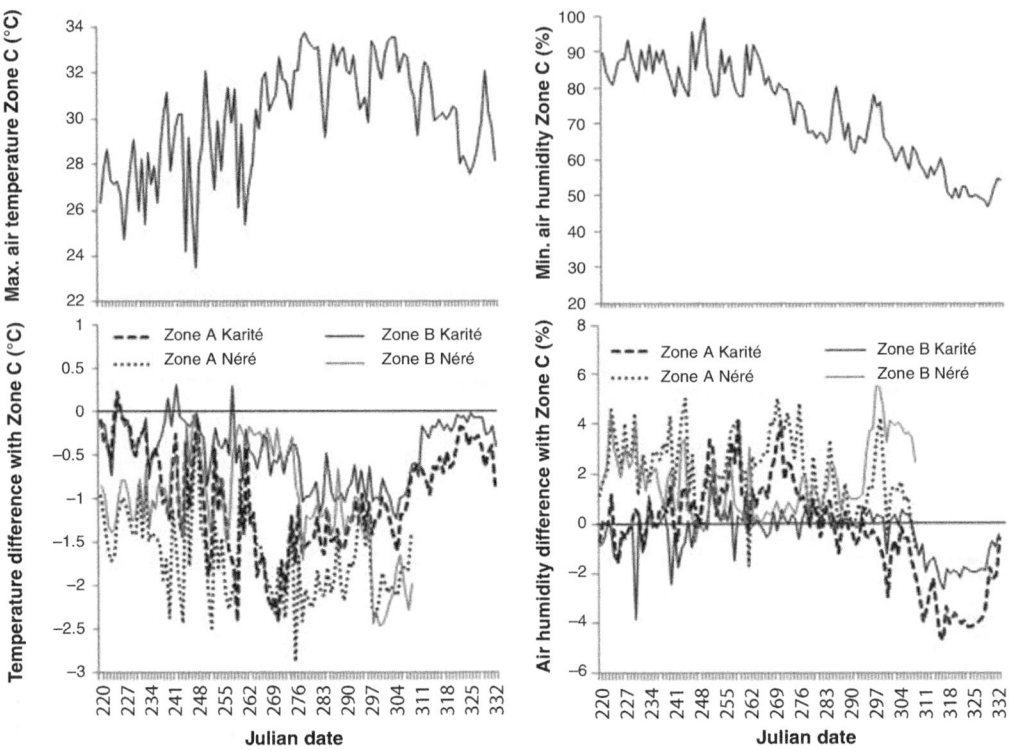

Fig. 14.3. Effect of position relative to a karité (*Vitellaria paradoxa*) or a néré (*Parkia biglobosa*) tree on maximum daily temperature at crop level (left panels) or minimum air humidity (right panels) for zones A (under the tree) and B (edge of tree canopy) compared to zone C (in between trees) in the parkland landscape of Sapone (Burkina Faso). (Data: Bayala et al., 2013a.)

mean maximum meristem temperature for maize in agroforestry systems was 5–6.0°C lower than for sole maize, depending on season (Lott et al., 2009).

Large effects of trees on temperature were reported in a study of urban trees in Bangalore (India), where street trees provided a maximum reduction in afternoon ambient air temperatures of 5.6°C, and of tarmac road surface temperatures of 27.5°C (Vailshery et al., 2013). Similar data can now be obtained directly from satellite spectral information. Lai et al. (2012) discussed the relation between near-ground air temperature (T_a) as measured in climate stations and land surface temperature (T_s) as derived from the moderate resolution imaging spectroradiometer (MODIS) instruments installed on the Aqua and Terra Earth observation satellites. Correlation coefficients of 0.91–0.96 were obtained, and the standard deviations of the differences between the two sets were 1.2–1.8°C. However, differences in spatial resolution, different sensitivity to soil moisture buffering and technical challenges in the correct interpretation of the satellite data remain. Cassidy et al. (2013) used MODIS satellite data to derive a daily pattern of surface temperatures over different land cover types in the Mekong region in Southeast Asia. They estimated a 15°C daily amplitude for an ambient air temperature and 10, 22, 25, 30 and 37°C for the surface temperature of secondary forest, a cassava crop (young stage), a fallow rice field grassy residue, bare land (ploughed field) and a tarmac road, respectively.

14.5 Trees Modifying Water Balance

14.5.1 Processes

Cannavo et al. (2011) considered the net effect of shade trees on the water balance of coffee production systems and found that a net increase in infiltration rate in the agroforestry version of coffee was linked to an effectively deeper soil system that provided a stronger buffer and reduced losses by leaching. Hydraulic lift, the transfer of water from relatively wet deeper layers to drier soil higher in the profile, has been described for both natural tree–grass mixtures and agroforestry systems (Richards and Caldwell, 1987; Dawson, 1993; Caldwell et al., 1998; Midwood et al., 1998; Norton and Hart, 1998; Zou et al., 2005; Hawkins et al., 2009). If the topsoil is re-wetted by rainfall, flow will be in the reverse direction (Hultine et al., 2004) and the term 'hydraulic redistribution' may describe the process irrespective of direction (Burgess et al., 1998; Smith et al., 1999; Brooks et al., 2002).

Nadezhdina et al. (2010) describe four types of hydraulic redistribution, including canopy capture of fog and transfer to drier soil layers. The physics are well understood, as the root–soil contact that allows the uptake of water to occur where the water potential inside the roots is more negative than that in soil cannot prevent water to enter the soil where the gradient in water potential is the reverse. Root shrinkage, reducing root–soil contact, and dieback can reduce water loss for the plant, but needs to be reversed if the soil re-wets. Hydraulic redistribution may benefit the plant in two ways: it allows for higher daytime water uptake, using water that is brought to the well-rooted topsoil, and it facilitates nutrient uptake in situations where nutrients are in the topsoil but difficult to access because of low effective diffusion rates. These two benefits, however, are 'public goods' in the soil layer and other plants rooted in the re-wetted soil layers benefit as well.

Water vapour transport can play a dominant role in transport in the upper 15 cm, subject to strong day–night cycles in temperature, but root-based hydraulic redistribution likely dominates in deeper layers (Warren et al., 2011). Ludwig et al. (2003) found that hydraulic redistribution in *Acacia tortilis* could re-wet soil up to 10 m from the tree stem, but was absent in a dry year. The competitive effect of trees can outweigh the facilitation by hydraulic redistribution (Ludwig et al., 2004). Prieto et al. (2010) found that by restricting daytime tree transpiration, the net effect of hydraulic redistribution on soil water content could be increased. The total effects were found to depend on soil texture and associated soil physical properties. The role of sparse root systems in the subsoil in the overall water balance is much more than that expected; hydraulic equilibration effectively provides the roots with an opportunity to function for 24 h day^{-1} rather than only during transpiration peak demand (Amenu and Kumar, 2008).

Sekiya et al. (2010) found that aboveground removal (pruning) of stems could enhance greatly the net positive effects on companion crops. Burgess (2011) described attempts to use such effects in designed agricultural intercropping patterns. Siriri et al. (2013) reported that shoot and root pruning of *Alnus acuminata* and *Sesbania sesban* on terrace risers increased water content in the cropping area. Wang et al. (2011) found that hydraulic redistribution could have positive effects on transpiration and plant growth during regular dry seasons but negative effects on net primary productivity (NPP) under extreme droughts, such as those during El Niño years in the Amazon forest. The latter is due to more rapid depletion of the groundwater reserves, reaching the permanent wilting point earlier. Climatic limits to the potential positive effect of hydraulic redistribution can thus be defined on the basis of the required build-up of deep groundwater reserves (potentially concentrated in groundwater flows and allowing use at some distance from the source), and the relevance of wetter topsoil to survive dry spells.

The bumper harvests that can be obtained after the pollarding of parkland trees have usually been interpreted as the effect of 'fertility islands' in an environment with strong nutrient limitations. It may well be, however, that the persistence of the tree roots plays a complementary role in supporting crop growth in these circumstances. Published estimates of the volume of water involved vary from 5% to 30% of potential daily evapotranspiration. Most publications so far indicate that hydraulic lift can mitigate, but not reverse, the drying of soil layers, postponing the emergence of water stress. Bayala et al. (2008), however, described a case where night-time re-wetting of topsoil layers exceeded daytime uptake after the harvest of a grain crop.

14.5.2 Essential and sufficient assumptions for modelling hydraulic equilibration

Current procedures for quantifying 'hydraulic lift' are based on the amplitude of the day–night cycle in either soil water potential or soil water content, often involving the fit of a sinusoidal relationship. At hourly, daily or weekly timescales, the mass balance for any layer of soil (i) prescribes:

$$\Delta W_s = HR_{S+R,i} - f_i \text{Uptake} + g_i \text{Rain} - h_i \text{Drain}$$

(Eqn 14.1)

where ΔW_s is the change in soil water content, HR_{S+R} indicates hydraulic redistribution through soil and roots, f, g and h are partitioning factors that satisfy $\Sigma f = \Sigma g = \Sigma h = 1$. The 'rainfall' term (Rain) includes irrigation, the balance of runoff and runoff and/or snowmelt under relevant conditions, the 'Drain' term both vertical and horizontal (net of outflow minus inflow) components, and 'Uptake' includes surface evaporation for the topsoil compartment.

In the absence of rainfall, this implies

$$HR_{S+R,i} = f_i \text{Uptake} - \Delta W_s + h_i \text{Drain}$$

(Eqn 14.2)

Night-time hydraulic redistribution can be derived from the night-time change in soil water content only under the assumption of zero uptake, or the absence of restoring water stocks in the tree trunk. Hydraulic redistribution on a daily timescale requires assessment of the uptake term, which is non-negligible at this timescale. The 'Drain' term is of some interest, as it relates to the soil component of HR_{S+R}.

14.5.3 Hydraulic equilibration in WaNuLCAS

A more detailed account of hydraulic equilibration through the root systems of crops or trees that connect relatively dry and relatively wet zones of the soil was incorporated into the model of water, nutrient and light capture in agroforestry systems (WaNuLCAS model) (van Noordwijk et al., 2011c). The process of 'hydraulic equilibration' is driven by the existence of differences in water potential among the layers (and zones) of a soil profile and the availability of conductors in the form of root systems that are connected to the soil.

Implementation requires the following steps:

- Estimation of equilibrium stem base water potential at zero flux, from the root-weighted average of the soil hydraulic potential in each cell; the proportionality factor consists of root-length density and the volume of the cell as other proportionality factors cancel out in the equation.
- Derivation of the equivalent equilibrium volumetric soil water content in each cell on the basis of this stem base potential for each tree or crop type and the parameters of the pedotransfer function.
- Calculation of the amount of water involved in the difference between current and equilibrium soil water content (positive differences as 'potential supply' of water, negative ones as 'demand').
- Derivation of the potential flux as the minimum of a 'cap' (HydEq fraction that relates to soil transport constraints that may have to be calibrated to actual data; the HydEq fraction default value is 0.1 day^{-1}; for a value of 1, the model becomes

a 1-pool soil model, for a value of 0 there is no hydraulic equilibration) of the difference between the target and the actual volumetric soil water content, and a potential flux that is in accordance with the potential difference, the hydraulic conductivity of the roots, the root diameter and root length density and the period of time available (based on the fraction of day that stomata are expected to be closed).

- Reductions on either the positive or the negative potential fluxes are to be in accordance with a zero-sum net process, by calculating the minimum of the total potential supply and the total potential demand, and scaling down the cell-specific differences such that total supply matches total demand.
- Implementing the resulting flux in or out of each cell on a daily time-step basis and checking the consistency of the water balance for errors or inconsistencies.

For a 'standard' case of parklands (with parameterization for a parkland system in Burkina Faso, as simulated by Bayala *et al.*, 2013b) the implementation leads to (Fig. 14.4):

- a total hydraulic equilibration flux through tree roots that is 64% of the tree transpiration
- slight increases for processes that depend on topsoil water content: runoff, soil evaporation
- a 9% increase in crop water uptake
- a 22% decrease of tree water uptake (and 10% decrease in canopy interception)
- a 15% decrease in vertical drainage.

These results are only moderately sensitive to the value (arbitrarily) selected for the HydEq Fraction; values above 0.2 may be unrealistic. In all situations, the tree + crop uses more water than the crop alone would have done.

14.6 Trees and Progressive Climate Change

Actions towards climate change adaptation, according to Vermeulen *et al.* (2012), fall into two broad overlapping areas:

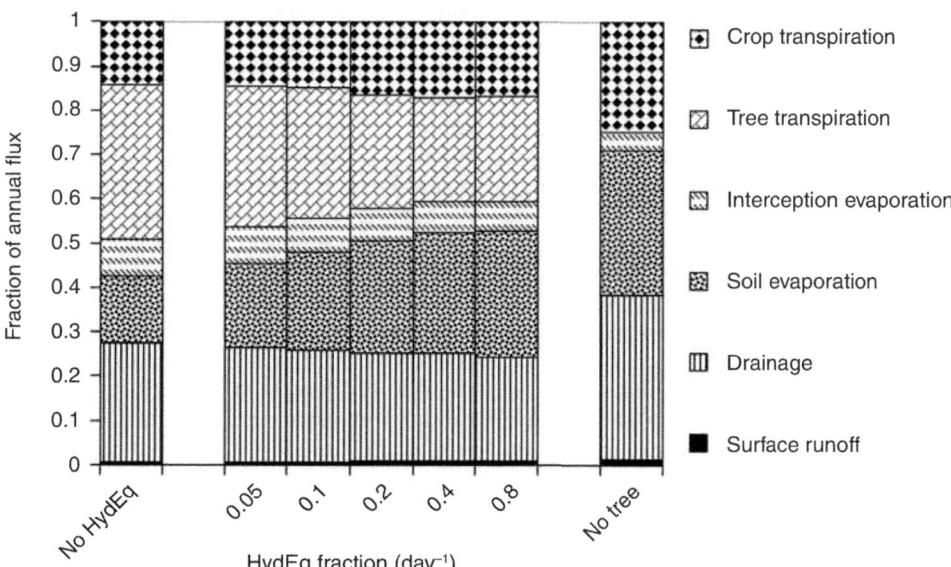

Fig. 14.4. Impacts on the water balance of a parkland system with a rainfall of approximately 750 mm year^{-1} of the presence of trees and the inclusion of hydraulic equilibration in the model, for a range of values of the (arbitrarily set) HydEqFraction parameter.

- better management of agricultural risks associated with increasing climate variability and extreme events
- accelerated adaptation to progressive climate change over decadal timescales; for example, integrated packages of technology, agronomy and policy options for farmers and food systems.

The buffering effects of trees in agricultural landscapes as discussed here and in van Noordwijk *et al.* (2011a) are part of the first approach but are not mentioned explicitly by Vermeulen *et al.* (2012).

Many trees have a considerable adaptive range, especially the slower-growing species. Gebrekirstos *et al.* (2011) describe the negative relationship between tree growth rates under favourable circumstances, facilitated by sapwood with large xylem vessels, and the performance under severe drought, when small xylem vessels and dense wood are of an advantage. As the rate of climate shifts, the lateral displacement of iso-climes over the land surface, may well exceed the intergenerational rate of tree seed dispersal, except for wind-borne pioneer trees, the tolerance to climate change will determine the fate of many tree species. Tree species of recognized human use will have to be helped in reaching the parts of the world that still suit them.

Many of the most widespread agroforestry species behave like invasive exotics in places where they have been introduced and have naturalized. For such trees, climate change is not likely to be a major challenge, but for other tree species, especially in locations currently on the edge of a distribution range, it may be relevant to seek tree germplasm in 'climate analogues', places where the current climate resembles what can, in future, be expected for the place where the trees are to be planted. Climate change may lead to some novel climate situations, without current analogues, but its primary effect will be shifting current climates to new places, so that analogues can be found in the current situation, allowing farmer-to-farmer communication, social exchange, value chain shifts and germplasm transfer to be explored.

Analysis of tree spread in response to warming after interglacial periods, or their expansion from mountain refuges after dry and hot geological climate episodes, has shown that tree species move at individually differentiated rates, rather than as assemblages (Hewitt, 1999). The presence of a suitable soil microflora with suppression of disease organisms probably is an underrated factor in the spread of tree species to new habitats, with little predictive skill in current science. The expansion of termites to temperate regions and urban environments without frost has been noted as a challenge to trees not adapted to these rhizovores (van Noordwijk *et al.*, 1998).

Biophysical buffering effects of trees on micro- or meso-climate, as discussed so far, will not be sufficient in the face of progressive climate change. Their interaction with the buffering of socio-economic processes related to social and human capitals will need further attention in a dynamic landscape context (Table 14.1).

14.7 Concluding Remarks

Further interdisciplinary research on the way dynamic landscapes provide buffering and other ecosystem services may benefit from a number of analytical steps to assess supply and demand, current sufficiency and trends of buffering across the multiple capitals of Table 14.1. It can, however, build on local ecological knowledge and experience in dealing with past shocks, as well as quantitative scenario studies of local socio-ecological systems and their linkages with the wider outside world.

The biggest stumbling blocks for realizing the full contributions agroforestry can make to the challenge of adapting our food production systems, multifunctional landscapes and rural livelihood systems probably still are: (i) the mindset of agricultural scientists trained to think that open-field agriculture is the norm and standard; (ii) climate scientists who have not even started serious downscaling of climate change predictions to include the effects of local land cover change on local temperature,

Table 14.1. Examples of buffering across natural (N), social (S), human (H), physical infrastructure (P) and financial (F) capital types. The letters 'AF' after each capital symbol refers to the contribution of agroforestry to the buffering function. IPR = intellectual property rights.

Capital type	Buffer functions in the face of external variability	Landscape and livelihood elements providing buffer	Threats to buffering/increasing buffer demand
N – AF	Protection from wave action	Coral reefs, coastal vegetation, mangrove (Bayas et al., 2011)	Destructive fisheries, shrimp ponds, mangrove decline
N – AF	Flow buffering, protection from floods and droughts, given climatic variability	Vegetated upper watersheds, riparian zone, wetlands (van Noordwijk et al., 2004b, 2011b; Ma et al., 2010; Verbist et al., 2010)	Forest conversion; increasing need in face of more extreme rainfall events alternating with prolonged droughts
N – AF	Protection from landslides and erosion, given extreme rainfall events	Trees on slopes, through deep and superficial rooting patterns (Reubens et al., 2007)	Increasing need in face of more extreme rainfall events
N – AF	Freshwater supplies as groundwater	Maximize rainwater infiltration (Hairiah et al., 2006)	Increasing need in face of more variable rainfall
N – AF	Nutrient supply between temporary excess and shortage	Organic matter-rich soil and synchrony management (van Noordwijk and Cadisch, 2002)	Decline of soil organic matter under intensified management
N – AF	Limit pest and disease outbreaks	Biological reservoirs of control agents (Jackson et al., 2012)	Loss of agrobiodiversity and ecological connectivity
N – AF	Specific response of crops and domesticated stock to abiotic and biotic variation	Agrobiodiversity with portfolio of multiple species reducing risk (Nguyen et al., 2012)	IPR rules restrict free exchange of germplasm; climate shifts increase needs
N/S – AF	Transitions and interactions between 'conservation' and 'development' functions	Integrated conservation and development programmes (Minang and van Noordwijk, 2013)	Land shortage heightens conflict between conservation managers and villagers
N/S/F	External appreciation for local ecosystem functions (ES) beyond their weight in land-use decisions	Co-investment in ecosystem services (van Noordwijk and Leimona, 2010; van Noordwijk et al., 2012)	Commoditization of ES can undermine local norms and create new dependencies
S	Internal social pressures	Social networks within extended families	Urbanization undermines safety network functions
S	External social pressures	Social networks across neighbouring villages or islands (Silvey and Elmhirst, 2003)	Competition for external resources and attention
S – AF	Shifting policies on forest–community interactions (de/re-centralization)	Village forest, community-based forest management agreements (Akiefnawati et al., 2010)	Landscape management requires clarity of access rules for all mosaic transitions
S/H	Response of flora and fauna to environmental variation and human management	Local ecological knowledge (LEK), its accumulation and reproduction	Loss of LEK transmission due to focus on formal schooling, changing practices
H	Fluctuating job opportunities	Education and diversification of skills within families	Rapid shift to urban lifestyles and expectations
P	Engineering structures in response to earthquakes, tsunamis, sea-level fluctuations	Engineering for coastal protection, water storage, marine + aerial transport	Sea-level rise and episodic high tides; siltation of reservoirs; landslides
F	Fluctuations in economic markets, exchange rates and credit supply	Traditional + modern savings and insurance systems, microcredit provision	Globalization and increased synchronization of risks in financial markets

humidity, wind speed and other parameters of direct human relevance; (iii) the makers and shapers of agricultural, forestry and land-use policies who treat forestry and agriculture as opposite sides of a coin that can only fall on either side of the institutional divide. The main supporters of the emergence of agroforestry as part of the solution are the farmers of the world who have defied the advice to oversimplify and overspecialize their farms and landscapes. Studies such as that by Nguyen *et al.* (2012) start to document the ways farmers perceive that trees substantially reduce their exposure to climate risk, and part of the research community is picking up this challenge – but as other chapters in this book will probably show, most of the analysis of climate change adaptation in agriculture is not thinking outside of the specialization box.

Acknowledgements

ICRAF's research is financed as part of the CGIAR Research Program on Forests, Trees and Agroforestry and that on Climate Change Agriculture and Food Security. We thank Dr Ric Coe for comments on an earlier draft.

References

Akiefnawati, R., Villamor, G.B., Zulfikar, F., Budisetiawan, I., Mulyoutami, E., Ayat, A., *et al.* (2010) Stewardship agreement to reduce emissions from deforestation and degradation (REDD): case study from Lubuk Beringin's Hutan Desa, Jambi Province, Sumatra, Indonesia. *International Forestry Review* 12, 349–360.

Amenu, G.G. and Kumar, P. (2008) A model for hydraulic redistribution incorporating coupled soil–root moisture transport. *Hydrology and Earth System Sciences* 12, 55–74.

Arnfield, A.J. (2003) Two decades of urban climate research: a review of turbulence, exchanges of energy and water, and the urban heat island. *International Journal of Climatology* 23, 1–26.

Bayala, J., Heng, L.K., van Noordwijk, M. and Ouedraogo, S.J. (2008) Hydraulic redistribution study in two native tree species of agroforestry parklands of West African dry savanna. *Acta Oecologica* 34, 370–378.

Bayala, J., Sanou, J., Bazié, P. and van Noordwijk, M. (2013a) Empirical data collection of tree effects on temperature and humidity at crop level. CRP 7 Activity Report. World Agroforestry Centre, Nairobi.

Bayala, J., Bazié, H.R. and Sanou, J. (2013b) Competition and facilitation-related factors impacts on crop performance in an agro-forestry parkland system in Burkina Faso. *African Journal of Agricultural Research* 8, 5307–5314.

Bayas, J.L., Marohn, C., Dercon, G., Dewi, S., Piepho, H., Joshi, L., *et al.* (2011) Influence of coastal vegetation on the 2004 tsunami wave impact in West Aceh. *Proceedings of the National Academy of Sciences* 108, 18612–18617.

Bebbington, A. (1999) Capitals and capabilities: a framework for analyzing peasant viability, rural livelihoods and poverty. *World Development* 27, 2021–2044.

Beddington, J., Asaduzzaman, M., Fernandez, A., Clark, M., Guillou, M., Jahn, M., *et al.* (2011) *Achieving Food Security in the Face of Climate Change.* Final Report from the Commission on Sustainable Agriculture and Climate Change. CGIAR Research Program Climate Change, Agriculture and Food Security (CCAFS), Copenhagen (http://ccafs.cgiar.org/commission/reports, accessed 6 December 2013).

Beer, J. (1987) Advantages, disadvantages and desirable characteristics of shade trees for coffee, cacao and tea. *Agroforestry Systems* 5, 3–13.

Beer, J., Muschler, R., Kass, D. and Somarriba, E. (1997) Shade management in coffee and cacao plantations. *Agroforestry Systems* 38, 139–164.

Boffa, J.M. (1999) Agroforestry parklands in sub-Saharan Africa. *FAO Conservation Guides 34.* FAO, Rome.

Bonan, G.B. (2002) *Ecological Climatology: Concepts and Applications.* Cambridge University Press, Cambridge, UK.

Bonan, G.B. (2008) Forests and climate change: forcings, feedbacks, and the climate benefits of forests. *Science* 320, 1444–1449.

Breman, H. and Kessler, J.J. (1997) The potential benefits of agroforestry in the Sahel and other semi-arid regions. *European Journal of Agronomy* 7, 25–33.

Brenner, A.J. (1996) Microclimatic modifications in agroforestry. In: Ong, C.K. and Huxley, P. (eds) *Tree–Crop Interactions: A Physiological Approach.* CAB International (in association with ICRAF), Wallingford, UK, pp. 159–187.

Brooks, J.R., Meinzer, F.C., Coulombe, R. and Gregg, J. (2002) Hydraulic redistribution of soil water during summer drought in two contrasting

Pacific Northwest coniferous forests. *Tree Physiology* 22, 1107–1117.
Burgess, S.S.O. (2011) Can hydraulic redistribution put bread on our table? *Plant and Soil* 341, 25–29.
Burgess, S.S.O., Adams, M.A., Turner, N.C. and Ong, C.K. (1998) The redistribution of soil water by tree root systems. *Oecologia* 115, 306–311.
Caborn, J.M. (1965) Shelterbelts and windbreaks. *Scottish Forestry* 19, 265–286.
Cairns, M.F. (ed.) (2007) *Voices from the Forest: Integrating Indigenous Knowledge into Sustainable Upland Farming*. Resources for the Future Press, Washington, DC.
Caldwell, M.M., Dawson, T.E. and Richards, J.H. (1998) Hydraulic lift: consequences of water efflux from the roots of plants. *Oecologia* 113, 151–161.
Cannavo, P., Sansoulet, J., Harmand, J.-M., Siles, P., Dreyer, E. and Vaast, P. (2011) Agroforestry associating coffee and Inga densiflora results in complementarity for water uptake and decreases deep drainage in Costa Rica. *Agriculture, Ecosystems and Environment* 140, 1–13.
Cassidy, L., Southworth, J., Gibbes, C. and Binford, M. (2013) Beyond classifications: combining continuous and discrete approaches to better understand land-cover change within the lower Mekong River region. *Applied Geography* 39, 26–45.
Cleugh, H.A. (1998) Effects of windbreaks on airflow, microclimates and crop yields *Agroforestry Systems* 41, 55–84.
Dawson, T.E. (1993) Hydraulic lift and water use by plants: implications for water balance, performance and plant–plant interactions. *Oecologia* 95, 565–574.
Eichhorn, M.P., Paris, P., Herzog, F., Incoll, L.D., Liagre, F., Mantzanas, K., *et al.* (2006) Silvoarable systems in Europe – past, present and future prospects. *Agroforestry Systems* 67, 29–50.
Foresta, H. de, Somarriba, E., Temu, A., Boulanger, D., Feuilly, H. and Gauthier, M. (2013) Towards the assessment of trees outside forests. *Resources Assessment Working Paper 183*. FAO, Rome.
Forman, R.T. (1990) Ecologically sustainable landscapes: the role of spatial configuration. In: Zonneveld, I.S. and Forman, R.T.T. (eds) *Changing Landscapes: An Ecological Perspective*. Springer, New York, pp. 261–278.
Gebrekirstos, A., van Noordwijk, M., Neufeldt, H. and Mitlöhner, R. (2011) Relationships of stable carbon isotopes, plant water potential and growth: an approach to assess water use efficiency and growth strategies of dry land agroforestry species. *TREES* 25, 95–102.
Gold, M.A. and Garrett, H.E. (2009) Agroforestry nomenclature, concepts, and practices. In: Gene Garrett, H.E. (ed.) *North American Agroforestry: An Integrated Science and Practice*, 2nd edn. American Society of Agronomy, Auburn, Alabama, pp. 45–56.
Gordon, A.M. and Newman, S.M. (1997) *Temperate Agroforestry Systems*. CAB International, Wallingford, UK.
Hairiah, K., Sulistyani, H., Suprayogo, D., Widianto, Purnomosidhi, P., Widodo, R.H. and van Noordwijk, M. (2006) Litter layer residence time in forest and coffee agroforestry systems in Sumberjaya, West Lampung. *Forest Ecology and Management* 224, 45–57.
Harvey, C.A., Villanueva, C., Esquivel, H., Gómez, R., Ibrahim, M., Lopez, M., *et al.* (2011) Conservation value of dispersed tree cover threatened by pasture management. *Forest Ecology and Management* 261, 1664–1674.
Hawkins, H.-J., Hettasch, H., West, A.G. and Cramer, M.D. (2009) Hydraulic redistribution by Protea 'Sylvia' (Proteaceae) facilitates soil water replenishment and water acquisition by an understorey grass and shrub. *Functional Plant Biology* 36, 752–760.
Hewitt, G M. (1999) Post-glacial re-colonization of European biota. *Biological Journal of the Linnean Society* 68, 87–112.
Hultine, K.R., Scott, R.L., Cable, W.L., Goodrich, D.C. and Williams, D.G. (2004) Hydraulic redistribution by a dominant, warm-desert phreatophyte: seasonal patterns and response to precipitation pulses. *Functional Ecology* 18, 530–538.
Jackson, L.E., van Noordwijk, M., Bengtsson, J., Foster, W., Lipper, L., Pulleman, M., *et al.* (2010) Biodiversity and agricultural sustainability: from assessment to adaptive management. *Current Opinion in Environmental Sustainability* 2, 80–87.
Jackson, L.E., Pulleman, M.M., Brussaard, L., Bawa, K.S., Brown, G., Cardoso, I.M., *et al.* (2012) Social-ecological and regional adaptation of agrobiodiversity management across a global set of research regions. *Global Environmental Change* 22, 623–639.
Jackson, R.B., Randerson, J.T., Canadell, J.G., Anderson, R.G., Avissar, R., Baldocchi, D.D., *et al.* (2008) Protecting climate with forests. *Environmental Research Letters* 3, 044006.
Jonsson, K., Ong, C.K. and Odongos, J.C.W. (1999) Influence of scattered néré and karité on microclimate, soil fertility and millet yield in Burkina Faso. *Experimental Agriculture* 35, 39–53.
Kenney, W.A. (1987) A method for estimating windbreak porosity using digitized photographic silhouettes. *Agricultural and Forest Meteorology* 39, 91–94.

Kho, R.M. (2000) On crop production and the balance of available resources. *Agriculture, Ecosystems and Environment* 80, 71–85.

Kho, R.M., Yacouba, B., Yayé, M., Katkoré, B., Moussa, A., Iktam, A., et al. (2001) Separating the effects of trees on crops: the case of *Faidherbia albida* and millet in Niger. *Agroforestry Systems* 52, 219–238.

King, K.F.S. (1979) Agroforestry and the utilisation of fragile ecosystems. *Forest Ecology and Management* 2, 161–168.

Kohli, A. and Saini, B.C. (2003) Microclimate modification and response of wheat planted under trees in a fan design in northern India. *Agroforestry Systems* 58, 109–118.

Lai, Y.J., Li, C.F., Lin, P.H., Wey, T.H. and Chang, C.S. (2012) Comparison of MODIS land surface temperature and ground-based observed air temperature in complex topography. *International Journal of Remote Sensing* 33, 7685–7702.

Larcher, F. and Baudry, J. (2013) Landscape grammar: a method to analyse and design hedgerows and networks. *Agroforestry Systems* 87, 181–192.

Lin, B.B. (2007) Agroforestry management as an adaptive strategy against potential microclimate extremes in coffee agriculture. *Agricultural and Forest Meteorology* 144, 85–94.

Lott, J.E., Ong, C.K. and Black, C.R. (2009) Understorey microclimate and crop performance in a *Grevillea robusta*-based agroforestry system in semi-arid Kenya. *Agricultural and Forest Meteorology* 149, 1140–1151.

Ludwig, F., Dawson, T.E., Kroon, H., Berendse, F. and Prins, H.H.T. (2003) Hydraulic lift in *Acacia tortilis* trees on an East African savanna. *Oecologia* 134, 293–300.

Ludwig, F., Dawson, T.E., Prins, H.H.T., Berendse, F. and De Kroon, H. (2004) Below-ground competition between trees and grasses may overwhelm the facilitative effects of hydraulic lift. *Ecology Letters* 7, 623–631.

Ma, X., Xu, J. and van Noordwijk, M. (2010) Sensitivity of streamflow from a Himalayan catchment to plausible changes in land-cover and climate. *Hydrological Processes* 24, 1379–1390.

McCarthy, N., Lipper, L. and Branca, G. (2011) Climate smart agriculture: smallholder adoption and implications for climate change adaptation and mitigation. Mitigation of Climate Change in Agriculture Working Paper. FAO, Rome.

Martius, C., Höfer, H., Garcia, M.V., Römbke, J., Förster, B. and Hanagarth, W. (2004) Microclimate in agroforestry systems in central Amazonia: does canopy closure matter to soil organisms? *Agroforestry Systems* 60, 291–304.

Matocha, J., Schroth, G., Hills, T. and Hole, D. (2012) Integrating climate change adaptation and mitigation through agroforestry and ecosystem conservation. In: *Agroforestry – The Future of Global Land Use*. Springer, the Netherlands, pp. 105–126.

Midwood, A.J., Boutton, T.W., Archer, S.R. and Watts, S.E. (1998) Water use by woody plants on contrasting soils in a savanna parkland: assessment with δ^2H and $\delta^{18}O$. *Plant and Soil* 205, 13–24.

Minang, P.A. and van Noordwijk, M. (2013) Design challenges for achieving reduced emissions from deforestation and forest degradation through conservation: leveraging multiple paradigms at the tropical forest margins. *Land Use Policy* 31, 61–70.

Monteith, J.L., Ong, C.K. and Corlett, J.E. (1991) Microclimatic interactions in agroforestry systems. *Forest Ecology and Management* 45, 31–44.

Munang, R., Thiaw, I., Alverson, K., Goumandakoye, M., Mebratu, D. and Liu, J. (2013) Using ecosystem-based adaptation actions to tackle food insecurity. *Environment: Science and Policy for Sustainable Development* 55, 29–35.

Nadezhdina, N., David, T.S., David, J.S., Ferreira, M.I., Dohnal, M., Tesař, M., et al. (2010) Trees never rest: the multiple facets of hydraulic redistribution. *Ecohydrology* 3, 431–444.

Nguyen, Q., Hoang, M.H., Öborn, I. and van Noordwijk, M. (2012) Multipurpose agroforestry as a climate change adaptation option for farmers – an example of local adaptation in Vietnam. *Climatic Change* 117, 241–257.

Norton, J.L. and Hart, S.C. (1998) Hydraulic lift: a potentially important ecosystem process. *Trends in Ecology and Evolution* 13, 232–235.

Nuberg, I. and Bennell, M. (2009) Trees protecting dryland crops and soil. In: Nuberg, I., George, B. and Reid, R. (eds) *Agroforestry for Natural Resource Management*. CSIRO Publishing, Collingwood, Australia, pp. 69–85.

Ong, C.K. and Huxley, P. (1996) *Tree–Crop Interactions – A Physiological Approach*. CAB International, Wallingford, UK.

Ong, C.K., Corlett, J.E., Singh, R.P. and Black, C.R. (1991) Above and below ground interactions in agroforestry systems. *Forest Ecology and Management* 45, 45–57.

Ong, C.K., Black, C.R., Wallace, J.S., Khan, A.A.H., Lott, J.E., Jackson, N.A., et al. (2000) Productivity, microclimate and water use in *Grevillea robusta*-based agroforestry systems on hillslopes in semi-arid Kenya. *Agriculture, Ecosystems and Environment* 80, 121–141.

Pielke, R.A., Adegoke, J.O., Chase, T.N., Marshall,

C.H., Matsui, T. and Niyogi, D. (2007) A new paradigm for assessing the role of agriculture in the climate system and in climate change. *Agricultural and Forest Meteorology* 142, 234–254.

Prieto, I., Martínez-Tillería, K., Martínez-Manchego, L., Montecinos, S., Pugnaire, F.I. and Squeo, F.A. (2010) Hydraulic lift through transpiration suppression in shrubs from two arid ecosystems: patterns and control mechanisms. *Oecologia* 163, 855–865.

Rao, M.R., Nair, P.K.R. and Ong, C.K. (1997) Biophysical interactions in tropical agroforestry systems. *Agroforestry Systems* 38, 3–50.

Reubens, B., Poesen, J., Danjon, F., Geudens, G. and Muys, B. (2007) The role of fine and coarse roots in shallow slope stability and soil erosion control with a focus on root system architecture: a review. *Trees* 21, 385–402.

Richards, J.H. and Caldwell, M.M. (1987) Hydraulic lift: substantial nocturnal water transport between soil layers by *Artemisia tridentata* roots. *Oecologia* 73, 486–489.

Sánchez, I.A., Lassaletta, L., McCollin, D. and Bunce, R.G. (2010) The effect of hedgerow loss on microclimate in the Mediterranean region: an investigation in Central Spain. *Agroforestry Systems* 78, 13–25.

Santos-Martin, F. and van Noordwijk, M. (2011) Is native timber tree intercropping an economically feasible alternative for smallholder farmers in the Philippines? *Australian Journal of Agricultural and Resource Economics* 55, 257–272.

Schoeneberger, M., Bentrup, G., de Gooijer, H., Soolanayakanahally, R., Sauer, T., Brandle, J., *et al.* (2012) Branching out: agroforestry as a climate change mitigation and adaptation tool for agriculture. *Journal of Soil and Water Conservation* 67, 128A–136A.

Sekiya, N., Araki, H. and Yano, K. (2010) Applying hydraulic lift in an agroecosystem: forage plants with shoots removed supply water to neighboring vegetable crops. *Plant and Soil* 341, 39–50.

Silvey, R. and Elmhirst, R. (2003) Engendering social capital: women workers and rural urban networks in Indonesia's crisis. *World Development* 31, 865–879.

Siriri, D., Wilson, J., Coe, R., Tenywa, M.M., Bekunda, M.A., Ong, C.K., *et al.* (2013) Trees improve water storage and reduce soil evaporation in agroforestry systems on bench terraces in SW Uganda. *Agroforestry Systems* 87, 45–58.

Smith, D.M., Jackson, N.A., Roberts, J.M. and Ong, C.K. (1999) Reverse flow of sap in tree roots and downward siphoning of water by *Grevillea robusta. Functional Ecology* 13, 256–264.

Swann, A.L., Fung, I.Y. and Chiang, J.C. (2012) Mid-latitude afforestation shifts general circulation and tropical precipitation. *Proceedings of the National Academy of Sciences* 109, 712–716.

Takimoto, A., Nair, P.K.R. and Nair, V.D. (2008) Carbon stock and sequestration potential of traditional and improved agroforestry systems in the West African Sahel. *Agriculture, Ecosystems and Environment* 125, 159–166.

Tamang, B., Andreu, M.G. and Rockwood, D.L. (2010) Microclimate patterns on the leeside of single-row tree windbreaks during different weather conditions in Florida farms: implications for improved crop production. *Agroforestry Systems* 79, 111–122.

Thornton, P.K., Van de Steeg, J., Notenbaert, A. and Herrero, M. (2009) The impacts of climate change on livestock and livestock systems in developing countries: a review of what we know and what we need to know. *Agricultural Systems* 101, 113–127.

Tscharntke, T., Clough, Y., Bhagwat, S.A., Buchori, D., Faust, H., Hertel, D., *et al.* (2011) Multifunctional shade-tree management in tropical agroforestry landscapes – a review. *Journal of Applied Ecology* 48, 619–629.

Vailshery, L.S., Jaganmohan, M. and Nagendra, H. (2013) The impact of street trees in combating air pollution and mitigating microclimate in a tropical city. *Urban Forestry and Urban Greening* 12, 408–415.

van Noordwijk, M. and Cadisch, G. (2002) Access and excess problems in plant nutrition. *Plant and Soil* 247, 25–40.

van Noordwijk, M. and Leimona, B. (2010) Principles for fairness and efficiency in enhancing environmental services in Asia: payments, compensation, or co-investment? *Ecology and Society* 15, 17 (http://www.ecologyandsociety.org/vol15/iss4/art17/, accessed 6 December 2013).

van Noordwijk, M., Dijksterhuis, G. and Van Keulen, H. (1994) Risk management in crop production and fertilizer use with uncertain rainfall: how many eggs in which baskets. *Netherlands Journal of Agricultural Science* 42, 249–269.

van Noordwijk, M., Martikainen, P., Bottner, P., Cuevas, E., Rouland, C. and Dhillion, S.S. (1998) Global change and root function. *Global Change Biology* 4, 759–772.

van Noordwijk, M., Cadisch, G. and Ong, C.K. (eds) (2004a) *Belowground Interactions in Tropical Agroecosystems.* CAB International, Wallingford, UK.

van Noordwijk, M., Poulsen, J. and Ericksen, P. (2004b) Filters, flows and fallacies: quantifying off-site effects of land use change. *Agriculture, Ecosystems and Environment* 104, 19–34.

van Noordwijk, M., Hoang, M.H., Neufeldt, H., Öborn, I. and Yatich, T. (eds) (2011a) *How Trees and People Can Co-adapt to Climate Change: Reducing Vulnerability through Multifunctional Agroforestry Landscapes*. World Agroforestry Centre (ICRAF), Nairobi.

van Noordwijk, M., Widodo, R.H., Farida, A., Suyamto, D., Lusiana, B., Tanika, L., et al. (2011b) *GenRiver and FlowPer: Generic River and Flow Persistence Models*. User Manual Version 2.0. Bogor, Indonesia: World Agroforestry Centre (ICRAF) Southeast Asia Regional Program (http://www.worldagroforestry.org/sea/publication?do=view_pub_detail&pub_no=MN0048-11, accessed 6 December 2013).

van Noordwijk, M., Lusiana, B., Khasanah, N. and Mulia, R. (2011c) *WaNuLCAS version 4.0, Background on a Model of Water Nutrient and Light Capture in Agroforestry Systems*. World Agroforestry Centre (ICRAF), Bogor, Indonesia.

van Noordwijk, M., Leimona, B., Jindal, R., Villamor, G.B., Vardhan, M., Namirembe, S., et al. (2012) Payments for environmental services: evolution towards efficient and fair incentives for multifunctional landscapes. *Annual Review of Environment and Resources* 37, 389–420.

Verbist, B., Poesen, J., van Noordwijk, M., Widianto, Suprayogo, D., Agus, F. and Deckers, J. (2010) Factors affecting soil loss at plot scale and sediment yield at catchment scale in a tropical volcanic agroforestry landscape. *Catena* 80, 34–46.

Verchot, L.V., van Noordwijk, M., Kandji, S., Tomich, T.P., Ong, C.K., Albrecht, A., et al. (2007) Climate change: linking adaptation and mitigation through agroforestry. *Mitigation and Adaptation Strategies for Global Change* 12, 901–918.

Vermeulen, S.J., Aggarwal, P.K., Ainslie, A., Angelone, C., Campbell, B.M., Challinor, A.J., et al. (2012) Options for support to agriculture and food security under climate change. *Environmental Science and Policy* 15, 136–144.

Wang, G., Alo, C., Mei, R. and Sun, S. (2011) Droughts, hydraulic redistribution, and their impact on vegetation composition in the Amazon forest. *Plant Ecology* 212, 663–673.

Warren, J.M., Brooks, J.R., Dragila, M.I. and Meinzer, F.C. (2011) *In situ* separation of root hydraulic redistribution of soil water from liquid and vapor transport. *Oecologia* 166, 899–911.

Watson, R.T., Noble, I., Bolin, B., Ravindranath, N.H., Verardo, D.J. and Dokken, D.J. (2000) *Land Use, Land-use Change, and Forestry: A Special Report of the Intergovernmental Panel on Climate Change*. Cambridge University Press, Cambridge, UK.

Williams, M. (2006) *Deforesting the Earth, from Prehistory to Global Crisis: An Abridgement*. University of Chicago Press, Chicago, Illinois.

Yeh, C.P., Tsai, C.H. and Yang, R.J. (2010) An investigation into the sheltering performance of porous windbreaks under various wind directions. *Journal of Wind Engineering and Industrial Aerodynamics* 98, 520–532.

Zomer, R.J., Trabucco, A., Coe, R. and Place, F. (2009) Trees on Farm: Analysis of Global Extent and Geographical Patterns of Agroforestry. ICRAF Working Paper 89. World Agroforestry Centre (ICRAF), Nairobi.

Zou, C.B., Barnes, P.W., Archer, S. and McMurtry, C.R. (2005) Soil moisture redistribution as a mechanism of facilitation in savanna tree-shrub clusters. *Oecologia* 145, 32–40.

15 Channelling the Future? The Use of Seasonal Climate Forecasts in Climate Adaptation

Lauren Rickards,[1] Mark Howden[2] and Steven Crimp[3]

[1]*Melbourne Sustainable Society Institute, The University of Melbourne, Carlton, Australia;* [2]*CSIRO Climate Adaptation Flagship and Ecosystem Sciences, Canberra, Australia*

15.1 Introduction

More than ever before, contemporary life is characterized by growing recognition of the day-to-day risks associated with everyday living and a preoccupation with managing this risk (Giddens, 1991; Beck, 1992, 2008). Underpinning this awareness is the reported expansion, propagation and intensification of the risks associated with anthropogenic climate change. New fragilities are emerging as modern systems become more complex; causes and effects are cascading as feedback loops lengthen and tighten (Rockström et al., 2009; Galaz et al., 2012). Our concern with the risk-saturated nature of our modern lives is also increasing as a result of our growing awareness and knowledge of it. While many feedbacks undoubtedly remain undetected in the way that the existence of anthropogenic climate change did for decades, our understanding of the factors and relationships involved is deepening as new 'big data' science and other efforts put the pieces together and report on previously unknown threats and unwelcome trajectories (Lynch, 2008; Frew and Dozier, 2012). More than simply increasing our understanding, such efforts are directed at improving our *management* of risks, including an attenuated but enduring interest in risk as opportunity and risk management as the basis of business competition and business decision making through risk versus return relationships. This 'practice turn' in research, development and extension (RD&E) has revitalized the Enlightenment ideal of guiding human decisions with rational knowledge, with research agendas increasingly targeted at topics judged by parts of the science community to be socially relevant (e.g. Lubchenco, 1998; Cash et al., 2003; Moser, 2010).

'Information' about the future is especially valued as a decision-making guide (Sarewitz et al., 2000; Anderson, 2010). Its predicted instrumental value means that huge levels of resources are being expended in both creating such information and promoting and enabling its utilization by decision makers. Irrespective of scale, decisions of all kinds are expected increasingly to incorporate not only a deep understanding of their general exposure to risk and the need to manage it profitably but also a mounting array of 'specific' risk information. As argued below, however, much of this remains poorly contextualized and its utility is open to debate.

This chapter contributes to understanding this context by bringing together a diversity of literatures pertinent to the use of seasonal climate forecasts (SCFs). In particular, it examines the role of such forecasts in assisting agricultural producers to adapt to climate variability and change. SCFs are projections of the trajectory of major climatic variables (notably precipitation) over a timescale of weeks to months (Luo et al., 2011). They allow some revision of the probability of occurrence of specific climate conditions when compared with historic experience (i.e. climatology). Information about upcoming weather and expected climate has always been highly valued in society – at least in temperate areas (Strauss and Orlove, 2003) – because of their pervasive influence. It is especially pertinent in the context of anthropogenic climate change. This is not only because climate change is already posited as disrupting familiar patterns of climate variability and extremes, including seasonal trends, leading to the need for decision makers to be more watchful and responsive (Verdon-Kidd and Kiem, 2009). It is also because, increasingly, such forecasts are perceived as a stimulus and eventual route to improving decision makers' acceptance of, knowledge of and adaptation to longer-term climate change (McKeon et al., 1993).

Defined in various ways, climate change adaptation is about managing the multi-faceted challenges of climate change in a way that continuously maintains or improves one's capacity to do so. According to Moser and Ekstrom, climate change adaptation:

> ... involves changes in social–ecological systems in response to actual and expected impacts of climate change in the context of interacting non-climatic changes. Adaptation strategies and actions can range from short-term coping to longer-term, deeper transformations, aim to meet more than climate change goals alone, and may or may not succeed in moderating harm or exploiting beneficial opportunities.
>
> (Moser and Ekstrom, 2010)

Climate change adaptation is considered to be particularly pertinent to agriculture. Not only is agriculture a major contributor to climate change (e.g. Smith et al., 2008) but also it is unusually exposed and sensitive to the resultant risks (Easterling et al., 2007). Assessments of how capable agriculture will be in adapting to the risks posed by climate change are varied, dynamic and subjective. In some ways, the sector's long experience in explicitly managing climate variability is considered to give it an advantage of awareness and knowledge, especially in comparison to sectors such as health or financial services that are often relatively unfamiliar with explicit climate risk. However, agriculture's exposure and sensitivity to current climate risk also threatens to reduce its adaptive capacity, to the extent that near-term climate variability erodes its resilience and/or embeds knowledge, habits and systems that are inflexible and optimized to current levels of variation. It may be because farmers are so familiar with climate's changeability and risk that a disproportionate number of them deny the data supporting anthropogenic climate change, at the same time that a high proportion report actively adapting to observed climate changes. For example, in Australia, only 39% of farmers surveyed agreed that climate change was happening and was human induced (Leviston et al., 2011), while in a different survey, 75% of farmers stated that they had changed their management practices in response to perceived climate change (ABS, 2009).

Central to work on climate change adaptation is the notion of adaptive capacity: the ability to mobilize resources to adapt to anticipated or observed climate change impacts (Smit and Wandel, 2006; Moser and Ekstrom, 2010; Marshall et al., 2012). Engle argues that:

> Increasing adaptive capacity improves the opportunity of systems to manage varying ranges and magnitudes of climate impacts, while allowing for flexibility to rework approaches if deemed at a later date to be on an undesirable trajectory.
>
> (Engle, 2011)

Adaptive capacity involves the ability to draw on diverse assets at a particular point in time (e.g. skills, infrastructure, soil quality, finance and, notably, information) in order to manage concurrently near-term climate variability and long-term climate change (Nelson et al., 2010a,b). Informed by resilience thinking (cf. Folke et al., 2010), adaptive capacity needs to balance the demand for continuity and stability with that for flexibility and transformation. It stands in contrast to the older, related concept of 'coping capacity', which refers more specifically to the ability to survive near-term climate risks and other disturbances. While managing near-term climate variability and extremes remains a key task for adaptive capacity, the latter refers to the ability to achieve higher-level and longer-term success. How adaptive capacity relates to coping capacity is poorly understood (Thomas et al., 2007; Berman et al., 2012). But increasingly it is agreed that genuine adaptive capacity requires critical reappraisal of current practices, including coping strategies (e.g. Davies, 1993; Eriksen et al., 2005; Adger et al., 2009; Heltberg et al., 2009). Empirical work indicates that the process of assessing one's current adaptive capacity (e.g. by systematically considering all of the natural and human-made forms of 'capital' one has to draw upon) can be a powerful route to such critical awareness (e.g. Brown et al., 2010; Nelson et al., 2010a,b; Crimp et al., 2010). Furthermore, critical re-evaluation of actors' circumstances in light of future risks may reveal the need and/or opportunity for transformational change (e.g. a fundamental change in activity, approach, identity or goals) (Rickards and Howden, 2012). Recent work with Australian graziers suggests that, in general, their capacity to adapt to long-term climate change is limited by their lack of 'transformational capacity' – that is, the specific ability to manage a transformational shift in what they do (including, for example, a relocation or change in occupation) (Marshall et al., 2012). Overall, adaptive capacity needs to incorporate both the ability to cope with near-term risks and manage longer-term transformations, but the trade-offs involved remain unexamined.

As a valued form of information, SCFs have long been represented as an important tool in improving actors' coping capacity in light of near-term climate risks, helping to buffer them from the negative impacts of climatic extremes (e.g. Ziervogel and Calder, 2003; Ziervogel, 2004; Ash et al., 2007). In addition, some scholars have also long identified their potential role in improving actors' capacity to adapt to climate change (McKeon et al., 1993). Indeed, their use is now often assumed to be a climate change adaptation in its own right, with adoption rates of forecasts used in assessments of the extent to which producers are adapting to climate change (e.g. Marshall et al., 2011). The imagined role of seasonal forecasts in climate change adaptation is multifaceted. On the one hand, they are valued as a relatively familiar, non-threatening tool for facilitating producers' adaptation to climate variability by informing their management decisions at the seasonal scale. Increasingly viewed as vital to producers' coping capacity, they are considered to benefit adaptation by helping to identify critical periods when the climate poses threats or opportunities requiring especially decisive action if acutely negative outcomes are to be avoided or positive outcomes are to be seized (e.g. Anderson, 2003; Meinke and Stone, 2005; Marshall et al., 2011). On the other hand, SCFs also seem to be presumed to perform a more fundamental role in climate change adaptation, which is to serve as a potentially catalytic role in improving producers' climate literacy and interest in and acceptance of the reality of anthropogenic climate change. For, with their focus on climate, uncertainty and the pragmatic need to manage risk, seasonal forecasts represent a kind of 'adaptation boundary object'. A 'boundary object' refers to a shared but ambiguous focal point for discussion and practice between different groups and perspectives (Star and Griesemer, 1989). Forecasts are imagined to act in this capacity in two ways: first by providing a bridge or shared focal point between climate change-doubting producers and the researchers and policy makers working to support climate change adaptation, as authors such as Cash et al. (2006) have implied; and second by providing a kind of set of stepping stones between managing near-term and long-term climate risk (McKeon et al., 1993; McGray et al., 2007; Tubiello et al., 2007).

The original proposition about the utility of SCFs as an element of climate change adaptation developed by McKeon et al. (1993) was that *if* such forecasts were available and conveyed robust information, and if they were used appropriately, then they might be *one among several ways* of helping producers gradually, year-by-year, to adjust to climate change, because those who used them would be adjusting autonomously to changes in the frequency of climate extremes. We think this basic proposition retains value after 20 years. At the same time, we argue that since this idea was first presented, research into producers' experiences with SCFs has identified limitations to, as well as support for, this proposition.

This chapter critically appraises the perceived benefit of SCFs for climate change adaptation. Focused on a discussion of agricultural production in developed countries, the chapter speculates about the potential disadvantages of producers relying overly strongly on SCFs. It does so not in order to critique SCFs *per se*, but in order to reflect critically on what may be seen as their currently privileged position within climate change adaptation and to help protect against the risk of complacency. Some general characteristics of seasonal forecasts and how they have been used in agricultural decision making are addressed first, and then aspects of the use of SCFs as a route to climate change adaptation are critiqued.

15.2 Types of Seasonal Climate Forecasts

There are two broad types of SCFs that differ somewhat in the advantages and disadvantages they present. The first, oldest and still dominant type is statistical seasonal forecasting, which develops relationships between the variable of interest and past and present climate and oceanic observations. Such approaches range from simple statistical relationships (e.g. lagged rainfall relationship with the Southern Oscillation Index (SOI)) to complex ones (e.g. time-varying covariate relationships with several predictors). In general, they provide a degree of predictability as a function of the long-established relative stability of external forcing factors such as sea surface temperatures and temperature gradients and regional wind-fields (Charney and Shukla, 1981). Despite this relative stability, empirical forecast systems are reliant on continuously being updated with recent climate data and the predictor indices such as the SOI that rely on these. Not unexpectedly, these relationships are usually more robust for near-term forecasts (1–3 months), with the skill of the forecasts often dropping off as the timescale of the forecast extends, with a general limit in terms of utility of 6 months. In addition, the increasing non-stationarity of the climate needs to be addressed (Doblas-Reyes *et al.*, 2013), as the relationship between specific predictors and the climate variables of interest in the places of interest may change. Historically, changes in the relationship between specific predictors (e.g. SOI) and rainfall have challenged the assumption of stationarity that more traditional statistical models are built upon (Lough, 1991; Allan and Haylock, 1993; Opoku-Ankomah and Cordery, 1993; Nicholls *et al.*, 1996, 1999). The emerging signal of climate change in the present day (Zwiers and Hegerl, 2008; Stott *et al.*, 2010; Lewis and Karoly, 2013; Zwiers *et al.*, 2013) can increase the uncertainty in empirical forecasts, potentially requiring new approaches that can include such changing relationships (e.g. Kokic *et al.*, 2011, 2013).

The second type of seasonal climate forecasting uses process-based/dynamical modelling to understand the climate system. Dynamical seasonal prediction systems are based on atmospheric models or coupled global models, which are based in turn on current scientific understanding of the underlying physical and chemical phenomena related to the conservation of energy in the atmosphere and oceans, as well as on land. Technically complex and computationally- and data-demanding, ensemble means from different models and model runs are increasingly used, as they offer more robust outcomes than individual models. To generate these ensembles, both significant storage and processing infrastructure is required. In addition, the interpretation of the vast quantities of data produced through these approaches also requires considerable resources. Beginning with information about current starting conditions, model simulations are used to predict what climatic conditions will emerge, and multiple model runs establish ensembles of results, which then provide some indication of the probability of specific climate events occurring. In general, dynamical forecasts are better able to predict temperature than precipitation. They suffer from relatively large errors in their representation of mean climate, climate variability and their interaction, leading to often overly confident (under-dispersive) forecasts (Doblas-Reyes et al., 2013). The decline in forecast skill with longer-term forecasts is similar to that of statistical SCF approaches. They also tend to be more limited in their ability to produce information at the local and regional scale due to their reliance on global-scale modelling (Doblas-Reyes et al., 2013). In contrast, empirical forecasts are generally able to provide more locally relevant information that is also more transparent in its origins, more easily interpretable in the local and historical context and easier for users to run (Hansen, 2005; Stone and Meinke, 2005; Newman, 2013). This last point is particularly important in the developing world context.

One major potential advantage of dynamical forecasts over statistical approaches is that they are thought to be able to incorporate implicitly the effects of rising anthropogenic greenhouse gas forcing (as well as its existing effect on climate). With the exception of some new-generation statistical forecasts (Cressie and Wilkie, 2011; Kokic et al., 2011; Bakar and Sahu, 2013), this is something current statistical forecasts cannot do (Liniger et al., 2007; Luo et al., 2011; Doblas-Reyes et al., 2013). At the same time, the ability of dynamical forecasts to incorporate climate change also remains severely constrained. In relation to the key forecasting variable of inter-annual rainfall variability, for example, forecasting skill is limited by the failure of global circulation models (GCMs) to simulate the changes already occurring in core climate drivers such as the subtropical ridge, cyclones, El Niño Southern Oscillation (ENSO), land-surface feedbacks and enhancement of the hydrological cycle (Christensen et al., 2007; Hegerl et al., 2007; Kent et al., 2011; Durack et al., 2012; Guilyardi et al., 2012; Ramirez-Villegas et al., 2013). By contributing to systematic underestimation of climate variability (Ramirez-Villegas et al., 2013), these blind spots in GCMs collectively add significant uncertainty to both future climate projections and SCFs (Hallegatte, 2009; Wilby et al., 2009), with little likelihood of resolution in the near term (Hallegatte, 2008, 2009). These limitations have refocused attention back on statistical forecasts, which recent work indicates can be modified relatively simply to include a trend component (e.g. Palmer et al., 2008; Solomon et al., 2011). Hybrid forecasts that use the current state of multiple known climate drivers with time-varying relationships between them hold some promise as a more advanced statistical option (e.g. Kokic et al., 2011), while Bayesian approaches to develop a rigorous statistical consensus model bringing in results from statistical and dynamic models seems a logical next step that builds on the strengths of different approaches. Overall, dynamical and statistical forecasts present both advantages and disadvantages, and an over-reliance on either seems unwise. At present, they provide an approximately equal level of forecasting skill (e.g. Feser et al., 2011; Kerr, 2011; Kar et al., 2012), but differ in the types of knowledge they produce and how (Table 15.1).

15.3 Benefits of Seasonal Climate Forecasts

The prime rationale for the use of SCFs in climate risk management is that they help to inform, focus and thus improve decision making about the upcoming forecast period by providing a probabilistic indication of what rainfall and temperature conditions may emerge over the forthcoming 1–6 months. Research suggests that relative to a reliance on local experiential or historical

Table 15.1. Overview of main relative benefits of the two types of SCFs.

Relative benefits of statistical forecasts	Relative benefits of dynamical forecasts
Basis in historical record	Basis in physical and chemical laws
Simpler, cheaper and familiar to run	Greater systemic accuracy due to ability to incorporate climate change
Well-established pathway of integration with other environmental models	Forecasts of multiple climate variables at one time
More transparent origins	Based on the principle of non-stationarity and so accounts for waxing and waning of predictors
Finer-grained information possible	
Development of multi-scale forecasts is relatively simple	Able to initialize from a number of different starting conditions and so provide tailored outputs based on seasonal conditions

knowledge alone, SCF information can improve decision making by helping to correct for various biases that humans tend to fall prone to when thinking about the future (Keogh et al., 2004; Hayman and Cox, 2005). This includes availability bias, which is the common tendency to overestimate the reoccurrence of recently or vividly experienced conditions because of the impression such conditions have left on our memories (Tversky and Kahneman, 1974).

In general, improved decision making in the context of managing climate variability and change refers to the ability to minimize losses and maximize gains in the face of risk. Research into the use of SCFs in agriculture suggests that these improved decisions can lead to a range of positive outcomes. As Hayman et al. describe:

> Although far from solving the problem of climate variability, SCFs are seen as an innovation that will improve farm profitability (Hammer et al., 1996, Marshall et al., 1996, McIntosh et al., 2005), minimize land degradation (McKeon et al., 2004), assist with self-reliance and drought preparation (Drought Review Panel, 2004), and play an important part in reducing vulnerability to future climate change (McInnes et al., 2002, Meinke and Stone, 2005).
>
> (Hayman et al., 2007)

The authors note that such is the importance some commentators attach to SCFs based on ENSO (e.g. Cusack, 1983) that the technology has been viewed as comparable to the New Green Revolution or the discovery of DNA.

Use of SCFs can appeal to producers because, relative to some other forms of management advice, the information they provide is apparently of clear pertinence to producer decisions in terms of being about variables of known importance (notably rainfall) over timeframes of decision-making relevance (notably the coming season). For a growing number of producers, the SCFs available are also now relatively familiar, low cost and varied, lowering the risk they represent as a form of innovation.

To assist producers' decision making, seasonal climate information is often converted into factors of more direct interest, such as yields or economic benefit through a climate-oriented decision support system that systematically combines variables to assist with multifactorial decisions of the sort producers need to make. Many such systems have been developed for agriculture, often with high hopes and considerable effort (Jakku and Thorburn, 2010). Although these aspirations for decision support systems (DSS) have largely been unmet (Hayman and Easdown, 2002; Matthews et al., 2008), there is evidence that DSS can serve as valuable boundary objects between producers and researchers, particularly by facilitating interpersonal interactions between researchers and producers in cases where producers are involved in developing and evaluating DSS (e.g. Cash et al., 2003, 2006; Jakku and Thorburn, 2010; Dilling and Lemos, 2011; Leith, 2011; Moser and Dilling, 2011; Duru et al., 2012).

Some evaluations of producers' use of DSS over time suggest that its value to these producers may be primarily pedagogical rather than informative. In contrast to conventional notions of what DSS 'adoption' involves, McCown (2012) and McCown et al. (2012) found that producers in northern Australia used the DSS as an exploratory tool to refine their management heuristics, and then discontinued its use to focus on applying their new 'rules of thumb' – a process found even for the very first decision support system (SIRATAC) in the 1980s (Hearn and Bange, 2002). Such an outcome highlights that producer decision making is framed strongly in terms of local knowledge and pragmatism, not detailed climate information.

Using SCFs may also appeal to producers for socio-psychological reasons. Sociology of risk literature suggests that a common motivation for people to adopt formal risk management tools is to gain some sense of control over an uncertain situation by giving them a basis for decision making and allowing them to 'do something' (Reith, 2004). The psychological security that use of formal risk management tools can impart is especially effective if the tool is recognized by outside observers as evidence of responsible risk management, providing the decision maker with a form of 'insurance' against social criticism (and self-flagellation) in the event that they suffer a poor decision outcome. As Reith writes of decision makers using formal risk management tools:

> They may not know what the future holds, but by following expert advice, they know they are 'doing the right thing': armed with appropriate information, individuals can feel secure as they go about their business, knowing that they are taking active steps to protect their well-being and shape their future. It may be impossible to know 'what will happen', but at least if things do go wrong, they have the comfort of knowing that they acted 'correctly' in ways that are consonant with agreed principles of risk management.
>
> (Reith, 2004)

The value of SCFs as a source of comfort (independent of their detailed content) aligns with the emphasis some capacity-building programmes place on producers maintaining confidence in their decision making despite setbacks (e.g. Grain and Graze, see Rickards and Price, 2009). In turn, producer confidence is promoted as part of the ideal of self-reliance, both generally within agricultural and rural culture (Herbert-Cheshire and Higgins, 2004; Rickards, 2006), and more specifically in efforts to enhance producers' capacity to adapt to climate change (Hayman and Rickards, 2013).

As we discussed above, SCFs hold appeal to those interested in promoting adaptation to climate change for two further reasons: as a means of adapting to the near-term climate variability that inevitably continues to be part of climate change; and as a tool for engaging producers on the topic of climate change adaptation. In terms of the first, SCFs have been promoted by numerous programmes in Australia and overseas on the basis that they help producers to avoid the downside risks associated with difficult climatic conditions such as drought, and thus help them to avoid disaster and its negative repercussions for society, national economies and the natural environment (Hayman et al., 2007). SCFs are also promoted as a tool for enabling producers to better anticipate and exploit favourable conditions. In Australia, this opportunistic strategy is an increasingly popular response to the increased climate variability under climate change. It is based on the idea that under climate change it is likely that producers will experience fewer good climatic years, and so those that occur need to deliver maximal production (and financial) outcomes to see businesses through (Hansen et al., 2007). As discussed below, this approach is not without its critics.

The idea that SCFs represent a means for engaging producers on climate change reflects the desire to find ways of working around the ongoing climate scepticism that dominates within the Australian agricultural sector (see above). The appeal of SCFs is that, with their focus on near-term climate variability, they do not require people to 'buy in' to climate change in order to use them. Indeed, to the extent that producers' belief in longer-term climate change is perceived to be in competition with a belief in (natural) climate variability, SCFs may hold extra appeal by seeming to support and represent the latter. At the same time, by engaging producers on the topic of climate risk and encouraging them to use technology to access scientific accounts of the future (even if of the relative near term), SCFs hold promise as a means of helping producers to 'transition' to engagement on climate change proper. For example, they likely prompt producers to learn about climate systems and improve their basic climate literacy independent of specific content about climate change. Discussed above in relation to the role of SCFs as boundary objects, this possible 'developmental role' for SCFs is illustrated by Patt's idea that using SCFs is about 'learning to crawl':

> Applying SCFs is like a baby learning to crawl. Eventually what a baby will need to do is learn to walk... But crawling can help to develop the necessary coordination that walking will later demand. And crawling can offer its own rewards... So too can seasonal forecast application be a valuable way to improve adaptive capacity and accelerate future adaptation. Forecast application can also offer immediate rewards, even if the actions taken today are quite different from what will be taken in the future.
>
> (Patt, 2010)

While appealing, the idea that SCF use is valuable for improving adaptive capacity is open to challenge. We turn now to consider six critiques that such a suggestion faces.

15.4 Critiques of SCFs as a Route to Climate Change Adaptation

The extent to which SCFs ultimately help or hinder adaptation to climate change is debatable. While SCFs provide potentially valuable information, they are also open to question on six interrelated grounds: being uncertain; being insufficient; being distracting; being misleading; being unnecessary; and being divisive. We discuss these in turn before putting them into the context of the benefits forecasts can provide.

15.4.1 Seasonal climate forecasts as uncertain

The value proposition of using forecasts as an adaptation tool is fundamentally dependent on the forecasts having some 'predictive skill' (accuracy and precision) at a confidence level that the users find acceptable. Ongoing constraints on how well SCFs achieve this goal mean that the longest-standing critique of SCFs is that they do not reduce uncertainty about the future satisfactorily, despite providing better-quality information than that attainable through extrapolation of historic records.

Uncertainties are introduced into forecasts through diverse avenues, from inadequate available data to limitations in model simulations, systematic biases in model design and the inherently chaotic nature of the climate system. The latter non-linearity means that forecasts are highly sensitive to small changes in initial conditions, as first demonstrated in the 1960s by Lorenz, the father of chaos theory (Slingo and Palmer, 2011). Knowledge of this fact has stimulated the development of ensemble modelling and probabilistic forecasting noted above. Great effort has been expended trying to find ways of increasing the spread or dispersion of forecast results in order to enable them to capture more accurately the full range of possible states of the real atmosphere (Slingo and Palmer, 2011; Doblas-Reyes et al., 2013). This has in itself introduced more challenges in communicating the forecast outcomes to individuals who may not have a strong grasp of the concepts associated with probabilities and ensemble means.

How skilful SCFs are is context specific and subjective. For producers, what matters is how skilful SCFs are at their main spatial scale of concern: the local scale. Recent studies have evaluated farmers' needs for climate information in terms of scale (Dunn et al., 2012). They found that information at too fine a scale could present an unnecessary cost to producers, as regional-scale information is often adequate (at least for more strategic decisions). But many SCFs remain more spatially coarse grained than producers feel is relevant to their decision making, adding to the uncertainty SCFs pose for them (Crane et al., 2010).

From the decision maker point of view, uncertainty about SCF information can be exacerbated by their uncertainty about how uncertain the forecasts are. Forecast skill tends to alter according to when the forecast is produced (e.g. what calendar month a forecast is produced in and what type of climate system (e.g. ENSO) is dominant at the time) and the period that is being forecast (e.g. 3 months hence versus 12 months hence) (Hammer et al., 2000). One important consideration is that it needs to be communicated with clarity when there is no skill in the forecast and hence that users may need to rely on climatology or some other information such as current soil moisture status for decision making. Unfortunately, not all operational forecast systems identify such periods clearly, potentially increasing uncertainty and eroding trust in the system. Uncertainty is also generated by ambiguity in the communication of forecast uncertainty through the use of probabilities (e.g. Buizer et al., 2010; Millner and Washington, 2011; Millner, 2012). Social research indicates that users interpret probabilistic information in diverse ways, as do modellers themselves (Handmer and Proudley, 2007). Potential forecast users also differ in the technical skill they attribute to any given forecast and on what they base such assessments (Crane et al., 2010). A well-known influence is the level of trust potential users place in the individuals and groups who develop and deliver forecasts, reflecting the important role of trust as an influence on producers' adoption of new technologies and practices overall (Pannell et al., 2006; Hayman et al., 2007; Pannell and Vanclay, 2011).

From an adaptation perspective, uncertainties associated with SCFs present the risk of producer overconfidence in the accuracy or determinism of the forecast information or the appropriateness of their management decisions (e.g. believing 'this year *will* be wetter than average' or 'this is a good year to grow canola') (Lemos and Rood, 2010). Little is actually known about how SCF use or specific forecast information affects producers' decisions or their cumulative, collective or distal outcomes. Many studies (particularly with dynamic forecasts) indicate very marginal benefit from their use (e.g. Kar et al., 2012). At the same time, most evaluations of SCF use tend to be snapshots rather than longitudinal and focused on production effects at the individual business level (Hammer et al., 2000). Especially little is known about the medium- to long-term consequences of SCF-informed decisions. This means that how SCF use affects climate change adaptation is unknown. An overly deterministic interpretation of SCFs could lead to producers building 'specific resilience' at the expense of 'generic resilience' (that is, resilience to a limited array of outcomes (e.g. drought) at the expense of the capacity to manage any outcome (e.g. flood, fire, price hike, market access closure)) (Walker and Salt, 2006). While attributing any decision to any one piece of information, or any outcome to any one decision, is very difficult, longitudinal empirical and modelling work is needed to explore the cumulative adaptation implications of ongoing forecast use, as well as the implications of any radical decisions (e.g. de-stocking) made in response to specific forecasts.

15.4.2 SCFs as insufficient

The insufficiency critique of SCFs stems from the understanding that access to information about climate risk is necessary but not sufficient for adaptive capacity. As a large number of authors recognize (e.g. Vogel and O'Brien, 2006; Tribbia and Moser, 2008; McCown et al., 2009; Patt, 2010), information about climate risks – as provided by SCFs – is only one of many sources of information and knowledge that actors need to adapt to climate change. This contextual critique of SCFs focuses attention on the way climate information is or is not integrated with other considerations and forms of knowledge pertinent to decision making, including dependent and independent non-climatic variables, local knowledge and intuition, and critical assessment of the resource condition and the outcomes of past management approaches. Decision making in agriculture is especially complex and, in contrast to the climate-centrism of SCFs, producers' decision making explicitly and implicitly takes into account a broad array of factors (Crane et al., 2010; Carof et al., 2013). To the extent that they do take into account climate, the factors of most interest (e.g. extreme rather than average conditions) may not be provided by SCFs (Lemos and Rood, 2010).

Efforts to 'educate' producers about climate change futures are often guided by the idea that the experiential knowledge that producers have relied on traditionally is of declining utility, because climate change is ushering in new conditions (Speranza et al., 2010; Kalanda-Joshua et al., 2011; McDowell and Hess, 2012). But personal, experiential knowledge is not 'replaceable'. Thus, what is needed are ways to 'recalibrate' such intuitive understanding to become more sensitive to the novelty that may emerge. This means that the exploratory and educational role of SCF (discussed above) may be one of its most important aspects and should be reframed as not a failure of adoption but a valuable form of adoption and learning (e.g. Hearn and Bange, 2002). At the same time, if producers engage only briefly with SCFs and proceed to manage without them on the assumption that they have an improved understanding of climate dynamics, this could lead to a maladaptive overconfidence.

15.4.3 SCFs as distracting

Another possible but neglected source of knowledge about the future that could complement existing experiential knowledge and climate information is local analysis and imagining of possible futures (Fig. 15.1). The profound uncertainty that exists about the future requires a strong devolution of expertise about what that future might hold. As imagination is needed to conceive of possibilities (Yusoff and Gabrys, 2011), local people have a valuable role in providing both a diversity of views and values as well as a sense of the internal coherence and plausibility of possible trajectories and outcomes. For this reason, participatory exploration of possible futures using scenario planning and action learning with local communities is being used increasingly to generate valuable ideas about what might lie ahead in their region (e.g. Bardsley and Sweeney, 2010; Bardsley and Rogers, 2011). In addition, new local knowledge based on adaptive management to new conditions, reflection on past approaches and con-

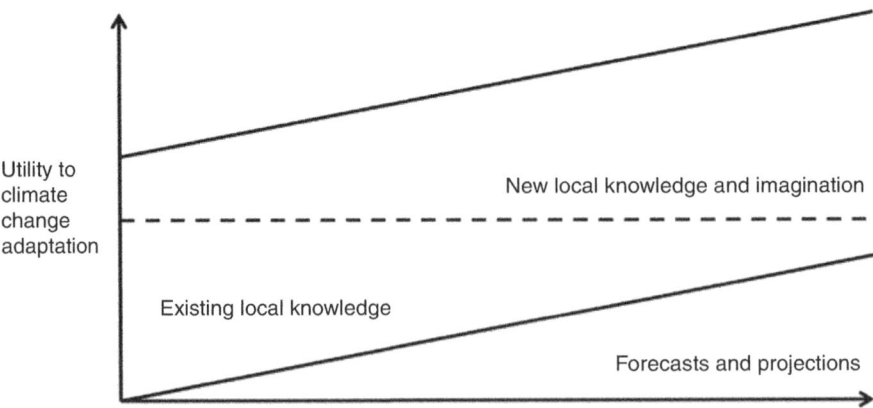

Fig. 15.1. Schematic of the possible relative importance of different forms of knowledge in understanding the future.

sideration of future options is likely to be a crucial source of new insights.

More generally, 'scenario thinking' is being encouraged increasingly as an alternative way of thinking about the future under climate change. In such a framework, the emphasis is on opening up the range of possibilities considered. It therefore stands in contrast to forecasting where the objective is to narrow down the range of possibilities under consideration. Indeed, for proponents of scenario thinking (e.g. Chermack, 2004; Frittaion et al., 2010; Smith et al., 2011), the continued use of forecasting approaches and their reliance on linear cause and effect relationships is a barrier to the type of radical openness of mind that they believe climate change adaptation requires. They focus on the risk of adaptation falling prey to the low probability but high consequence risks that climate change may intensify, or missing the opportunity to transform systems positively as part of the adaptation effort (Rickards and Wiseman, 2011; Rickards et al., 2013). While SCFs are not the target of this critique *per se*, as part of the prediction paradigm and climate-centric view of the future, they are part of the dominant approach that commentators such as O'Brien et al. (2007), Dessai et al. (2009a) and Jasanoff (2010) criticize for distracting society from both the downside and upside risks that climate change presents.

Whether seen as distracting from scenario thinking or climate change projections, SCFs are more generally critiqued as an exemplar of a limited techno-rationalist approach to climate change adaptation, which some commentators also consider formal risk management in general to be part of (e.g. Fünfgeld and McEvoy, 2011, 2012). A techno-rationalist framing of climate change adaptation is viewed critically from numerous perspectives, including those who advocate for more transformational changes (e.g. Allison and Hobbs, 2004; Pelling, 2011). Aligned with the need for transformational change are those who critique techno-rationalist options as a distraction from the ethical issues exposed by climate change. In particular, there is a strong movement to focus adaptation efforts on existing 'adaptation deficits' and vulnerabilities, including the need to improve how current climate risks are managed and related threats to ecological sustainability (McGray et al., 2007; Burton, 2009). As Lemos and Rood argue:

> [an] over-reliance on science and technical solutions might crowd out the moral imperative to do what is needed to improve livelihoods and to guarantee ecosystems' long-term sustainability.
>
> (Lemos and Rood, 2010)

Although SCF use by individual producers can actively contribute to these objectives, this critique suggests that SCFs need to be used in such a way that allows for such success, which requires that they are embedded within much larger systemic and collective efforts and repeatedly reviewed in light of the changing nature of climate interactions.

15.4.4 SCFs as misleading

Not only do SCFs potentially distract from other important perspectives on the future and decision options but also the outlook they provide may be skewed systematically towards conditions that are no longer plausible. As discussed above, anthropogenic climate change affects numerous sources of forecasting uncertainty, including present-day conditions used for indices, the relationship of these to the variable of interest and, in the case of dynamical approaches, the discrepancies it exposes in model design and fundamental climatological knowledge. As illustrated by Slingo and Palmer (2011), forecasts that fail to accommodate adequately the changes being wrought by climate change run the risk of producing results that not only fail to represent the full range of possibilities but that are directed systematically at an option space that only partially overlaps with the actual future. Such skewedness is especially a problem for common statistical forecasts because, as discussed above, they generally do not incorporate the implications of climate change.

15.4.5 SCFs as unnecessary

SCFs are also open to the critique – more often levelled at climate projections – that such information is not a hurdle requirement for decision making about the future. Work by Dessai et al. (2009a,b) and others argues that the promise of useful insights into the future posed by predictions of various types can inadvertently encourage decision makers to delay their decision making as they wait for better information to become available. That is, the very existence of forecasts can imply that the information they provide is required for responsible decision making. This can encourage practitioners to become psychologically and institutionally reliant on forecast information, even if the actual effect of such guidance on their decisions is not clear (Lemos and Rood, 2010). This risk may be especially acute in the case of SCFs, in that SCF skill tends to improve across a farming season due to shifting climate dynamics. This means that early on in a season, when producers ideally make a large number of strategic decisions about the season ahead, skilful forecasts tend to be unavailable. If they were to wait until the forecasts became more skilful, they are likely to have actually missed the main window of opportunity to act on such information. Thus, what producers need in addition to (or instead of) SCFs are insights into how best to make decisions under uncertainty. One increasingly popular general approach is the idea of 'robust decision making': making decisions that produce favourable outcomes relative to alternatives independent of what climate conditions emerge, similar to the idea of building generic resilience (Lempert and Schlesinger, 2000) (see Webber et al., Chapter 11, this volume). Hallegatte (2009) suggests related approaches, including focusing on 'no regrets' options, maintaining flexibility and building in 'safety margins'. An alternative approach applies simplified 'real options' ideas, which encourage management actions that allow for future decisions while cultivating better potential outcomes (Nelson et al., 2013).

15.4.6 SCFs as divisive

Finally, SCFs may inadvertently erode adaptive capacity at the community level if care is not taken to ensure that their availability does not weaken social capital by entrenching existing social inequities. Many adoption programmes are oriented to SCF adoption by individual producers for use in increasing their business competitiveness. If access to the forecasts or the capacity to use them is distributed unevenly across the farming population, an individualistic approach to SCF adoption and use can intensify existing differences between producers (Lemos and Rood, 2010). As Pelling (2011) notes, the poor can suffer the double burden of being not only highly exposed and sensitive to climatic impacts but also blinded to their probable occurrence by their relative lack of access to helpful technologies. For example, if some producers within a community do not have access to information forecasting a high probability of upcoming extreme conditions and then suffer severely negative consequences as a result of not amending their management decisions, their suffering stresses the bonds between producers and places a moral and potentially practical burden upon those who were forewarned. The risk management framing of climate change adaptation tends to encourage a focus on human capital and individual well-being, and it is from this angle that SCFs are generally promoted and adopted. However, it is increasingly well recognized that climate change adaptation requires social as well as human capital (e.g. Nelson et al., 2010a,b), and coordinated as well as individual actions. Thus, SCFs need to be promoted in a way that does not weaken adaptive capacity at the community and sectoral as well as individual levels. That said, there is potential that widespread access to SCF information could lead to perverse outcomes as convergent producer behaviour intersects with market forces (e.g. a large number of producers make similar management decisions in response to a certain forecast, triggering oversupply and low prices). Research is needed to help guide producers in how to negotiate the intersection of climatic and market forces, and how to use SCFs to foster social capital, collective adaptive capacity and synergistic adaptation actions. In essence, the use of SCFs must be done in such a way as to encourage routine evaluation and adaptive management in order to ensure that continual adaptation occurs.

15.5 Conclusions: Pros and Cons of SCFs as a Route to Climate Change Adaptation

As addressed in the above two sections and in Table 15.2 below, SCFs pose a range of potential advantages and disadvantages from a climate change adaptation perceptive. As a risk management tool, they provide a valuable but partial 'telescope' on to the future, exemplifying the way technology is exposing previously invisible risks in society.

Their effective use as an adaptation tool requires recognition of the risks they pose as well as expose. In particular, over-reliance on SCFs at the expense of other forms of knowledge and more systemic perspectives poses the risk of maladaptation: 'action taken ostensibly to avoid or reduce vulnerability to climate change that impacts adversely on, or increases the vulnerability of other systems, sectors or social groups' (Barnett and O'Neill, 2010).

How adaptive, or maladaptive, SCF use is demands attention across a range of timescales. Research is needed to understand how efforts to adapt to short-term climate variability affect efforts to adapt to long-term climate change, both in the case of SCFs and more broadly. More specifically, the extent to which SCFs act as an effective boundary object within agricultural adaptation efforts demands serious attention, complementing ongoing work on the effects of SCF use on producers' and communities'

Table 15.2. Summary of pros and possible cons of SCF use as a climate change adaptation tool in agriculture.

Possible pros	Possible cons
Improves producers' ability to prepare for and thus cope with or exploit particularly poor or favourable conditions	Uncertainty about the robustness of forecasts as the climate changes and the ability of consistent forecast use to improve outcomes over different time frames
Provides climate information on a timescale of more relevance to producers than that provided by climate change projections	If the SCFs use multi-model ensemble runs, they can be complex and costly to produce, and difficult to communicate and train producers in
Provides information about pertinent variables such as rainfall	Producers who lack basic knowledge in how to use probabilities may use SCF information in an overly deterministic and potentially maladaptive manner
Allows producers to engage with climate risk management using relatively familiar and low-cost technologies	Requires basic knowledge of the climate and how it is changing
Are often presented within decision support systems that integrate a range of information, convert climate variables into relevant factors such as yields and help develop their managerial capacity	Risk that SCF use distracts from other important sources of information (e.g. non-climatic factors, outcomes of past strategies, longer-term future)
Engages producers on climate risk issues without requiring them to 'buy in' to climate change as an issue	Risk that SCF use implies that forecasting information is necessary for immediate climate change adaptation action, delaying decisions
Using SCFs can enhance producers' sense of control and thus decision-making confidence	Risk that SCF use overly focuses attention on particular incremental changes, obscuring the full range of incremental and transformational options
Using SCFs can help to 'recalibrate' producers' intuition, helping them to develop 'rules of thumb' that better take into account shifting climatic conditions	Risk that SCF use increases inequalities within communities and erodes social capital and collective adaptive capacity

capacity to cope with and adapt to climate variability and extremes. Understanding the ways in which SCFs inadvertently and implicitly may undermine the steps needed to adapt to climate change in the longer term is needed in order to ensure their considerable benefits come to fruition. As climate change progresses, the types of adaptations needed are likely to be increasingly intensive and intentional. How to use SCFs in a strategic rather than an uncritical manner will be key to this progression.

Acknowledgements

We would like to thank the CSIRO Climate Adaptation Flagship for support in writing this chapter. Lauren would also like to acknowledge the support of the Melbourne Sustainable Society Institute (University of Melbourne); and Serena Schroeter for her assistance with the references and with chapter formatting.

References

ABS (Australian Bureau of Statistics) (2009) *Environmental Views and Behaviour, 2007–2008*. Category No 4626.0.55.001. ABS, Canberra.

Adger, W., Dessai, S., Goulden, M., Hulme, M., Lorenzoni, I., Nelson, D., *et al.* (2009) Are there social limits to adaptation to climate change? *Climatic Change* 93, 335–354.

Allan, R.J. and Haylock, M. (1993) Circulation features associated with the winter rainfall decrease in south-western Australia. *Australian Journal of Climate* 6, 1356–1367.

Allison, H.E. and Hobbs, R.J. (2004) Resilience, adaptive capacity, and the 'lock-in trap' of the Western Australian agricultural region. *Ecology and Society* 9, 3 (http://www.ecologyandsociety.org/vol9/iss1/art3/, accessed 16 December 2013).

Anderson, B. (2010) Preemption, precaution, preparedness: anticipatory action and future geographies. *Progress in Human Geography* 34, 777–798.

Anderson, J.R. (2003) Risk in rural development: challenges for managers and policy makers. *Agricultural Systems* 75, 161–197.

Ash, A., Mcintosh, P., Cullen, B., Carberry, P. and Smith, M.S. (2007) Constraints and opportunities in applying seasonal climate forecasts in agriculture. *Australian Journal of Agricultural Research* 58, 952–965.

Bakar, K.S. and Sahu, S.K. (2013) Spatio-temporal Bayesian modelling using R. Technical Report. (http://www.personal.soton.ac.uk/sks/research/papers/spTimeRpaper.pdf, accessed 9 December 2013).

Bardsley, D. and Rogers, G. (2011) Prioritizing engagement for sustainable adaptation to climate change: an example from natural resource management in South Australia. *Society and Natural Resources* 24, 1–17.

Bardsley, D. and Sweeney, S. (2010) Guiding climate change adaptation within vulnerable natural resource management systems. *Environmental Management* 45, 1127–1141.

Barnett, J. and O'Neill, S. (2010) Maladaptation. *Global Environmental Change–Human and Policy Dimensions* 20, 211–213.

Beck, U. (1992) *Risk Society: Towards a New Modernit.* Sage, London.

Beck, U. (2008) *World at Risk*. Polity Press, Cambridge, UK.

Berman, R., Quinn, C. and Paavola, J. (2012) The role of institutions in the transformation of coping capacity to sustainable adaptive capacity. *Environmental Development* 2, 86–100.

Brown, P.R., Nelson, R., Jacopbs, B., Kokic, P., Tracey, J., Ahmend, M., *et al.* (2010) Enabling natural resource managers to self-assess their adaptive capacity. *Agricultural Systems* 103, 562–568.

Buizer, J., Jacobs, K. and Cash, D. (2010) Making short-term climate forecasts useful: linking science and action. *Proceedings of the National Academy of Sciences* doi:10.1073/pnas.0900518107 (http://www.pnas.org/content/early/2012/01/18/0900518107.short, accessed 9 December 2013).

Burton, I. (2009) Climate change and the adaptation deficit. In: Schipper, L. and Burton, I. (eds) *The Earthscan Reader on Adaptation to Climate Change*. Earthscan, London, pp. 89–98.

Carof, M., Colomb, B. and Aveline, A. (2013) A guide for choosing the most appropriate method for multi-criteria assessment of agricultural systems according to decision-makers' expectations. *Agricultural Systems* 115, 51–62.

Cash, D.W., Clark, W.C., Alcock, F., Dickinson, N.M., Eckley, N., Guston, D.H., *et al.* (2003) Knowledge systems for sustainable development. *Proceedings of the National Academy of Sciences* 100, 8086–8091.

Cash, D.W., Borck, J.C. and Patt, A.G. (2006)

Countering the loading-dock approach to linking science and decision making – comparative analysis of El Nino/Southern Oscillation (ENSO) forecasting systems. *Science Technology and Human Values* 31, 465–494.

Charney, J.G. and Shukla, J. (1981) Predictability of monsoons. In: Lighthill, S.J. and Pearce, R.P. (eds) *Monsoon Dynamics*. Cambridge University Press, Cambridge, UK, pp. 99–109.

Chermack, T.J. (2004) Improving decision-making with scenario planning. *Futures* 36, 295–309.

Christensen, J.H., Hewitson, B., Busuioc, A., Chen, A., Gao, X., Held, I., et al. (2007) Regional climate projections. In: Solomon, S., Qin, D., Manning, M., Chen, Z., Marquis, M., Averyt, K.B., et al. (eds) *Climate Change 2007: The Physical Science Basis. Contribution of Working Group I to the Fourth Assessment Report of the Intergovernmental Panel on Climate Change*. Cambridge University Press, Cambridge, UK, and New York, pp. SM11.1–SM11.46.

Crane, T.A., Roncoli, C., Paz, J., Breuer, N., Broad, K., Ingram, K.T., et al. (2010) Forecast skill and farmers' skills: seasonal climate forecasts and agricultural risk management in the south-eastern United States. *Weather, Climate, and Society* 2, 44–59.

Cressie, N.A.C. and Wikle, C.K. (2011) *Statistics for Spatio-temporal Data*. Wiley, Hoboken, New Jersey.

Crimp, S.J., Stokes, C.J., Howden, S.M., Moore, A.M., Jacobs, B., Brown, P.R., et al. (2010) Managing MDB livestock production systems in a variable and changing climate: challenges and opportunities. *Rangelands Journal* 32, 293–304.

Cusack, D.F. (1983) Introduction: reviving the Green Revolution. In: Cusack, D.F. (ed.) *Agroclimate Information for Development: Reviving the Green Revolution*. Westview Press, Boulder, Colorado, pp. xii–xvi.

Davies, S. (1993) Are coping stragies a cop out? *Institute of Development Studies Bulletin* 24, 60–72.

Dessai, S., Hulme, M., Lempert, R. and Pielke, R. Jr (2009a) Climate prediction: a limit to adaptation? In: Adger, W.N., Lorenzoni, I. and O'Brien, K. (eds) *Adapting to Climate Change: Thresholds, Values, Governance*. Cambridge University Press, Cambridge, pp. 64–78.

Dessai, S., Hulme, M., Lempert, R. and Pielke, R. Jr (2009b) Do we need better predictions to adapt to a changing climate? *Eos* 90, 111–112.

Dilling, L. and Lemos, M.C. (2011) Creating usable science: opportunities and constraints for climate knowledge use and their implications for science policy. *Global Environmental Change* 21, 680–689.

Doblas-Reyes, F.J., Garcia-Serrano, J., Lienert, F., Biescas, A.P. and Rodrigues, L.R.L. (2013) Seasonal climate predictability and forecasting: status and prospects. *Wiley Interdisciplinary Reviews: Climate Change* 4, 245–268.

Dunn, M., Howden, M. and Lindsay, J. (2012) Quantifying user needs for future climate information in the wine-grape sector. National Climate Change Adaptation Research Facility (NCCARF) Conference 2013, 25–28 June 2013, Sydney, Australia (http://www.nccarf.edu.au/conference2013/wp-content/uploads/2013/06/NCCARF-2013-Abstracts-book_web.pdf, accessed 16 December 2013).

Durack, P.J., Wijffels, S.E. and Matear, R.J. (2012) Ocean salinities reveal strong global water cycle intensification during 1950 to 2000. *Science* 336, 455–458.

Duru, M., Felten, B., Theau, J.P. and Martin, G. (2012) A modelling and participatory approach for enhancing learning about adaptation of grassland-based livestock systems to climate change. *Regional Environmental Change* 12, 739–750.

Easterling, W.E., Aggarwal, P.K., Batima, P., Brander, K.M., Erda, L., Howden, S.M., et al. (2007) Food, fibre and forest products. In: Parry, M.L., Canziani, O.F., Palutikof, J.P., van der Linden, P.J. and Hanson, C.E. (eds) *Climate Change 2007 Impacts, Adaptation and Vulnerability. Contribution of Working Group II to the Fourth Assessment Report of the Intergovernmental Panel on Climate Change*. Cambridge University Press, Cambridge, UK, pp. 273–313.

Engle, N.L. (2011) Adaptive capacity and its assessment. *Global Environmental Change* 21, 647–656.

Eriksen, S.H., Brown, K. and Kelly, P.M. (2005) The dynamics of vulnerability: locating coping strategies in Kenya and Tanzania. *Geographical Journal* 171, 287–305.

Feser, F., Rockel, B., Von Storch, H., Winterfeldt, J. and Zahn, M. (2011) Regional climate models add value to global model data – a review and selected examples. *Bulletin of the American Meteorological Society* 92, 1181–1192.

Folke, C., Carpenter, S.R., Walker, B., Scheffer, M., Chapin, T. and Rockström, J. (2010) Resilience thinking: integrating resilience, adaptability and transformability. *Ecology and Society* 15, 9.

Frew, J.E. and Dozier, J. (2012) Environmental informatics. *Annual Review of Environment and Resources* 37, 449–472.

Frittaion, C.M., Duinker, P.N. and Grant, J.L. (2010) Narratives of the future: suspending disbelief in forest-sector scenarios. *Futures* 42, 1156–1165.

Fünfgeld, H. and McEvoy, D. (2011) *Framing Climate Change Adaptation in Policy and Practice*. Working Paper 1, Framing Adaptation in the Victorian Context Project. Victorian Centre for Climate Change Adaptation Research, and RMIT University Climate Change Adaptation Program, Melbourne (http://www.climateaccess.org/sites/default/files/Funfgeld_Framing%20Climate%20Adaptation%20in%20Policy%20and%20Practice.pdf, accessed 9 December 2013).

Fünfgeld, H. and McEvoy, D. (2012) Resilience as a useful concept for climate change adaptation? *Planning Theory and Practice* 13, 324–328.

Galaz, V., Biermann, F., Crona, B., Loorbach, D., Folke, C., Olsson, P., et al. (2012) 'Planetary boundaries': exploring the challenges for global environmental governance. *Current Opinion in Environmental Sustainability* 4, 80–87.

Giddens, A. (1991) *Modernity and Self-identity: Self and Society in the Late Modern Age*. Stanford University Press, Stanford, California.

Guilyardi, E., Bellenger, H., Collins, M., Ferrett, S., Cai, W. and Wittenberg, A. (2012) A first look at ENSO in CMIP5. *Clivar Exchanges* 58, 29–32.

Hallegatte, S. (2008) Adaptation to climate change: do not count on climate scientists to do your work. Regulation2point0, Working paper 458 (http://ideas.repec.org/p/reg/wpaper/458.html, accessed 9 December 2013).

Hallegatte, S. (2009) Strategies to adapt to an uncertain climate change. *Global Environmental Change* 19, 240–247.

Hammer, G.L., Nicholls, N. and Mitchell, C. (2000) *Applications of Seasonal Climate Forecasting in Agricultural and Natural Systems: The Australian Experience*. Kluwer Academic Publishers, Dordrecht, the Netherlands.

Handmer, J. and Proudley, B. (2007) Communicating uncertainty via probabilities: the case of weather forecasts. *Environmental Hazards* 7, 79–87.

Hansen, J.D., Liebig, M.A., Merrill, S.D., Tanaka, D.L., Krupinsky, J.M. and Stott, D.E. (2007) Dynamic cropping systems: increasing adaptability amid an uncertain future. *Journal of Agronomy* 99, 939–943.

Hansen, J.W. (2005) Integrating seasonal climate prediction and agricultural models for insights into agricultural practice. *Philosophical Transactions of the Royal Society of Biological Sciences* 360, 2037–2047.

Hayman, P. and Cox, P. (2005) Drought risk as a negotiated construct. In: Botterill, L. and Wilhite, D. (eds) *From Disaster Response to Risk Management: Australia's National Drought Policy*. Springer, Dordrecht, the Netherlands, pp. 113–126.

Hayman, P.T and Easdown, W.J. (2002) An ecology of a DSS: reflections on managing wheat crops in the northeastern Australian grains industry with WHEATMAN. *Agricultural Systems* 74, 57–78.

Hayman, P. and Rickards, L.A. (2013) Drought, climate change, farming and science: the interaction of four privileged topics. In: Botterill, L. and Cockfield, G. (eds) *Drought, Risk Management, and Policy: Decision-Making Under Uncertainty*. Taylor and Francis, Abingdon, Oxfordshire, pp. 45–68.

Hayman, P.J.C., Parton, K. and Mullen, J. (2007) How do seasonal climate forecasts compare to other innovations that farmers are encouraged to adopt. *Australian Journal of Agricultural Research* 58, 975–984.

Hearn, A.B. and Bange, M.P. (2002) SIRATAC and CottonLOGIC: persevering with DSSs in the Australian cotton industry. *Agricultural Systems* 74, 27–56.

Hegerl, G.C., Zwiers, F.W., Braconnot, P., Gillett, N.P., Luo, Y., Marengo Orsini, J.A., et al. (eds) (2007) *Climate Change 2007: The Physical Science Basis. Contribution of Working Group I to the Fourth Assessment Report of the Intergovernmental Panel on Climate Change*. Cambridge University Press, Cambridge, UK, and New York, pp. 663–745.

Heltberg, R., Siegel, P.B. and Jorgensen, S.L. (2009) Addressing human vulnerability to climate change: toward a 'no-regrets' approach. *Global Environmental Change* 19, 89–99.

Herbert-Cheshire, L. and Higgins, V. (2004) From risky to responsible: expert knowledge and the governing of community-led rural development. *Journal of Rural Studies* 20, 289–302.

Jakku, E. and Thorburn, P.J. (2010) A conceptual framework for guiding the participatory development of agricultural decision support systems. *Agricultural Systems* 103, 675–682.

Jasanoff, S. (2010) A new climate for society. *Theory, Culture and Society* 27, 233–253.

Kalanda-Joshua, M., Ngongondo, C., Chipeta, L. and Mpembeka, F. (2011) Integrating indigenous knowledge with conventional science: enhancing localised climate and weather forecasts in Nessa, Mulanje, Malawi. *Physics and Chemistry of the Earth, Parts A/B/C* 36(14–15), 996–1003.

Kar, S.C., Acharya, N., Mohanty, U.C. and Kulkarni, M.A. (2012) Skill of monthly rainfall forecasts over India using multi-model ensemble schemes. *International Journal of Climatology* 32, 1271–1286.

Kent, D.M., Kirono, D.G.C., Timbal, B. and Chiew, F.S. (2011) Representation of the Australian sub-tropical ridge in the CMIP3 models. *International Journal of Climatology* 33.1, 48–57.

Keogh, D.U., Abawi, G.Y., Dutta, S.C., Crane, A.J., Ritchie, J.W., Harris, T.R., et al. (2004) Context evaluation: a profile of irrigator climate knowledge, needs and practices in the northern Murray–Darling Basin to aid development of climate-based decision support tools and information and dissemination of research. *Australian Journal of Experimental Agriculture* 44, 247–257.

Kerr, R.A. (2011) Vital details of global warming are eluding forecasters. *Science* 334(6053), 173–174.

Kokic, P., Crimp, S. and Howden, M. (2011) Forecasting climate variables using a mixed-effect state-space model. *Environmetrics* 22, 409–419.

Kokic, P., Jin, H. and Crimp, S. (2013) Improved point scale climate projections using a block bootstrap simulation and quantile matching method. *Climate Dynamics* 41, 853–866.

Leith, P. (2011) Public engagement with climate adaptation: an imperative for (and driver of) institutional reform? In: Whitmarsh, L., O'Neill, S. and Lorenzoni, I. (eds) *Engaging the Public with Climate Change: Behaviour Change and Communication.* Earthscan, London, pp. 100–119.

Lemos, M.C. and Rood, R.B. (2010) Climate projections and their impact on policy and practice. *WIREs Climate Change* 1, 670–682.

Lempert, R.J. and Schlesinger, M.E. (2000) Robust strategies for abating climate change. *Climatic Change* 45, 387–401.

Leviston, Z., Leitch, A., Greenhill, M., Leonard, R. and Walker, I. (2011) *Australian's Views of Climate Change.* CSIRO Report. Commonwealth Scientific and Industrial Research Organisation (CSIRO), Canberra.

Lewis, S.C. and Karoly, D.J. (2013) Anthropogenic contributions to Australia's record summer temperatures of 2013. *Geophysical Research Letters* 40, 1–5.

Liniger, M., Mathis, H., Appenzeller, C. and Doblas-Reyes, F. (2007) Realistic greenhouse gas forcing and seasonal forecasts. *Geophysical Research Letters* 34, L04705.

Lough, J.M. (1991) Rainfall variations in Queensland, Australia: 1891–1986. *International Journal of Climatology* 11, 745–768.

Lubchenco, J. (1998) Entering the century of the environment: a new social contract for Science. *Science* 279, 491–497.

Luo, J.-J., Behera, S.K., Masumoto, Y. and Yamagata, T. (2011) Impact of global ocean surface warming on seasonal-to-interannual climate prediction. *Journal of Climate* 24, 1626–1646.

Lynch, C. (2008) Big data: How do your data grow? *Nature* 455, 28–29.

McCown, R.L. (2012) A cognitive systems framework to inform delivery of analytic support for farmers' intuitive management under seasonal climatic variability. *Agricultural Systems* 105, 7–20.

McCown, R.L., Carberry, P.S., Hochman, Z., Dalgliesh, N.P. and Foale, M.A. (2009) Reinventing model-based decision support with Australian dryland farmers. 1. Changing intervention concepts during 17 years of action research. *Crop and Pasture Science* 60, 1017–1030.

McCown, R.L., Carberry, P.S., Dalgliesh, N.P., Foale, M.A. and Hochman, Z. (2012) Farmers use intuition to reinvent analytic decision support for managing seasonal climatic variability. *Agricultural Systems* 106, 33–45.

McDowell, J.Z. and Hess, J.J. (2012) Accessing adaptation: multiple stressors on livelihoods in the Bolivian highlands under a changing climate. *Global Environmental Change* 22, 342–352.

McGray, H., Hammill, H., Bradley, R., Schipper, E.L. and Parry, J.E. (2007) *Weathering the Storm: Options for Framing Adaptation and Development.* World Resources Institute, Washington, DC.

McKeon, G.M., Howden, S.M., Abel, N.O.J. and King, J.M. (1993) Climate change: adapting tropical and subtropical grasslands. In: M.J. Baker (ed.) *Grasslands for Our World.* SIR Publishing, Wellington.

Marshall, N., Gordon, I. and Ash, A. (2011) The reluctance of resource-users to adopt seasonal climate forecasts to enhance resilience to climate variability on the rangelands. *Climatic Change* 107, 511–529.

Marshall, N., Park, S., Adger, W., Brown, K. and Howden, S. (2012) Transformational capacity and the influence of place and identity. *Environmental Research Letters* 7, 034022, doi:10.1088/1748-9326/7/3/034022.

Matthews, K., Schwarz, G., Rivington, M. and Miller, D. (2008) Wither agricultural DSS? *Computers and Electronics in Agriculture* 61, 149–159.

Meinke, H. and Stone, R. (2005) Seasonal and inter-annual climate forecasting: the new tool for increasing preparedness to climate variability and change in agricultural planning and operations. *Climatic Change* 70, 221–253.

Millner, A. (2012) Climate prediction for adaptation: Who needs what? *Climatic Change* 110, 143–167.

Millner, A. and Washington, R. (2011) What determines perceived value of seasonal climate forecasts? A theoretical analysis. *Global Environmental Change* 21, 209–218.

Moser, S.C. (2010) Now more than ever: the need for more societally relevant research on vulnerability and adaptation to climate change. *Applied Geography* 30, 464–474.

Moser, S.C. and Dilling, L. (2011) Communicating climate change: closing the science–action gap. In: Dryzek, J.S., Norgaard, R.B. and Schlosberg, D. (eds) *The Oxford Handbook of Climate Change and Society*. Oxford University Press, Oxford, UK, pp. 161–174.

Moser, S.C. and Ekstrom, J.A. (2010) A framework to diagnose barriers to climate change adaptation. *Proceedings of the National Academy of Sciences* 107, 22026–22031.

Nelson, R., Kokic, P., Crimp, S., Martin, P., Meinke, H., Howden, S.M., et al. (2010a) The vulnerability of Australian rural communities to climate variability and change: Part II – Integrating impacts with adaptive capacity. *Environmental Science and Policy* 13, 18–27.

Nelson, R., Kokic, P., Crimp, S., Meinke, H. and Howden, S.M. (2010b) The vulnerability of Australian rural communities to climate variability and change: Part I – Conceptualising and measuring vulnerability. *Environmental Science and Policy* 13, 8–17.

Nelson, R., Howden, M. and Hayman, P. (2013) Placing the power of real options analysis into the hands of natural resource managers – taking the next step. *Journal of Environmental Management* 124, 128–136.

Newman, M. (2013) An empirical benchmark for decadal forecasts of global surface temperature anomalies. *Journal of Climate* 26, 5260–5269.

Nicholls, N., Lavery, B., Frederiksen, C. and Drosdowsky, W. (1996) Recent apparent changes in relationships between the El Niño – southern oscillation and Australian rainfall and temperature. *Geophysical Reserach Letters* 23, 3357–3360.

Nicholls, N., Chambers, L., Haylock, M., Frederiksen, C., Jones, D. and Drosdowsky, W. (1999) Climate variability and predictability for south-west Western Australia. In: *Towards Understanding Climate Variability in South Western Australia*, Research reports on the First Phase of the Indian Ocean Climate Initiative, October 1999. Indian Ocean Climate Initiative Panel, East Perth, Western Australia, pp. 1–52 (http://www.ioci.org.au/publications/doc_download/1-first-research-report.html, accessed 16 December 2013).

O'Brien, K., Eriksen, S., Nygaard, L.P. and Schjolden, A. (2007) Why different interpretations of vulnerability matter in climate change discourses. *Climate Policy* 7, 73–88.

Opoku-Ankomah, Y. and Cordery, I. (1993) Temporal variation of relations between New South Wales rainfall and the southern oscillation. *International Journal of Climatology* 13, 51–64.

Palmer, T.N., Doblas-Reyes, F.J., Weisheimer, A. and Rodwell, M.J. (2008) Toward seamless prediction: calibration of climate change projections using seasonal forecasts. *Bulletin of the American Meteorological Society* 89, 459–470.

Pannell, D. and Vanclay, F. (eds) (2011) *Changing Land Management: Adoption of New Practices by Rural Landholders*. CSIRO, Melbourne, Australia.

Pannell, D., Marshall, G., Barr, N., Curtis, A., Vanclay, F. and Wilkinson, R. (2006) Understanding and promoting adoption of conservation practices by rural landholders. *Australian Journal of Experimental Agriculture* 46, 1407–1424.

Patt, A.G. (2010) Learning to crawl: how to use seasonal climate forecasts to build adaptive capacity. In: Adger, W.N., Lorenzoni, I. and O'Brien, K. (eds) *Adapting to Climate Change: Thresholds, Values, Governance*. Cambridge University Press, Cambridge, pp. 79–95.

Pelling, M. (2011) *Adaptation to Climate Change: From Resilience to Transformation*. Routledge, London.

Ramirez-Villegas, J., Challinor, A.J., Thornton, P.K. and Jarvis, A. (2013) Implications of regional improvement in global climate models for agricultural impact research. *Environmental Research Letters* 8, 024018, doi:10.1088/1748-9326/8/2/024018.

Reith, G. (2004) Uncertain times: the notion of 'risk' and the development of modernity. *Time and Society* 13, 383–402.

Rickards, L. (2006) Capable, enlightened and masculine: constructing English agriculturalist ideals in formal agricultural education, 1845–2003. DPhil thesis, University of Oxford, UK.

Rickards, L. and Howden, M. (2012) Transformational adaptation: agriculture and climate change. *Crop and Pasture Science* 63, 240–250.

Rickards, L. and Price, R.J. (2009) Cultural dimensions of a large-scale mixed-farming program: competing narratives of stakeholder actors. *Animal Production Science* 49, 956–965.

Rickards, L. and Wiseman, J. (2011) From

probability to possibility: using scenarios to get our heads around climate change. *The Conversation*, 10 June 2011 (https://theconversation.com/from-probability-to-possibility-using-scenarios-to-get-our-heads-around-climate-change-348, accessed 9 December 2013).

Rickards, L., Wiseman, J., Edwards, T. and Biggs, C. (2013) The problem of fit: scenario planning and climate change adaptation in the public sector. *Environment and Planning C: Government and Policy,* in press.

Rockström, J., Steffen, W., Noone, K., Persson, Å., Chapin, F.S., Lambin, E., *et al.* (2009) Planetary boundaries: exploring the safe operating space for humanity. *Ecology and Society*, 14(2), 32 (http://www.ecologyandsociety.org/vol14/iss2/art32/, accessed 9 December 2013).

Sarewitz, D.R., Pielke, R.A. and Byerly, R. (2000) *Prediction: Science, Decision Making, and the Future of Nature.* Island Press, Washington, DC.

Slingo, J. and Palmer, T. (2011) Uncertainty in weather and climate prediction. *Philosophical Transactions of the Royal Society A: Mathematical, Physical and Engineering Sciences* 369, 4751–4767.

Smit, B. and Wandel, J. (2006) Adaptation, adaptive capacity and vulnerability. *Global Environmental Change–Human and Policy Dimensions* 16, 282–292.

Smith, J.-A., Mulligan, M. and Nadarajah, Y. (2011) Scenarios for engaging a rural Australian community in climate change adaptation work. In: Ford, J.D. and Berrang-Ford, L. (eds) *Climate Change Adaptation in Developed Nations*, Advances in Global Change Research 42. Springer, the Netherlands, pp. 413–422.

Smith, P., Martino, D., Cai, Z., Gwary, D., Janzen, H.H., Kimar, P., *et al.* (2008) Greenhouse gas mitigation in agriculture. *Philosophical Transactions of the Royal Society London Biological Sciences* 363, 789–813.

Solomon, A., Goddard, L., Kumar, A., Carton, J., Deser, C., Fukumori, I., *et al.* (2011) Distinguishing the roles of natural and anthropogenically forced decadal clmate variability. *Bulletin of the American Meteorological Society* 92, 141–156.

Speranza, C.I., Kiteme, B., Ambenje, P., Wiesmann, U. and Makali, S. (2010) Indigenous knowledge related to climate variability and change: insights from droughts in semi-arid areas of former Makueni District, Kenya. *Climatic Change* 100, 295–315.

Star, S.L. and Griesemer, J.R. (1989) Institutional ecology, translations and boundary objects: amateurs and professionals in Berkeley's Museum of Vertebrate Zoology, 1907–1939. *Social Studies of Science* 19, 387–420.

Stone, R.C. and Meinke, H. (2005) Operational seasonal forecasting of crop performance. *Philosophical Transactions of the Royal Society of Biological Sciences* 360, 2109–2124.

Stott, P.A., Gillett, N.P., Hegerl, G.C., Karoly, D.J., Stone, D.A., Zhang, X., *et al.* (2010) Detection and attribution of climate change: a regional perspective. *Wiley Interdisciplinary Reviews: Climate Change* 1, 192–211.

Strauss, S. and Orlove, B. (2003) Up in the air: the anthropology of weather and climate. In: Strauss, S. and Orlove, B. (eds) *Weather, Climate, Culture.* Berg, Oxford, UK, pp. 3–14.

Thomas, D., Twyman, C., Osbahr, H. and Hewitson, B. (2007) Adaptation to climate change and variability: farmer responses to intra-seasonal precipitation trends in South Africa. *Climatic Change* 83, 301–322.

Tribbia, J. and Moser, S.C. (2008) More than information: what coastal managers need to plan for climate change. *Environmental Science and Policy* 11, 315–328.

Tubiello, F.N., Soussana, J.-F. and Howden, S.M. (2007) Crop and pasture response to climate change. *Proceedings of the National Academy of Sciences* 104, 19686–19690.

Tversky, A. and Kahneman, D. (1974) Judgement under uncertainty: heuristics and biases. *Science* 185, 1124–1131.

Verdon-Kidd, D.C. and Kiem, A.S. (2009) Nature and causes of protracted droughts in southeast Australia: comparison between the federation, WWII, and big dry droughts. *Geophysical Research Letters* 36, L22707, doi: 10.1029/2009GL041067.

Vogel, C.H. and O'Brien, K. (2006) Who can eat information? Examining the effectiveness of seasonal climate forecasts and regional climate-risk management strategies. *Climate Research* 33, 111–122.

Walker, B. and Salt, D. (2006) *Resilience Thinking: Sustaining Ecosystems and People in a Changing World.* Island Press, New York.

Wilby, R.L., Troni, J., Biot, Y., Tedd, L., Hewitson, B.C., Smith, D.G., *et al.* (2009) A review of climate risk information for adaptation and development planning. *International Journal of Climatology* 29, 1193–1215.

Yusoff, K. and Gabrys, J. (2011) Climate change and the imagination. *Wiley Interdisciplinary Reviews: Climate Change* 2, 516–534.

Ziervogel, G. (2004) Targeting seasonal climate forecasts for integration into household level decisions: the case of smallholder farmers in Lesotho. *Geographical Journal* 170, 6–21.

Ziervogel, G. and Calder, R. (2003) Climate variability and rural livelihoods: assessing the impact of seasonal climate forecasts in Lesotho. *Area* 35, 403–417.

Zwiers, F. and Hegerl, G. (2008) Climate change – attributing cause and effect. *Nature* 453, 296–297.

Zwiers, F., Alexander, L., Hegerl, G., Knutson, T., Kossin, J., Naveau, P., *et al.* (2013) Climate extremes: challenges in estimating and understanding recent changes in the frequency and intensity of extreme climate and weather events. In: Asrar, G.R. and Hurrell, J.W. (eds) *Climate Science for Serving Society.* Springer, the Netherlands, pp. 339–369.

16 Agricultural Adaptation to Climate Change: New Approaches to Knowledge and Learning

Julie Ingram

Countryside and Community Research Institute, University of Gloucestershire, Gloucester, UK

16.1 Introduction

A better understanding of how agricultural adaptation can be supported is needed if practices and systems such as those proposed in the preceding chapters are to be utilized effectively. In particular, it is important to understand how the flow of knowledge can be enhanced to support and enable adaptation. It has become clear that complex problems like climate change, and the uncertainty associated with them, require adaptive solutions and a focus on resilience at the farm level. The centrality of knowledge in formulating these solutions is apparent, both with respect to providing technological solutions and valid scientific information and to facilitating farmer learning and strengthening the adaptive capacity of farmers, institutes and communities (IAASTD, 2009).

This chapter focuses on knowledge needs for farm-level agricultural adaptation. Although there is a large literature exploring the many dimensions of, and barriers to, climate change adaptation, there has been little analysis of knowledge requirements for agricultural adaptation. This chapter therefore draws on relevant work in areas such as extension and innovation science and farmer decision making and behavioural research in the wider context of sustainable agriculture and rural development. The term 'knowledge' is used here in its broadest sense, incorporating information and advice. However, these represent qualitatively different concepts. Information comprises facts and interpretations, knowledge refers to how people understand and attribute meaning to this information, while advice implies a recommendation based on acquired knowledge.

Agricultural adaptation options are multiple and diverse and this presents challenges for the Agricultural Knowledge System (AKS), the set of agricultural actors, organizations and their linkages engaged with supporting farmer decision making. A number of approaches to providing farmers with the knowledge and means to adapt and innovate have emerged. These operate against a backdrop of transition in the AKS, with a shift away from pushing the *adoption* of technological solutions towards enabling *adaptation* through facilitation, learning and innovation. This has been in response to a growing recognition that challenges such as adapting to more sustainable and resilient systems requires new approaches to engaging farmers. Farmers' utilization of knowledge is influenced by a number of factors, including the quality and perceived relevance and credibility of the information; while their ability to use knowledge depends on institutional contexts and household status. Thus, simply supplying knowledge is not enough to achieve adaptation, wider institutional changes are also required.

Among these is the need to strengthen the AKS. This chapter therefore focuses on individual adaptations at the farm level but places the discussion within the wider context of agricultural knowledge and innovation systems where potentially planned adaptations can occur.

16.2 Knowledge Needs for Agricultural Adaptation to Climate Change

16.2.1 Adaptation to climate change – complex knowledge needs

Adaptation is generally described as those responses by individuals, groups and governments to climatic change or other stimuli that are used to reduce their vulnerability to adverse impacts (Bradshaw et al., 2004). It refers to changes in processes, practices and structures in response to the actual or perceived threat of climate change, as well as changes in social and institutional structures and technical options (Howden et al., 2007). Adaptive capacity refers to the potential or capability of a system or an individual to make these adjustments. Studies indicate that individual adaptations tend to be incremental and ad hoc, to take multiple forms, to be in response to multiple stimuli and to be constrained by economic, social, technological, institutional and political conditions (Smit and Pilifosova, 2001). Consequently, knowledge needs for farm-level individual adaptations will be complex.

16.2.2 The multiple forms of adaptations have implications for the provision of knowledge

Agricultural adaptation to climate change is a complex, multidimensional and multi-scale process that takes on a number of forms (see Webber et al., Chapter 11, this volume). A wide variety of agricultural adaptations to climatic variability and change are available (Bryant et al., 2000; Smit and Skinner, 2002; Bryan et al., 2009).

Adaptations in agriculture have been characterized according to a number of attributes, such as the form they take (technological or behavioural); purpose (to sustain current activities or to develop new ones); intent (spontaneous or planned); timing (reactive, concurrent or anticipatory); duration (short or long term); spatial extent (localized or widespread); responsibility (e.g. government, producers, etc.); and to individuals' choice options (Smit and Pilifosova, 2001). At the farm level, possible adaptations to climatic variability and change therefore are multiple and can be implemented at different levels of intensity and duration. Tactical and strategic actions are distinguished. For example, farmers can adapt tactically, or fine-tune at a micro scale, to climate change conditions by changing the timing of planting, input use and harvesting (de Loe et al., 2001), while they can also adapt strategically by altering soil management practices such as tillage (Dumanski et al., 1986) or their selection of crop types/varieties (Mendelsohn, 2000), by diversifying their farm enterprise (Kelly and Adger, 2000), or purchasing crop insurance (Smit, 1994). Such multiple forms of adaptation have implications for the provision of knowledge, tailored, place-specific advice being more appropriate than a 'one size fits all' approach.

16.2.3 Behavioural adaptation is qualitatively different from incremental adjustments

Some commentators refer to actions at the managerial level which are short-lived and consistent with existing management practices as adjustments, and to actions which result in a more fundamental change in the system as behavioural adaptation (Bryant et al., 2000). The latter goes beyond technological fixes and focuses more on long-term change to different kinds of activities and restructuring. These are evident in more dramatic ways, such as broad-scale shifts in systems, either at the farm system or community scale (Smithers and Smit, 1997). This behavioural adaptation is qualitatively different from incremental adjustments, and as such, requires different ways of providing knowledge where facilitation, rather than promoting adoption of innovations, is believed to be most appropriate. Farmers, and their supporting institutions, also need to change established frames of reference for thinking and acting. Pelling (2010) highlights the role of social learning in enabling such transformational adaptation which builds on alternative values and norms.

16.2.4 Adaptations are hard to identify

Farmers are adapting continuously to a number of stresses and changes, and their adaptation actions in response to climatic events often revolve around the complex interplay of other non-climate factors such as market forces (cost/price ratios, consumer demands, etc.), institutional factors, etc. As such, adaptation actions tend to constitute 'on-going processes, reflecting many factors or stresses, rather than discrete measures to address climate change specifically' (IPCC, 2007). Thus, not only are there multiple adaptation options available to farmers but also diverse and unpredictable adaptation responses. Because of this, there is some debate as to whether suitable adaptive strategies are actually discernible. While adaptations that are generally suitable for managing climate-related risks can be identified, some argue that it is not so easy to identify adaptations that are suitable for managing multiple risks such as downturns in commodity markets and changes to government policy and support (Dolan et al., 2001; Bradshaw et al., 2004). In this sense, adaptations are hard to 'pin down' and prescribe; formulating the right sort of knowledge to support them is equally difficult.

16.2.5 Adaptation not adoption

Although a variety of technological innovations have emerged from science to assist farmers in managing climate change, simply supplying interventions does not provide the certainty of successful adaptation; ultimately, this rests on the adoption and successful implementation of specific strategies. Considerable effort has been spent on examining the constraints to, and opportunities for, the adoption of innovations, although this 'supply-led' focus is only one approach. Adaptation is a process, and rather than relying on farmers to adopt 'bolt-on' technologies, there should be more emphasis on the adaptation of principles to place. As with other adaptive processes or systemic changes in farming, such as a shift from conventional to agroecological/organic farming or from plough to conservation tillage systems, there is no blueprint. Evidence from studies of other agricultural transitions has shown that locally adapted solutions are knowledge intensive, complex and non-prescriptive, requiring incremental learning as well as a good understanding of the local agroecosystem (Kroma, 2005). Supporting learning, rather than teaching, should therefore, it is argued, underpin approaches supporting adaptation.

16.2.6 Maximizing production is often no longer the main goal

Increasing the resilience of the system to cope with change is one of the cornerstones of climate change adaptation. Resilience is defined by the IPCC (2007) as the ability of a social or ecological system to absorb disturbances while retaining the same basic structure and ways of functioning, the capacity for self-organization and the ability to adapt to stress and change. The context in which farmers manage their farms is continually in a state of flux in response to changing circumstances. Climate change, the increasing frequency of extreme climatic events and the long-term prospect of climate variability, is a key source of uncertainty; however, it is not the only one, with the biophysical environment, markets, resources and supplies also changing unpredictably. Given such turbulence and uncertainty, farmers are seeking resilient farming systems that can be sustained despite large fluctuation in yields, prices, diseases, climate, etc. (Darnhofer et al., 2010). As a consequence, maximizing production is often no longer the main goal, and this has implications for knowledge provision.

16.2.7 Communicating risk is a key feature of climate change adaptation

Communicating information about climate variability and extremes and the associated risks requires a particular approach and specific types of information. The focus needs to shift from formulating options and knowledge for average conditions to supporting learning for uncertainty and variability. As such, rather than promoting field crops whose spatial range has been extended for average climatic conditions, there is a need now to promote broadly adapted species that are tolerant of inter-annual variations. This distinction is important, for it is these deviations from so-called normal conditions that may well define the experience of climate change (Smit et al., 1999). Also, given the context of multiple risks (e.g. market fluctuations), some argue that there is a need to consider all of the key sources of risk to provide effective decision making and learning for farmers, and thus improve decision makers' 'climate knowledge' overall (Meinke et al., 2006).

Thus, agricultural adaptation to climate change is a complex process, leading to multiple adaptation options and choices, from minor adjustments to fundamental shifts in farm systems and farmer behaviour. It is an ongoing process reflecting many factors or stresses, where risk is a central element and resilience is often the goal. Farmers operating in this challenging arena need the appropriate knowledge and support if they are to develop effective adaptation solutions. A discussion of these needs can be framed within the debates concerning the transition of the AKS. This provides a useful framework in which to understand the communication activities of many actors and institutions operating in agriculture.

16.3 Agricultural Knowledge Systems and Adaptation to Climate Change

16.3.1 Agricultural Knowledge Systems in transition

In line with the emerging challenges and transformations in agriculture associated with large-scale threats such as food security, water insecurity and resource degradation, and the new ecosystem services required from agriculture, there has been an evolution of ideas about knowledge and innovation (EU SCAR, 2012). The original AKS, understood as the 'triangle' of agricultural research, education and extension (advisory service) establishments (Rivera and Sulaiman, 2009), has made great contributions to the development of food provision and rural development. However, the need for transition in the AKS to make it fit for purpose in a new agricultural context has been widely articulated (Knickel et al., 2009). Earlier linear models of knowledge transfer from science to practice failed to represent the increasingly pluralistic and fragmented arrangements of actors, institutions, structures and multiple sources of knowledge. Instead, the notion of AKS evolved to describe a complex set of agricultural organizations and/or persons, and the links, networks and interactions between them, engaged in all knowledge processes with the purpose of supporting decision making and innovation in agriculture (Röling and Engel, 1991). In AKS theory, the debate has moved on from a concentration on the interaction between farmers and technologies/science to incorporate wider perspectives of institutional change and innovation systems perspectives (Hall et al., 2003). These theoretical developments in the AKS frame our understanding of the flows of knowledge between actors and institutions and of the approaches used to provide farmers with information, knowledge and advice.

In the agricultural extension literature, it is possible to document an evolution in theory and practice from persuasive 'knowledge transfer' approaches to more facilitative 'human development' perspectives (Röling and Jiggins, 1994). Theory and methodology has traditionally been predicated on the promotion of technological innovations with a reliance on the top-down, unilinear model of transfer from science to practice (the knowledge transfer model). This notion of a 'one-way' path was developed and adapted by a number of authors, the most pervasive being Roger's diffusion of innovation theory (Rogers, 1995) and the technology transfer (TOT) model which has underpinned the activities of many extension services and development activities.

However, this supply-driven paradigm has been criticized for reflecting what interventions are available rather than the needs of farmers in their local context, and specifically for failing to equip them with appropriate knowledge to meet the multiple challenges they are facing (Röling and Jiggins, 1994). Assumptions that farmers are a homogeneous group and as such respond uniformly to one-size-fits-all technologies, that non-adopting farmers are 'dumb', that they just need the 'right' information or technology to respond and that farmers do not contribute any knowledge of their own to the process have been shown to be flawed. In response to these criticisms, a range of 'human development' approaches emerged incorporating the principles of the participation and facilitation of farmer learning drawing on their experiences and knowledge. Here, the implication is that, given the right conditions, information, mutual interaction and opportunity, land managers will use their own knowledge and develop their own appropriate solutions to their problems. However, as Black (2000) points out, belief in a 'participation fix' may be as naïve as in a 'technological fix'. Thus, while commentators often conceptualize the top-down technology transfer based on scientific knowledge and the bottom-up participatory approaches drawing on local knowledge as two ends of a spectrum, the territory in between is of most interest.

16.3.2 Climate change adaptation and Agricultural Knowledge Systems

In agriculture, among the most frequently advocated strategies for climate adaptation is technology research and development. The development of technologies to assist farmers in managing the vagaries of weather has been an important focus of agricultural research over the past several decades, and many examples exist of previous technological innovations that have provided farmers with the means to respond to climatic limits and possibilities, both with respect to new crop varieties and better agronomic practices (see Chhetri et al., 2012, for a review). For example, developing technological capability to help improve the efficiency and resilience of the agriculture sector to enable it to respond to climate change is recommended as a future strategy in the UK (Committee on Climate Change, 2013). As such, there is an abiding belief in the ability of technology to continue to provide farmers with the needed strategic and tactical options for handling future weather-related uncertainties, although the alternative view that questions whether climate innovations in agriculture will flow when needed has been voiced with reference to a number of institutional and economic constraints (Smithers and Blay-Palmer, 2001).

Beyond the issue of technological innovation there are further critiques of this technology-centred discourse, mainly with respect to assumptions concerning on-the-ground adaptation through agricultural extension and farmer adoption of specified strategies. The supply-led approach assumes that to achieve effective responses to climate, the 'problem' lies not so much in the ability to develop innovative solutions but in the farmers' ability to adopt them, and therefore the challenge of developing effective strategies for adaptation lies with technology transfer, that is, persuading farmers to adopt (Smithers and Blay-Palmer, 2001). The acceptance and implementation of new practices at the farm level is a fundamental element of technology-related climate adaptation in agriculture, and there has been extensive behavioural research looking at socio-economic determinants that influence farmer decision making and the probability of uptake of adaptation measures (e.g. Bryan et al., 2009). These have explored, for example, farmer perceptions of risk and related farm management decisions, including those related to the use of selected technologies (Brklacich et al., 1997). However, while this research has provided insights into the process of innovation adoption, there has been less emphasis on the role of farmer knowledge in 'adapting' technologies to local conditions and on their experience of managing past climatic risks (Christoplos, 2010). Significantly, this approach fails to address fully the complex knowledge needs of adaptation which, as noted earlier, requires the adaptation of principles to place and learning rather than the adoption of technological fixes (Darnhofer et al., 2010). It is argued that adaptation in increasingly complex situations requires that the emphasis be placed on empowering individuals and groups to engage in ongoing learning. In particular, a context of uncertainty and unpredictability requires continuous learning processes that incorporate new information and experiences and individual experimentation (Funtowicz and Ravetz, 1993; Folke et al., 2003).

As reflected in the theoretical discussion of knowledge transfer, the middle ground between these two extremes, technological innovation and farmer learning, is arguably most suited to formulating approaches to supporting farm-level adaptation. While learning and drawing on local knowledge are important, new complex environmental problems need innovation, scientific input and technical know-how. It is considered that it is unfair to expect farmer groups to solve difficult and complex problems alone and that often farmers' knowledge and their ability to learn and cope unsupported are over-romanticized; this is particularly the case when considering the inherent risks associated with climate variability and extremes. Also importantly, farmers have individual learning styles; they learn about innovations in different ways and therefore need different levels of engagement. A combination of scientific and human development solutions, therefore, can be regarded as an effective model, and in practice a mix of approaches to providing knowledge for climate change adaptation are likely to be employed.

Furthermore, there has been increasing recognition that technological innovations in agriculture come from multiple sources, including public institutions, private firms and farmers. Countries that have been successful in developing location-specific technologies have been able to 'socialize' the process of technological innovation; that is, to increase interactions between farmers and their supporting institutions (Hayami and Ruttan, 1985). Such innovation of technologies at the local level is crucial for enhancing adaptive capacity of farmers. This draws on both local knowledge from farmers operating under specific climatic conditions and on scientific knowledge embedded in the institutions that are designed to minimize uncertainties at the decision level (Chhetri et al., 2012).

16.4 Approaches to Providing Knowledge for Agricultural Adaptation to Climate Change

Many options for policy-based adaptation to climate change have been identified for agriculture (Agrawal, 2008). Actions associated with building adaptive capacity, including financial incentives, developing infrastructure and capacity building in the broader user community and institutions, have been proposed by a number of commentators (see Adger, 2003, for example). With respect to knowledge, government activities typically comprise communicating climate change information, specifically improving the state of weather forecasting and building awareness of future scenarios and potential impacts; and providing information about farm-level adaptations. Extension systems have always been a key component of the AKS, and public extension (rural advisory) services have a central role in delivering policy measures. These services are critical to dealing with national food security, providing objective information, reaching disadvantaged groups and enabling farmers to deal better with risk. However, although public extension services are important, in an increasingly pluralistic and demand-driven AKS, farmer organizations, NGOs, commercial companies and public–private partnerships are also involved in providing information and support to farmers and as such will have a role in supporting climate change adaptation.

A suite of strategies and mechanisms operating within the AKS have developed at a number of scales and governance levels (local, regional, national and international) to provide farmers with the information, the tools and the means to make both short-term adjustments and longer-term, more systemic change adaptations to climate change. The nature and extent of these will reflect the varying agricultural contexts and needs of the farmers, the perceived level of vulnerability and adaptive capacity, as well as the market opportunities, institutional resource settings, policy objectives and state and function of extension services. These approaches will be aligned to, and often integrated with, approaches used to address other large-scale issues such as food security, sustainable agriculture, a range of rural development goals, which share many of the same knowledge needs as adaptation. However, climate change knowledge and advice will need specifically to take into account the risks associated with climate variability and extremes which affect vulnerable farmers (Christoplos, 2010). The following subsections describe some of the approaches.

16.4.1 Providing climate information

Communicating information about climate change to raise awareness is a key area of activity for national agencies. This is done using a range of media. Newspapers and radio have been used extensively to reach rural communities in developing countries, with digital technologies increasingly being harnessed as more people gain access to them. Community radio has played an important role in disseminating climate change information (Myers, 2008). For example, in Malawi, where 90% of households are engaged in agriculture, community radio is being used as a catalyst in communicating food security issues caused by climate change.

Improving the state of weather forecasting is a central part of many national adaptation strategies (Bradshaw et al., 2004) (see Rickards et al., Chapter 15, this volume). This involves the provision of information about climate variability and change to help reduce unpredictability associated with climate events and trends. The Kenya Meteorological Department, for example, releases seasonal forecasts on local radio with a view to helping farmers' cropping decisions. The use of web-based tools and initiatives is widespread, with investment in forecasting and early warning capacity increasing in a number of African countries (see www.africa-adapt.net).

However, to achieve its potential, rural climate change information needs to be accurate, accessible to and useful for farmers. The usefulness of climate information has been shown to be a key determinant of adaptation (Roncoli et al., 2002; Deressa et al., 2009). The quality of the climate forecasts is important. Accurate climate forecasts have been found to improve household well-being, while poor forecast information has been shown to actually be harmful to poor farmers (Ziervogel et al., 2005). Farmers' responses to forecasts also depend largely on their own experiences and observations. Under conditions of climate risk and uncertainty, farmer decisions can be based as much on personal experience (e.g. of extreme events; rainfall frequency, timing and intensity; and early or late frosts) as on forecast information, often giving greater weight to recent events (Vogel and O'Brien, 2006).

Projections of climate change over a range of time frames from short-term tactical to long-term strategic are also available to help inform farmers (as well as agribusiness and policy makers) about the implications of a changing climate. However, it is considered important to align the scales (spatial, temporal and sectoral) and reliability of the information with the scale and nature of the decision. Long-term projections of climate may not be that helpful for farm-level decision making, given the high uncertainties at the finer spatial and temporal scales (Howden et al., 2007; Newsham et al., 2011).

Partial understanding of climate impacts and uncertainty about the benefits of adaptation has been identified as a barrier to adaptation (Hammill and Tanner, 2011). Climate change information is often difficult to communicate beyond the scientific community, due to its inherent uncertainty and complexity, so that providing end users with information in a format that is appropriate to them is a challenge. The importance of creating a dialogue between those producing and those using information, often through a brokerage organization, has been highlighted. For example, in southern Africa, a brokerage exercise helped to reveal that scientists' concern with improving the confidence in predicting the start of the rainy season did not match the farmers' interest, which was on distribution throughout the season (Davis, 2012). The role of extension is important. Although extension services have always helped farmers adapt to changing climatic conditions using, for example, study circles for farmers to 'talk about the weather', these discussions now need to be scaled up and better informed through increased attention to uncertainty and vulnerability. At the same time, more effective ways of 'downscaling' climate forecasts to make them relevant to specific agrometeorological zones is needed (Christoplos, 2010).

Farmers also need to be convinced that projected climate changes are real and are likely to continue before they make any adaptive changes. In the same way, farmers also need to be confident that the projected changes will impact their enterprise significantly. Credibility of information is therefore important. This is backed up from studies in other contexts that have found that credibility, trust in the source and in the messenger is critical if it is to influence behavioural change (Hallam et al., 2012; Sutherland et al., 2012).

16.4.2 Promotion of climate change agricultural adaptation options

Providing information and advice about different adaptation options is central to government agricultural adaptation strategies (Brklacich et al., 2000; Smithers and Blay-Palmer, 2001). While governments acknowledge the value of human development approaches and locally derived adaptive solutions, they still need to draw on traditional extension models to achieve policy objectives. A large body of research has grown within the adoption–diffusion paradigm, showing the significance of information and examining the relative merits of different communication approaches (see Feder and Umali, 1993). More recent studies such as Maddison (2007) have looked at the relevance of this to climate change adaptation in Africa. Extensive research also provides insights into the determinants of farmer adaptation behaviours, as discussed later.

In the case of climate change, extension services provide information about different adaptation options and resources that might be available to help local actors adapt. Indeed, these services have been shown to be a key factor in determining farmers' decisions, facilitating adaptation and enhancing adaptive capacity both by increasing the likelihood of perceiving climate change and in encouraging a response to such a perception (Maddison, 2007; Nhemachena and Hassan, 2007; Bryan et al., 2009). However, the demands of adaptation, where the portfolio of adaptation strategies and options is so extensive, present new challenges for extension. Also, with rapid and unpredictable changes in local climates and in other factors such as markets, a new paradigm has been called for in extension that rejects blanket advice and favours tailoring advice and adaptation options to specific farmers in specific circumstances (GFRAS, 2012). It is considered that the modus operandi of many extension providers needs to change accordingly. Instead of supplying farmers with information and standard protocols about production based on average conditions, extension needs to provide a menu of options and relate this to information about seasonal weather forecasts and probabilities. According to Christoplos (2010), production maximization strategies based on producing a single variety which is expected to perform well in average weather conditions can bankrupt smallholders where increasing climate variability means that average years occur less frequently. This is a new and complex area of work for extension, where the emphasis needs to be more on resilience and less on achieving high production. Instead of encouraging farmers to specialize their production methods and adopt high-yielding varieties to be able to enter commercial markets, extension needs to provide advice on the different climate and market risks; this might entail retaining traditional production diversification strategies (agrobiodiversity) that might previously have been dismissed as irrational 'risk aversion' by extension agents. In particular, information about expected weather patterns needs to be combined with advice about what crops and varieties are appropriate in new and uncertain conditions. In addition, the need for extension agents to change their approaches from teaching to the promotion of learning with respect to climate change has been recognized (Christoplos, 2010).

16.4.3 Facilitating farmer collaboration, learning and adaptation

Participatory engagement and collective learning

Much recent climate action has concentrated on building local resilience through participatory techniques and community empowerment. As climate change affects the poorest populations and social groups most harshly, special attention has been directed to local structural inequalities and the voice and representation in decision making of these groups (Rodima-Taylor et al., 2012). In the context of farming, collaboration and participatory engagement approaches in climate change adaptation allow farmers' needs to be articulated, their practical knowledge to be considered and the values that are important to them to be recognized. Multi-level institutional partnerships have enabled the efficient transfer of agricultural technologies to farmers. Examining technological innovation in the context of agricultural adaptation to climate change in Nepal, Chhetri et al. (2012), for example, found that collaboration with farmers and NGOs was effective, as farmers were taken seriously, not only as end-users but also as active participants in the innovation of new technologies, and this led to more robust and enduring adaptation.

Participatory and collaborative approaches draw on substantial scientific knowledge of agricultural systems, but also enable the identification of a range of adaptations that scientists themselves might not explore. Involving farmers allows the assessment of the practicality, cost-effectiveness and acceptability of the options. The approach also enables solutions to be formulated that are sensitive to the complexity and variability of farmers' local production environments. Participatory approaches are also useful for developing step-wise mitigation and adaptation strategies against climate change, through systematic iterative assessment of the biophysical and the socio-economic aspects. Participatory research into climate change adaptation options can help agricultural decision makers realize that acting on the existing trends in climate now is likely to be to their advantage, as research assessing frost-free days in Australia has shown (Howden et al., 2003). The facilitation of collective or group learning has been applied in many contexts, from long-standing FAO Farmer-Field Schools, to group farmer learning and knowledge sharing in dealing with natural resource management, exemplified by the Landcare approach in

Australia. Farm monitoring discussion groups can also provide a collective learning environment. For example, in Scotland's Farming for a Better Climate programme, farmers and experts jointly formulate and assess mitigation options on Climate Change Focus Farms.

Tapping into farmer-to-farmer learning

The significance of communication within farmers' social networks, where individual members share and influence each other in a context of mutual trust and strong social capital, has been reported for a number of situations (Maddison, 2007). Often, variations in environmental perceptions and behaviour can be explained more by the character of social networks, interconnectedness and rule sharing than by demographic variables such as age and gender. Indeed, social networks, rather than the form and volume of information, have been identified as a key variable explaining whether people pay attention to climate change and enter into adaptive behavioural change (Rayner and Malone, 1998). It is suggested that networks that have already demonstrated an ability to adapt proactively to challenges might have the inherent capacity for further adaptation like climate change (Pelling et al., 2008). Studies have also highlighted the importance of social networks, social capital and relationships in facilitating or hindering adaptation in the wider community (Adger, 2003; Pelling and High, 2005; Agrawal, 2008). The need for extension efforts to target existing networks or groups with respect to messages about climate change has been recognized (Hallam et al., 2012). Another way in which farmers learn about what adaptations are appropriate is from observing and copying their neighbours. Maddison's (2007) review of the perception of, and adaptation to, climate change across 11 African countries suggests that strong neighbourhood or clustering effects of the adoption of certain technologies, on the basis of what they observe their neighbours doing, leads farmers to update their own prior beliefs.

Facilitating individual learning

Individual experiential learning has always been seen as an essential part of farmer innovation. This style of learning is considered particularly relevant to adaptation where uncertainty requires a continuous learning process that incorporates new information and experiences (Funtowicz and Ravetz, 1993; Berkes, 2009). This is the central tenet of the adaptive approach, which emphasizes the dynamic nature of the farming context. It considers that with societal and farm dynamics being uncertain, adaptability, resilience and flexibility become as important as maximizing production and income (Darnhofer et al., 2010). Learning also entails developing so-called adaptive competencies such as critical thinking, problem solving, futures thinking and hindsight, identification and control of variables affecting crops, openness to novelty and collaboration, as described by Pruneau et al. (2012) in a study of Canadian farmers' responses to climate change. Extension services can facilitate such learning through encouragement, providing a supportive environment and scientific input and verification where required.

16.4.4 Incorporating local experiences and knowledge

There is growing recognition that efforts to strengthen the resilience of farming systems needs to understand and build on local coping strategies (Eriksen et al., 2005) (see Deressa, Chapter 17, this volume). Studies have emphasized the hardy adaptive capacity that farmers display in responses to climate and other stresses, and their sophisticated strategies for coping with stress (Newsham and Thomas, 2011). These strategies include: diversified use of the landscape; mobility and access to multiple resources which increase the capacity to respond to environmental variability and change, including climate change; maintaining genetic and species diversity in fields and herds to provide a low-risk buffer in uncertain weather environments; and agricultural practices evolved in traditional farming systems (Chhetri et al., 2012; Nakashima et al., 2012). These experiences of handling climatic challenges in the past provide insights for current and future agricultural adaptation challenges.

Scholars claim that this resourcefulness and resilience demonstrated in the face of climate stresses is rooted in indigenous knowledge or traditional ecological knowledge. The value of such knowledge in dealing with climate change has been increasingly explored (see Nakashima et al., 2012, for a review). This recognition has led to the acknowledgement of indigenous knowledge in the Fourth Assessment Report (AR4) of the Intergovernmental Panel on Climate Change (IPCC) as 'an invaluable basis for developing adaptation and natural resource management strategies in response to environmental and other forms of change' (IPCC, 2007; Parry et al., 2007). This has been reaffirmed at the 32nd Session of the IPCC, which stated that 'indigenous or traditional knowledge may prove useful for understanding the potential of certain adaptation strategies that are cost-effective, participatory and sustainable' (IPCC, 2010).

The role of local knowledge(s) and capacities has long been a focus within development, 'farmer first' approaches to agricultural development, livelihoods and participation (Scoones and Thompson, 1994). Critics, however, point to the risks of mythologizing local knowledge and, where local farming practices are implicated in degradation, argue that attributing such knowledge to farmers is misguided. While there is agreement that human systems have evolved a wide range of strategies to cope with climatic risks, and that these strategies have potential applications to climate change vulnerabilities, for most systems and communities, changes in the mean condition commonly fall within the so-called 'coping range'. Many systems, however, are particularly vulnerable to changes in the frequency and magnitude of extreme events or conditions outside this coping range. Furthermore, some point to substantial losses from climatic variations and extremes which, they argue, indicate that autonomous adaptation has not been sufficient to offset the damages associated with temporal variations in climatic conditions (Smit and Pilifosova, 2001).

The debates about the relative merits of scientific and local knowledge with respect to supporting sustainable agriculture have been long running. There is now an acceptance that, where adaptation and resilience are the goal, agriculture needs to be supported by diverse knowledge systems that enable the co-production of different knowledges. Knowledge co-production is defined as 'the collaborative process of bringing a plurality of knowledge sources and types together to address a defined problem and build an integrated or systems-oriented understanding of that problem' (Armitage et al., 2011, p. 996). Giving validity to both scientific and local knowledge is thought more likely to lead to adaptive forms of environmental management and longer-lasting, more effective outcomes than relying only on one source of knowledge. Thus, while traditional knowledge, innovations and adaptation practices are seen to embody local adaptive management to the changing environment, it is the way in which they can complement scientific research, observations and monitoring that is of increasing interest (IIPFCC, 2009).

A number of examples exist of effective knowledge co-production. Indigenous observations and interpretations of meteorological phenomena can contribute to climate science by offering observations and interpretations at a much finer spatial scale and temporal depth than by climate scientists (Nakashima et al., 2012). In Africa, rainmakers in the Nganyi community of western Kenya (Guthiga and Newsham, 2011) and farmers in Nessa Village in southern Malawi (Kalanda-Joshua et al., 2011) have collaborated with meteorological scientists to produce integrated forecasts that are being disseminated by both indigenous and conventional methods to enhance community resilience. These are seen to be more intelligible, robust and locally useful seasonal forecasts, easier to understand and more relevant to the village level. Although, as Newsham et al. (2011) point out, there can be considerable cultural barriers to achieving such collaborations. In another case in north central Namibia, Ovambo farmers were found to have a sophisticated understanding of agroecological dynamics, which enabled the farming to be resilient to current climatic variability and impacts but not necessarily to future climate change impacts. Incorporating specific features of agricultural science such as the use of early maturing varieties of pearl millet, instead of traditional varieties, and fitting them in to existing patterns of land use was found to be an effective way of increasing adaptive capacity (Newsham and Thomas, 2009). Failure to engage with local farmers and incorporate their knowledge can mean that extension efforts have limited success (Nakashima et al., 2012).

16.5 Adaptive Capacity – Farmers' Ability to Engage with Knowledge

16.5.1 Enabling access to knowledge

Successful adaptation to climate change by farmers is not merely a question of providing climate information, promoting new adaptation technologies and facilitating learning; it depends on enabling access to these. Individual adaptation actions are constrained and enabled by a number of local and contextual determinants. Considerable attention has been devoted to the so-called determinants of adaptive capacity, which are the characteristics of communities, countries and regions that influence their propensity or ability to adapt, and hence their vulnerability to risks associated with climate change. The capacity to adapt to variability and change is seen to be dependent on underlying structures of vulnerability such as levels of poverty, property rights, entitlements to assets (Pelling and High, 2005); policy, institutional environment and regulatory structures (Agrawal, 2008); access to technology (Prno et al., 2011); availability of human and financial capital; the socio-economic position of the household (Adger et al., 2005; Ziervogel et al., 2006; Prno et al., 2011); access to agricultural services such as extension services and credit, electricity and markets, and tenure status (Maddison, 2007; Nhemachena and Hassan, 2007; Bryan et al., 2009); and weak market systems (Kabubo-Mariara, 2009). Put simply for individuals, their capacity to adapt to climate change, and their ability to engage with knowledge and information, 'is a function of their access to resources' (Adger, 2003, p. 29).

With respect to accessing knowledge for adaptation, studies have shown that access and the ability to respond to climate forecasts and the benefits obtained from their use are determined by a number of factors, including the policy and institutional environment and the socio-economic position of the household (Ziervogel et al., 2005; Vogel and O'Brien, 2006). Maddison's (2007) survey of 11 African countries also found that small-scale farmers' accessibility to agricultural innovations was often limited by socio-economic institutional deficiencies such as lack of credit or savings, land tenure issues and proximity to the market. In a household survey in Ethiopia and South Africa, Bryan et al. (2009) found that extension services, information on climate change and government aid facilitated adaptation among the poorest farmers, while wealthier farmers were more likely to adapt given access to land, credit and information about climate change. These studies show that information awareness raising and advice alone is insufficient. For the poorest farmers, they also require resources to implement adaptation options (Vogel and O'Brien, 2006). Although the provision of free extension advice may play a role in promoting adaptation, particularly with poor households, there are always some costs associated with acquiring knowledge, and as such, it is argued that larger farms will most likely be the first to utilize knowledge and adapt to climate change (Maddison, 2007) (see Deressa, Chapter 17, this volume). This suggests that strengthening extension alone is insufficient to ensure adaptation. Complementary activities are also required to enhance the institutional environment, particularly in the case of poorer farmers.

16.5.2 Capacity building – extension services

There is a recognized need for the mobilization of agricultural extension services to achieve a range of food security and rural development goals, and part of this is to enable farmers to understand, mitigate and adapt to new climate change challenges (Ozor and Nnaji, 2011; GFRAS, 2012). Policy makers have been urged to extend and improve adaptation extension services, ensuring that they reach small-scale subsistence farmers. However, few agricultural extension service providers can meet these challenges, as capacities are limited in terms of human resources, the effectiveness of organizations, funding and, most importantly, leadership and direction (GFRAS, 2012). Extension not only needs to provide information but also to synchronize and make accessible the materials, credit, training and information (at the right place, time and format) needed to ensure that innovations and adaptations are accessible and transaction costs minimized. Long-term institutional development is seen as key. New ways of accessing information (the Internet and mobile phones) about the weather, technological options, markets, etc., need to be anchored in a stable and coherent institutionalized extension infrastructure if they are to be effective. While a weather forecast may be helpful, it may only be useful if the farmer can discuss the implications of that forecast with respect to what to plant and how to access markets for any new varieties (Christopolos, 2010; GFRAS, 2012). The need to build capacity in delivering information and advice is not restricted to developing countries. Developed country governments have been called upon to renew their focus on disseminating climate change advice, research and technologies to farmers (Committee on Climate Change, 2013).

16.6 Understanding Farmers' Adaptation Responses

As well as understanding farmers' ability to adapt in terms of accessing resources, considerable attention has also been devoted to the personal determinants of farmer adaptive capacity. Studies of farm-level adaptation using household data sets have shown that the probability of uptake of adaptation measures to climate change is influenced by a range of personal, socio-economic attributes including: farmer education, age, farming experience and perceptions and awareness, and willingness, as well as farm factors such as size, farm assets and wealth factors (Smit et al., 1996; Brklacich et al., 1997; Bryant et al., 2000; Bradshaw et al., 2004; Maddison, 2007; Nhemachena and Hassan, 2007; Bryan et al., 2009; Below et al., 2012). This research, in aiming to predict farmer responses to prescribed adaptations and innovations, implicitly assumes a role for supply-led research.

The perception of climate change is one factor that increases the probability of adaptation (Bryant et al., 2000). However, a number of studies describe the disconnect between farmers' perceptions of climate change and actual adaptation (Smit et al., 1996). In Canada, for example, Brklacich et al. (1997) found that farmers, despite having perceived climate changes, did not adapt their farming practices. This was attributed to the declining relative importance of climate in relation to other factors influencing farm-level decision making as well as built-in resilience of the agriculture system. Similarly, Bryan et al.'s (2009) study of adaptation decisions based on household surveys conducted in Ethiopia and South Africa found that, despite having perceived changes in temperature and rainfall, a large share of farmers in both countries did not take any adaptive measures. Maddison (2007), from a study of 11 African countries, found that, whereas farming experience determined whether or not farmers perceived climate change, farmer education largely determined whether or not they adapted to it. These examples reveal the difficulty in understanding farmers' adaptive responses due to the influence of a number of competing factors. In other research, using data from over 15,000 Canadian prairie farms, Bradshaw et al. (2004) found that, rather than diversify their crops, an adaptation strategy which would reduce risks from climate change and variability, farmers in the region were actually becoming more specialized due to economic considerations, such as the high start-up costs and implications for achieving economies of scale.

The heterogeneity of human decision making and behaviour makes it hard to predict farmer responses to climate stimuli. In a study of Canadian farmers, Bryant et al. (2000) show that different agricultural systems and market systems in which farmers operate, and their different individual characteristics and contexts such as personal managerial style and entrepreneurial capacity and family circumstances, influence farmers' responses to climatic stimuli. As such, they are found to respond differently when faced with the same climate stimuli, even within the same geographic area. This accords with studies of farmer behaviour in other contexts where researchers have demonstrated the influence of different motivations, cultural norms, habits, identity, farming styles, values, goals and world views on farmers' environmental behaviour (Siebert et al., 2006). With respect to climate change adaptation, this was demonstrated in a study that found poor farmers were likely to take measures to ensure their survival, while wealthier farmers made decisions to maximize profits (Ziervogel et al., 2006). According to Rayner and Malone (1998), farmers rarely choose the best responses to climate change, that is, those that would reduce losses most effectively, often because of an established preference for, or aversion to, certain options. Given this heterogeneity and inherent variability in individual behaviour, the assumption that all farmers behave as rational economic decision makers has been shown to be untenable. This has implications for government extension programmes. It also leads some to question the widespread belief that adaptation strategies can generally be recognized by analysts and extension officers (Dolan et al., 2001).

16.7 Conclusion

Effective and resilient adaptation solutions to climate change require new approaches to knowledge and learning. A combination of scientific and human development solutions is considered most appropriate, drawing on the provision of climate information (communicating risk, weather forecasting, etc.), promotion of adaptation technologies, facilitation of farmer learning and the co-production of knowledge. In practice, the approaches used will be a function of intensity of the stress, the perceived level of vulnerability and adaptive capacity of the farmer, scale of activity, government resources, policy objectives and nature and capacity of the AKS.

The ability of governments and other bodies to implement communication approaches effectively depends largely on the capacity of the AKS. Strengthening the AKS to meet the challenges of adaptation to climate change can be directed to a number of areas, including improving the quality, credibility and usefulness of information (Howden et al., 2007; Deressa et al., 2009). Extension services are a central component of the AKS that need to be enhanced. Here, the challenge is not only communicating climate risk, providing portfolio adaptation options and supporting learning, but also ensuring that farmers can utilize this by providing support with respect to accessing markets, credit, etc. Another area for attention is addressing the science–action knowledge gap to achieve a more integrated AKS for climate change adaptation, both by enabling effective communication about climatic variability and risk between science and practice (Kristjanson et al., 2009) and by investing in applied research. The disconnection between climate science and policy which has led to a lack of use-inspired research has also been identified as a barrier to adaptation, as has the lack of adequate channels to enable farmer feedback into the innovation process.

Communication and engagement processes between individuals and institutions are an important consideration in the institutional and structural barriers to climate change adaptation, and the involvement of different actors provides the basis for the sharing of different forms of knowledge (Raymond and Robinson, 2013). Boundary organizations and extension services are seen to pay a key brokerage role in this respect (Christoplos, 2010). In the face of climate change risks and impacts that remain uncertain and unpredictable, the need for policies and action that foster such collaboration and co-production of knowledge, as well as the building of partnerships and alliances between farmers and their supporting institutions, is widely articulated (Newsham and Thomas, 2011; Chhetri et al., 2012).

A critical area of AKS development is the need to provide an enabling environment and strengthening the capacity of different actors, both to create, diffuse and use knowledge and also to access resources and services. In recognition of the wider institutional, political and commercial contexts in which farmers operate, the theoretical notion of AKS has evolved to Agricultural Innovation Systems (AIS) or Agricultural Knowledge and Innovation systems (AKIS) (Hall et al., 2003; EU SCAR, 2012). In line with these wider perspectives of the role of knowledge, there is a need to recognize that, as well as responding to climatic events, farmers are also continuously adapting to fluctuations in markets, policy, etc. Thus, to provide effective decision making and learning for farmers, the AKS needs to consider all of the key sources of risk (Meinke et al., 2006). By making agricultural adaptation measures and approaches consistent, or integrated, with other approaches and programmes that address non-climatic stresses and risks, there is a greater chance of effectiveness. As the goals underlying adaptive capacity are closely connected to wider agricultural and rural development issues, opportunities for achieving this alignment are present.

References

Adger, W.N. (2003) Social capital, collective action, and adaptation to climate change. *Economic Geography* 79, 387–404.

Adger, W.N., Arnell, N.W. and Tompkins, E.L. (2005) Successful adaptation to climate change across scales. *Global Environmental Change* 15, 77–86.

Agrawal, A. (2008) The role of local institutions in adaptation to climate change. Paper prepared for the social dimensions of climate change. Social Development Department, World Bank, Washington, DC.

Armitage, D., Berkes, F., Dale, A., Kocho-Schellenberg, E. and Patton, E. (2011) Co-management and the co-production of knowledge: learning to adapt in Canada's Arctic. *Global Environmental Change* 21, 995–1004.

Below, T.B., Mutabazi, K.D., Kirschke, D., Franke, C., Siebe, S., Siebert, R., et al. (2012) Can farmers' adaptation to climate change be explained by socio-economic household-level variables? *Global Environmental Change* 22, 223–235.

Berkes, F. (2009) Indigenous ways of knowing and the study of environmental change. *Journal of the Royal Society of New Zealand* 39, 151–156.

Black, A.W. (2000) Extension theory and practice: a review. *Australian Journal of Experimental Agriculture* 40, 493–502.

Bradshaw, B., Dolan, H. and Smit, B. (2004) Farm-level adaptation to climatic variability and change: crop diversification in the Canadian Prairies. *Climatic Change* 67, 119–141.

Brklacich, M., McNabb, D., Bryant, C. and Dumanski, I. (1997) Adaptability of agriculture systems to global climatic change: a Renfrew County, Ontario, Canada Pilot Study. In: Ilbery, B., Chiotti, Q. and Richard, T. (eds) *Agricultural Restructuring and Sustainability: A Geographical Perspective*. CAB International, Wallingford, UK, pp. 185–199.

Brklacich, M., Bryant, C., Veenhof, B. and Beauchesne, A. (2000) Agricultural adaptation to cimatic change: a comparative assessment of two types of farming in central Canada. In: Millward, H., Beesley, K., Ilbery, B. and Harrington, L. (eds) *Agricultural and Environmental Sustainability in the New Countryside*. Hignell Printing, Winnipeg, Canada, pp. 40–51.

Bryan, E., Deressa, T.T., Gbetibouo, G.A. and Ringler, C. (2009) Adaptation to climate change in Ethiopia and South Africa: options and constraints. *Environmental Science and Policy* 12, 413–426.

Bryant, C.R., Smit, B., Brklacich, M., Johnston, T., Smithers, J., Chiotti, Q., *et al.* (2000) Adaptation in Canadian agriculture to climatic variability and change. *Climate Change* 45, 181–201.

Chhetri, N., Chaudhary, P., Tiwari, P.R. and Yadaw, R.B. (2012) Institutional and technological innovation: understanding agricultural adaptation to climate change in Nepal. *Applied Geography* 33, 142–150.

Christoplos, I. (2010) *Mobilizing the Potential of Rural and Agricultural Extension.* FAO, Rome.

Committee on Climate Change (2013) Managing the land in a changing climate – Adaptation Sub-Committee progress report 2013 (http://www.theccc.org.uk/publication/managing-the-land-in-a-changing-climate/, accessed 11 December 2013).

Darnhofer, I., Bellon, S., Dedieu, B. and Milestad, R. (2010) Adaptiveness to enhance the sustainability of farming systems. A review. *Agronomy for Sustainable Development* 30, 545–555.

Davis, C. (2012) Bridging the gap: experiences of communicating climate information between producers and end-users in southern Africa. Presentation at Planet Under Pressure Conference, London 25–29 March 2012 (http://www.sarva.org.za/sadc/download/presentation_pup_2012.pdf, accessed 11 December 2013).

de Loe, R., Kreutzwiser, R. and Mararu, L. (2001) Adaptation options for the near term: climate change and the Canadian water sector. *Global Environmental Change* 11, 231–245.

Deressa, T.T., Hassan, R.M., Ringler, C., Alemu, T. and Yesuf, M. (2009) Determinants of farmers' choice of adaptation methods to climate change in the Nile Basin of Ethiopia. *Global Environmental Change* 19, 248–255.

Dolan, A.H., Smit, B., Skinner, M.W., Bradshaw, B. and Bryant, C.R. (2001) Adaptation to climate change in agriculture: evaluation of options. Occasional Paper 26. Department of Geography, University of Guelph, Canada.

Dumanski, J., Coote, D.R., Luciuk, G. and Lok, C. (1986) Soil conservation in Canada. *Journal of Soil and Water Conservation* 41, 204–210.

Eriksen, S.H., Brown, K. and Kelly, P.M. (2005) The dynamics of vulnerability: locating coping strategies in Kenya and Tanzania. *Geographical Journal* 171, 287–305.

EU SCAR (2012) Agricultural knowledge and innovation systems in transition – a reflection paper. European Commission, Standing Committee on Agricultural Research – Collaborative Working Group on Agricultural Knowledge and Innovation System (CWG AKIS), Brussels.

Feder, G. and Umali, D. (1993) The adoption of agricultural innovations: a review. *Technological Forecasting and Social Change* 43, 215–239.

Folke, C., Colding, J. and Berkes, F. (2003) Synthesis: building resilience and adaptive capacity in social-ecological systems. In: Berkes, F., Colding, J. and Folke, C. (eds) *Navigating Social-Ecological Systems. Building Resilience for Complexity and Change.* Cambridge University Press, Cambridge, UK, pp. 352–387.

Funtowicz, S. and Ravetz, J. (1993) Science for the post-normal age. *Futures* 25, 739–755.

GFRAS (2012) Building knowledge systems in agriculture. Five key areas for mobilising the potential of extension and advisory services. Global Forum for Rural Advisory Services (GFRAS) Position Paper, Summary June 2012.

Guthiga, P. and Newsham, A. (2011) Meteorologists meeting rainmakers: indigenous knowledge and climate policy processes in Kenya. *Institute of Development Studies Bulletin* 42, 104–109.

Hall, A., Sulaiman, R.V., Clark, N. and Yoganand, B. (2003) From measuring impact to learning institutional lessons: an innovation systems perspective on improving the management of international agricultural research. *Agricultural Systems* 78, 213–241.

Hallam, A., Bowden, A. and Kasprzyk, K. (2012) *Agriculture and Climate Change: Evidence on Influencing Farmers' Behaviours.* Rural Analytical Unit, Scottish Government Research Findings No 9/2012. Scottish Government Social Research, Scottish Government, UK.

Hammill, A. and Tanner, T. (2011) Harmonizing climate risk management: adaptation screening and assessment tools for development. OECD Environment Working Papers 36. OECD Publishing, Paris.

Hayami, Y. and Ruttan, V.W. (1985) *Agricultural Development: An International Perspective.* John Hopkins University Press, Baltimore, Maryland.

Howden, S.M., Meinke, H., Power, B. and McKeon, G.M. (2003) Risk management of wheat in a non-stationary climate: frost in Central Queensland. In: Post, D.A. (ed.) *Integrative Modelling of Biophysical, Social and Economic Systems for Resource Management Solutions.* Modelling and Simulation Society of Australia and New Zealand, Canberra, pp. 17–22.

Howden, S.M., Soussana, S.M.F., Tubiello, F.N., Chhetri, N., Dunlop, M. and Meinke, H. (2007) Adapting agriculture to climate change. *Proceedings of the National Academy of Sciences* 104, 19691–19696.

IAASTD (2009) *Agriculture at a Crossroads: The Synthesis Report.* International Assessment of Agricultural Knowledge, Science and Technology for Development, Washington, DC.

IIPFCC (International Indigenous Peoples Forum on Climate Change) (2009) Policy Paper on Climate Change (www.indigenousportal.com/Climate-Change/IIPFCC-Policy-Paper-on-Climate-Change-September-27-2009.html, accessed 11 December 2013).

IPCC (2007) *Summary for Policymakers, Fourth Assessment Report (AR4).* Cambridge University Press, New York.

IPCC (2010) *Review of the IPCC Processes and Procedures.* Report by the Inter Academy Council (IPCC-XXXII/Doc. 7), 32nd Session, Busan, Seoul, 11–14 October 2010. IPCC.

Kabubo-Mariara, J. (2009) Global warming and livestock husbandry in Kenya: impacts and adaptations. *Ecological Economics* 68, 1915–1924.

Kalanda-Joshua, M., Ngongondo, C., Chipeta, L. and Mpembeka, F. (2011) Integrating indigenous knowledge with conventional science: enhancing localised climate and weather forecasts in Nessa, Mulanje, Malawi. *Physics and Chemistry of the Earth* 36, 996–1003.

Kelly, P. and Adger, W. (2000) Theory and practice in assessing vulnerability to climate change and facilitating adaptation. *Climate Change* 47, 325–352.

Knickel, K., Brunori, G., Rand, S. and Proost, J. (2009) Towards a better conceptual framework for innovation processes in agriculture and rural development: from linear models to systemic approaches. *Journal of Agricultural Education and Extension* 15, 131–146.

Kristjanson, P., Robin, S., Reid, R.S., Dickson, N., Clark, W.C., Romney, D., *et al.* (2009) Linking international agricultural research knowledge with action for sustainable development. *Proceedings of the National Academy of Sciences* 106, 5047–5052.

Kroma, M.M. (2005) Organic farmer networks: facilitating learning and innovation for sustainable agriculture. *Journal of Sustainable Agriculture* 28, 5–28.

Maddison, D. (2007) The perception of and adaptation to climate change in Africa. *World Bank Policy Research Working Paper* 4308. World Bank, Washington, DC.

Meinke, H., Nelson, R., Kokic, P., Stone, R., Selvaraju, R. and Baethgen, W. (2006) Actionable climate knowledge: from analysis to synthesis. *Climate Research* 33, 101–110.

Mendelsohn, R. (2000) Efficient adaptation to climate change. *Climate Change* 45, 583–600.

Myers, M. (2008) Radio and Development in Africa: A Concept Paper. International Development Research Centre, Ottawa.

Nakashima, D.J., Galloway McLean, K., Thulstrup, H.D., Ramos Castillo, A. and Rubis, J.T. (2012) *Weathering Uncertainty: Traditional Knowledge for Climate Change Assessment and Adaptation.* UNESCO, Paris, and UNU, Darwin, Australia.

Newsham, A.J. and Thomas, D.S.G. (2009) Agricultural adaptation, local knowledge and livelihoods diversification in North-Central Namibia. *Tyndall Centre for Climate Change Research Working Paper* 140. Tyndall Centre Publications, University of East Anglia, Norwich, UK.

Newsham, A.J. and Thomas, D.S.G. (2011) Knowing, farming and climate change adaptation in North-Central Namibia. *Global Environmental Change* 21, 761–770.

Newsham A., Naess, L.O. and Guthiga, P. (2011) Farmers' knowledge and climate change adaptation: insights from policy processes in Kenya and Namibia. *Future Agricultures Policy Brief* 42, November 2011. Future Agricultures Consortium Secretariat, Institute of Development Studies, Brighton, UK.

Nhemachena, C. and Hassan, R. (2007) Micro-level analysis of farmers' adaptation to climate change in southern Africa. *IFPRI Discussion Paper* 00714. August 2007. Centre for Environmental Economics and Policy in Africa (CEEPA), Pretoria, Republic of South Africa.

Ozor, N. and Nnaji, C. (2011) The role of extension in agricultural adaptation to climate change in Enugu State, Nigeria. *Journal of Agricultural Extension and Rural Development* 3, 42–50.

Parry, M.L., Canziani, O.F., Palutikof, J.P., van der Linden, P.J. and Hanson, C.E. (2007) *Contribution of Working Group II to the Fourth Assessment Report of the Intergovernmental Panel on Climate Change.* Cambridge University Press, Cambridge, UK, and New York.

Pelling, M. (2010) *Adaptation to Climate Change: From Resilience to Transformation.* Taylor and Francis, New York.

Pelling, M. and High, C. (2005) Understanding adaptation: what can social capital offer assessments of adaptive capacity? *Global Environmental Change* 15, 308–319.

Pelling, M., High, C., Dearing, J. and Smith, D. (2008) Shadow spaces for social learning: a relational understanding of adaptive capacity to climate change within organisations. *Environment and Planning A* 40, 867–884.

Prno, J., Bradshaw, B., Wandel, J., Pearce, T., Smit, B. and Tozer, L. (2011) Community vulnerability to climate change in the context of other exposure-sensitivities in Kugluktuk, Nunavut. *Polar Research* 30, 73–63.

Pruneau, P., Kerry, J., Mallet, M., Freiman, V., Langis, J., Laroche, A., et al. (2012) The competencies demonstrated by farmers while adapting to climate change. *International Research in Geographical and Environmental Education* 21, 247–259.

Raymond, C.M. and Robinson, G.M. (2013) Factors affecting rural landholders' adaptation to climate change: insights from formal institutions and communities of practice. *Global Environmental Change* 23, 103–114.

Rayner, S. and Malone, E.L. (1998) *Human Choice and Climate Change*. Battelle Press, Columbus, Ohio.

Rivera, W. and Sulaiman, R.V. (2009) Extension: object of reform, engine for innovation. *Outlook on Agriculture* 38, 267–273.

Rodima-Taylor, D., Olwig, M.F. and Chhetri, N. (2012) Adaptation as innovation, innovation as adaptation: an institutional approach to climate change. *Applied Geography* 33, 107–111.

Rogers, E.M. (1995) *Diffusion of Innovations*, 4th edn. Free Press, New York.

Röling, N. and Engel, P. (1991) The development of the concept of agricultural knowledge and information systems. In: Rivera, W. and Gustafson, M. (eds) *Agricultural Extension: Worldwide Institutional Evolution and Forces for Change*. Elsevier, Amsterdam, pp. 125–137.

Röling, N.G. and Jiggins, J.L.S. (1994) Policy paradigm for sustainable farming. *European Journal of Agricultural Education and Extension* 1, 23–43.

Roncoli, C., Ingram, K. and Kirshen, P. (2002) Reading the rains: local knowledge and rainfall forecasting among farmers of Burkina Faso. *Society and Natural Resources* 15, 411–430.

Scoones, I. and Thompson, J. (1994) *Beyond Farmer First: Rural People's Knowledge, Agricultural Research and Extension Practice*. Intermediate Technology, London.

Siebert, R., Toogood, M. and Knierim, A. (2006) Factors affecting European farmers' participation in biodiversity policies. *Sociologia Ruralis* 46, 318–340.

Smit, B. (1994) Climate, compensation and adaptation. In: McCulloch, J. and Etkin, D. (eds) *Improving Responses to Atmospheric Extremes: The Role of Insurance and Compensation*. Environment Canada/Climate Institute, Toronto, Canada, pp. 229–273.

Smit, B. and Pilifosova, O. (2001) Adaptation to climate change in the context of sustainable development and equity. In: Mcarthy, J.J., Canziani, O.F., Learty, N.A., Dokken, D.J. and White, K.S. (eds) *Climate Change 2001: Impacts, Adaptation and Vulnerability, International Panel on Climate Change*. Cambridge University Press, Cambridge, UK, and New York, pp. 877–912.

Smit, B. and Skinner, M.W. (2002) Adaptation options in agriculture to climate change: a typology. *Mitigation and Adaptation Strategies for Global Change* 7, 85–114.

Smit, B., McNabb, D. and Smithers, J. (1996) Agricultural adaptation to climate change. *Climatic Change*, 33, 7–29.

Smit, B., Burton, I., Klein, R.J.T. and Street, R. (1999) The science of adaptation: a framework for assessment. *Mitigation and Adaptation Strategies for Global Change* 4, 199–213.

Smithers, J. and Blay-Palmer, A. (2001) Technology innovation as a strategy for climate change adaptation in agriculture. *Applied Geography* 21, 175–197.

Smithers, J. and Smit, B. (1997) Human adaptation to climatic variability and change. *Global Environmental Change* 7, 129–146.

Sutherland, L., Burton, R.J.F., Ingram, J., Blackstock, K., Slee, B. and Gotts, N. (2012) Triggering change: towards a conceptualisation of major change processes in farm decision-making. *Journal of Environmental Management* 104, 142–151.

Vogel, C. and O'Brien, K. (2006) Who can eat information? Examining the effectiveness of seasonal climate forecasts and regional climate-risk management strategies. *Climate Research* 33, 111–122.

Ziervogel, G., Bithell, M., Washington, R. and Downing, T. (2005) Agent-based social simulation: a method for assessing the impact of seasonal climate forecasts among smallholder farmers. *Agricultural Systems* 83, 1–26.

Ziervogel, G., Bharwani, S. and Downing, T.E. (2006) Adapting to climate variability: pumpkins, people and pumps. *Natural Resource Forum* 30, 294–305.

17 What are the Factors that Dictate the Choice of Coping Strategies for Extreme Climate Events? The Case of Farmers in the Nile Basin of Ethiopia

Temesgen Tadesse Deressa

Guest Scholar, Africa Growth Initiative, Global Economy and Development, Brookings Institute, Washington, DC, USA

17.1 Introduction

Droughts in Ethiopia can reduce household farm production by up to 90% of a normal year's output (World Bank, 2003) and lead to the death of livestock and humans. The recorded history of drought in Ethiopia dates back to 250 BC. Since then, droughts have occurred in different parts of the country at different times (Webb and von Braun, 1994). Studies show that the frequency of drought has increased over the past few decades, especially in the lowlands (Lautze et al., 2003; NMS, 2007). In addition to drought, floods and hailstorms also reduce yields significantly during excessively rainy seasons.

In response to these natural calamities, farmers in Ethiopia have developed different coping strategies. Several studies have identified the primary coping strategies employed by farmers during extreme climate events, especially drought. The country-level study conducted by the Ministry of Finance and Economic Development (MoFED, 2007) on the ability of farmers to cope with shocks revealed that the main coping strategies included the sale of animals, loans from relatives, the sale of crop outputs and cash savings. A study by Belay et al. (2005) revealed that arid and semi-arid pastoralists in Ethiopia temporarily migrated, adopted seasonal grazing patterns, diversified their herd and raided livestock to cope during harsh climatic conditions. Devereux and Guenther (2007) noted that productivity safety net programmes, household extension packages, voluntary resettlement and weather-indexed insurance programmes were some of the drought-coping strategies implemented by the Ethiopian government. Block and Webb (2001) revealed that households in the Wollo province of Ethiopia who survived famine were the ones with higher than average income and food consumption levels, and also had a more diversified income base as well as possessing more valuable assets, especially livestock.

Although these studies have contributed to expanding the knowledge base with regards to potential coping strategies, the factors that affect farmers' choice of a particular coping strategy, or combination thereof, have not been clearly identified. Other studies have analysed the choice of adaptation methods and climate change impact at a highly aggregated regional level (Africa), and hence have little ability to explain local-level coping strategies in the near term (Maddison, 2006; Hassan and Nhemachena, 2008; Kurukulasuriya and Mendelsohn, 2008; Seo and Mendelsohn, 2008). Some attempts have been made to study micro-level adaptation strategies for long-term climate change in Ethiopian agriculture (Deressa et al., 2009, 2011). Deressa et al. (2009) identified micro-level adaptation strategies and analysed the factors that affected the choice of the identified adaptation strategies for long-term climate change and found that different socio-economic and environmental factors affected the choice of these strategies. Deressa et al. (2011) studied how farmers in Ethiopia perceived long-term climate change and adapted accordingly. Although informative in terms of how farmers perceived long-term climate change and adapted, these studies did not indicate how farmers coped with short-term and unexpected extreme climate events and how farmers chose among different coping strategies.

This chapter argues that it is the differences in households' socio-economic and environmental characteristics that affect the choice of coping strategies rather than the type of extreme event. In other words, a household employs what is at its disposal to cope with an extreme climate event – whether that extreme might be drought, flood or hailstorm. This chapter identifies and analyses the factors affecting farmers' choice of coping mechanisms for extreme climate events in Ethiopia's Nile Basin without differentiating between coping strategies adopted by farmers against a particular event like drought, flood or hailstorm. In doing so, it argues that knowledge of the factors dictating farmers' selection of a particular coping method can assist in government policy interventions to reduce the harmful impacts of climate extremes.

17.2 Study Area and Context

This study was conducted in the Blue Nile Basin, situated in the north-western plateau of the country, which covers about 34% of the total geographic area of Ethiopia. The topography of the basin consists of highlands, hills, valleys and rock peaks ranging from 4000 m above sea level (m asl) in the north-western highlands to 700 m asl in the Sudan border (Sutcliffe and Parks, 1999). It comprises three major rivers in Ethiopia, namely the Abbay River, the Tekeze River and the Baro–Akobo rivers, with an estimated annual surface runoff of 80.83 billion m^3 $year^{-1}$, equivalent to nearly 74% of the runoff of Ethiopia's 12 river basins (MoWR, 1998). The basin provides about 62% of the flow reaching the Aswan Dam in Sudan (World Bank, 2006). The average annual rainfall in the basin ranges between 800 mm and 2200 mm $year^{-1}$. The rainfall is highly unpredictable in terms of both spatial and temporal distributions, which affects crop yield and sometimes causes total crop failure (Erkossa and Awulachew, 2009). About 40% of agricultural products and 45% of the surface water of the country are contributed by this basin (Erkossa and Awulachew, 2009). Of the nine regional states of Ethiopia, the basin covers different proportions of six of the regional states (Amhara, Oromia, Beneshangul-Gumuz, Gambela and Southern Nations, Nationalities and Peoples), with the highest proportion in Amhara (38%) and the least proportion in Southern Nations, Nationalities and People's Regional States (5%) (MoWR, 1998). Of the four traditional agroecological zones of Ethiopia; namely *bereha* (desert, below 500 m asl), *kola* (low land, 500–1500 m asl), *weynadega* (middle land, 1500–2500 m asl) and *dega* (highland, 2500–3500 m asl), the survey districts in the Blue Nile Basin selected for this study fall in the *dega*, *weynadega* and *kola* agroecological zones.

Rainfed mixed crop–livestock production systems with low levels of productivity characterize the basin. Agriculture productivity, and hence food security, is low in the basin, due mainly to unfavourable climatic conditions such as drought, poor land conservation practices and lack of advanced farming technology packages, lack of enabling infrastructure and institutional set-ups such as extension services (Erkossa and Awulachew, 2009). The basin is fully representative of Ethiopia, as all the economic, geographic and social diversities are found there, and hence the results from this study can be extrapolated to the rest of the country.

17.3 Methods of Data Collection and Analysis

A structured questionnaire was used to interview 1000 farmers in the Blue Nile Basin of Ethiopia during the 2004/05 production year. The selection of survey districts and peasant associations was based on purposive sampling, whereas the selection of farmers for interview was random. Purposive sampling in the selection of districts and peasant associations was adopted to include different environmental and social characteristics of farmers; a detailed description of data collection methods is available in Deressa et al. (2009). Among the data collected, this study focused on that collected on extreme climate events and their impacts over the past five years, the coping strategies adopted and the various socio-economic attributes of the households surveyed.

To know how farmers in the Nile Basin of Ethiopia choose from different coping options (strategies), the multinomial logit model (MNL) was adopted. MNL is suitable for analysing multi-category (more than two) discrete response variables.

The MNL model can be described as:

$$P(y=j|x) = \exp(x\beta_j) / \left[1 + \sum_{h=1}^{J} \exp(x\beta_h)\right], \ j=1,\ldots J$$

(Eqn 17.1)

where y is a random variable which takes the values $\{1, 2\ldots J\}$ for a positive integer J; y also represents coping options or categories and x is a $1 \times K$ vector of conditioning explanatory variables with first-element unity; x also represents different household, institutional and environmental attributes. The problem to be solved is how (keeping other factors constant) changes in the values of x alter the probabilities of response $P(y = j/x)$, $j = 1,2,\ldots J$. As the probabilities must sum up to one, $P(y = j/x)$ is determined once the probabilities B_j are known.

Farmers' responses on choice of coping strategies were used to construct the choice variable for this study. Seven response categories have been reported in a survey of farmers: doing nothing; selling livestock and borrowing from relatives; selling livestock and eating less; selling livestock and engaging in food-for-work; depending on food aid and liquidating other assets; and seeking off-farm income opportunities. The explanatory variables include household characteristics (i.e. education, gender, age of household head, household size, farm and non-farm income, livestock ownership, ownership of a radio, quality of houses and access to electricity); farm size, institutional factors (i.e. engagement with extension services on crop and livestock production, access to credit), social capital (which includes farmer-to-farmer extension services and the number of relatives in the *got* or village); local area and environmental characteristics, such as temperature, rainfall and agroecology (Table 17.1).

17.4 Results and Discussion

Survey results showed that the study area was drought prone, as many of the farmers surveyed had experienced drought during the past five years (Table 17.2).

A decline in crop yields coupled with loss of assets and income are the impacts most felt due to these events (Fig. 17.1). When faced with unfavourable climate extreme events that affect livelihoods in different forms, farmers in Ethiopia undertake different activities to cope. For instance, 51% of the respondents indicated that they did nothing to cope with extreme climate events. Those who did report attempting to cope with the adverse impacts indicated that they used a range of response measures. About 27% of the respondents indicated they only sold livestock to cope, whereas about 18% of the respondents combined the sale of livestock with other coping methods to withstand the calamities of severe climate events (Table 17.3).

Livestock ownership is an indicator of wealth in rural Africa, and wealthy households have higher capacity to cope with extreme events as they can liquidate their wealth to buffer against unfavourable conditions. This suggests that owning livestock serves not only as a source of

Table 17.1. Description of included explanatory variables.

Explanatory variables	Description
Education of household head	In number of years
Size of household	Number of people in the household
Gender of household head	Dummy, takes the value of 1 if male and 0 otherwise
Age of household head	In number of years
Farm income	Amount in Ethiopian birr
Non-farm income	Amount in Ethiopian birr
Livestock ownership	Dummy, takes the value of 1 if owned and 0 otherwise
Extension on crop and livestock	Dummy, takes the value of 1 if visited and 0 otherwise
Farmer-to-farmer extension	Dummy, takes the value of 1 if there is and 0 otherwise
Access to credit	Dummy, takes the value of 1 if there is access and 0 otherwise
Relatives in *got*	In number of people
Farm size	In hectares
Local agroecology *kola* (lowland)	Dummy, takes the value of 1 if *kola* and 0 otherwise
Local agroecology *weynadega* (mid-land)	Dummy, takes the value of 1 if *weynadega* and 0 otherwise
Local agroecology *dega* (highland)	Dummy, takes the value of 1 if *dega* and 0 otherwise
Temperature	Annual average for the 2004/05 survey period (°C)
Precipitation	Annual average for the 2004/05 survey period (mm)
Ownership of a radio	Dummy, takes the value of 1 if owned and 0 otherwise
Type of roof	Dummy, takes the value of 1 if roof is of corrugated iron sheets and 0 otherwise
Access to electricity	Dummy, takes the value of 1 if there is access and 0 otherwise

Table 17.2. Major extreme events encountered by surveyed farmers (out of 1000 respondents).

Shock	Per cent of farmers
Drought	38.0
Hailstorm	22.5
Flood	14.2
None	2.0
Other	21.3

labour for farming and manure for fertilizer (Yirga, 2007) but also as an insurance against shocks. In the absence of access to formal credit institutions, farmers borrow from each other (borrowed from relatives) to meet their needs when they cannot depend on savings. Similar findings were recorded by Gbetibouo *et al.* (2010), who indicated that farmers borrowed either from the bank or relatives to cope with drought in South Africa. Farmers are also forced to reduce their food consumption during harsh climatic conditions, to cope with food shortages (eating less). Ding *et al.* (2005) revealed that farmers in China reduced their food consumption by up to 31% to cope with drought. This coping strategy is generally observed among survey farmers with higher farm incomes. This could be due to the fact that farmers with high farm incomes can afford to consume more than the survival requirements during good weather conditions and do not necessarily need to depend on food aid or other means of survival as they can afford to reduce their consumption during extreme climate events. Income diversification by engaging in food-for-work and in off-farm jobs is the other coping strategy. Governmental and non-governmental organizations provide food (food-for-work) in exchange for activities such as constructing rural roads, water harvesting and soil conservation undertaken by farmers in many parts of rural Ethiopia. Farmers also engage in off-farm income-generating activities as daily labourers in suburban and urban areas, and the production of local farm tools and handwoven clothes to sell in local markets. Moreover, farmers also keep valuable assets such as gold and silver, which could be sold during

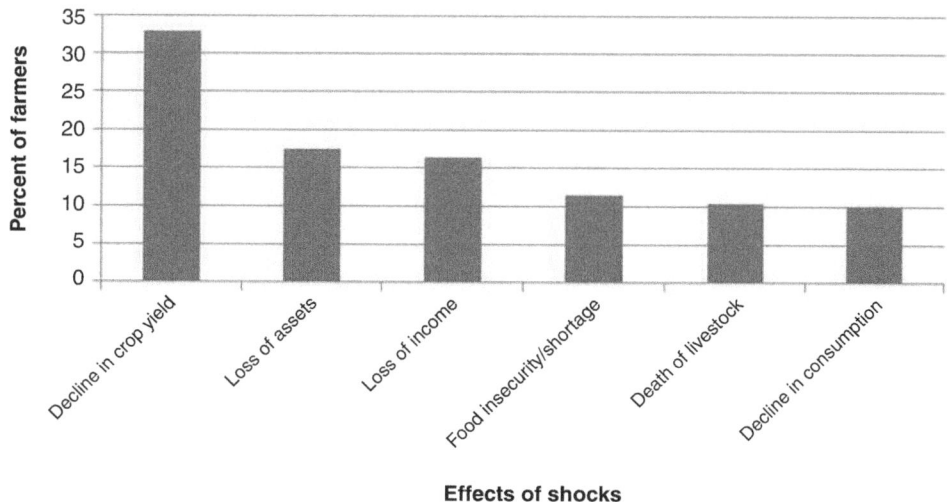

Fig. 17.1. Effects of shocks on surveyed farmers.

Table 17.3. Reported response strategies for coping with extreme climate.

Coping strategies	Number of farmers
Did nothing	503
Sold livestock	263
Sold livestock and borrowed from relatives	106
Sold livestock and ate less	35
Sold livestock and engaged in food-for-work	34
Depended on food aid and liquidated other assets	21
Sought off-farm opportunities	18

emergency situations. Block and Webb (2001) showed that households who survived famine in the Wollo region of Ethiopia had a more diversified income base and more valuable assets on hand, especially livestock. Finally, farmers also cope by depending on food aid being given to them, especially by international aid agencies.

The estimation of the MNL model for this study was undertaken by normalizing one category, which is normally referred to as the reference state or the base category. In this analysis, 'doing nothing' is the reference state. Estimation results of the MNL model's parameters are reported in Table 17.4. The estimated coefficients were compared with the base category of 'doing nothing' in response to extreme climate events. The parameter estimates reveal that different factors affect the choice of coping strategy.

The gender of the head of the household being male, the age of the head of the household, owning livestock and access to extension increase positively and significantly the likelihood of selling livestock as a coping mechanism (Table 17.4). The reason why male-headed households cope by selling livestock could be due to the fact that in low-income countries male-headed households are generally richer than female-headed households (Tenge and Hella, 2004), and livestock is one source of asset for wealthier rural households. Livestock ownership also facilitates coping by selling

Table 17.4. Parameter estimates of the multinomial logit model of climate coping strategies.

Explanatory variables	Sold livestock	Sold livestock and borrowed from relatives	Sold livestock and ate less	Sold livestock and engaged in food-for-work	Depended on food aid and liquidated other assets	Sought off-farm opportunities
Education of household head	1.057 (0.143)	0.991 (0.869)	0.909 (0.429)	0.970 (0.755)	0.968 (0.755)	1.065 (0.543)
Size of household	1.036 (0.433)	1.002 (0.976)	1.118 (0.286)	0.857 (0.225)	0.939 (0.590)	1.153 (0.285)
Gender of household head	3.066** (0.003)	2.740* (0.038)	1.604 (0.513)	6.605 (0.092)	2.098 (0.358)	0.753 (0.741)
Age of household head	1.024** (0.004)	1.013 (0.238)	1.012 (0.551)	0.959* (0.079)	1.004 (0.850)	1.006 (0.819)
Farm income	1.000 (0.194)	1.000 (0.679)	1.000** (0.021)	1.000** (0.034)	1.000 (0.219)	1.000 (0.851)
Non-farm income	1.000 (0.113)	1.000 (0.611)	1.000 (0.471)	0.999 (0.082)	0.998 (0.240)	1.000 (0.875)
Farm size	0.843 (0.054)	0.736* (0.023)	0.840 (0.430)	1.184 (0.365)	0.773 (0.302)	0.812 (0.488)
Livestock ownership	4.555*** (0.001)	2.988* (0.046)	2.401 (0.292)	1.970 (0.446)	0.924 (0.925)	0.420 (0.224)
Extension on crop and livestock	2.333*** (0.001)	3.141*** (0.001)	6.271** (0.009)	3.847* (0.037)	1.421 (0.560)	2.850 (0.146)
Access to credit	1.436 (0.112)	1.281 (0.425)	1.234 (0.694)	0.282 (0.127)	1.089 (0.888)	0.000 (1.000)
Farmer-to-farmer extension	2.508*** (0.000)	2.957** (0.002)	2.712 (0.082)	10.054*** (0.001)	2.157 (0.231)	2.140 (0.278)
Relatives in *got*	0.998 (0.715)	1.002 (0.789)	0.999 (0.899)	1.005 (0.613)	0.967 (0.294)	0.991 (0.594)
Local agroecology *kola*	0.596* (0.081)	0.079*** (0.000)	0.525 (0.412)	0.015** (0.003)	0.274 (0.147)	0.658 (0.699)
Local agroecology *weynadega*	0.935 (0.803)	0.604 (0.150)	0.437 (0.177)	0.142** (0.001)	0.468 (0.256)	0.959 (0.955)
Temperature	1.176** (0.050)	1.398* (0.011)	1.373 (0.130)	2.411*** (0.001)	1.577** (0.020)	0.600** (0.033)
Precipitation	−0.980** (0.040)	0.987 (0.954)	−0.961* (0.060)	−0.944** (0.018)	0.984 (0.173)	0.983 (0.107)
Ownership of a radio	0.999 (0.997)	1.797 (0.109)	0.890 (0.839)	1.597 (0.448)	0.581 (0.455)	0.280 (0.110)
Type of roof	1.094 (0.713)	0.940 (0.852)	3.030* (0.063)	1.103 (0.866)	1.571 (0.481)	0.836 (0.806)
Access to electricity	1.069 (0.761)	1.060 (0.846)	1.159 (0.760)	0.783 (0.627)	2.125 (0.227)	0.839 (0.776)
Constant	0.004*** (0.001)	0.000*** (0.000)	0.000* (0.028)	0.000** (0.005)	0.000* (0.015)	3051.708 (0.087)
Observations	791					
LI	−831.109					
Chi^2	410.318					

Notes: p-values in parentheses; *$p < 0.05$; **$p < 0.01$; ***$p < 0.001$.

livestock, as those who own livestock can sell that livestock to cope with extreme climate events. The reason why access to extension increases livestock selling as a coping strategy could be associated with the advice the farmers get from extension agents. Extension agents can obtain more information from the government on the timing of food aid delivery. If the delivery of food aid is late, farmers can be advised to sell their livestock earlier in order to be able to survive on the income from the sale. Additionally, extension agents could advise farmers to sell their livestock early to save for future consumption, as the value of livestock decreases the longer the extreme climate event continues, or even results in massive death of livestock, as has been witnessed in Ethiopia during drought. The reason why the selling of livestock increases with the age of the farmer could be associated with their experience of the effects of drought. For instance, older farmers may have experienced attempts earlier in their lives to retain livestock but ended up losing most or all of them. Conversely, older farmers could have more livestock (wealth) than younger and established farmers and, therefore, could afford to sell in order to cope.

The results from this study further reveal that male-headed households, asset ownership (land, livestock and better-quality houses), access to extension services and living in the *kola* agroecology zone are factors that increase the likelihood of farmers resorting to combining the sale of livestock with other response options to cope. Asset ownership or wealth gives high levels of resilience to shocks, due to insurance, social safety nets and entitlement programmes (Cutter et al., 2000). In low-income countries, especially rural Africa, the number of livestock owned, ownership of land and quality of residential homes are commonly used as indicators of wealth (Vyas and Kumaranayake, 2006). Farmers who have access to extension services are better positioned to borrow from relatives, reduce food consumption and engage in food-for-work, in addition to selling livestock as a coping mechanism. This could be associated with the advice they get from extension agents. For instance, extension agents can advise farmers to strengthen their social networks and develop more trust to fight hardships together by informally lending money to each other. Farmers may also be advised by extension agents to reduce their consumption to meet the challenges, as extension agents receive information from government and non-governmental organizations on relief programmes (most often, food aid arrives very late, after much damage has been experienced). Moreover, extension agents can guide and enrol farmers in food-for-work programmes, as they have information on the opportunities in rural areas. Farmer-to-farmer extension (obtaining information from other farmers) is a form of social capital (Deressa et al., 2009) in which farmers share information on the opportunities and constraints that they face. Farmers who are better positioned to share information with other farmers have more likelihood of selling livestock, borrowing from relatives and engaging in food for work, or combinations of these. More experienced farmers, and farmers who have access to information (extension), can advise other farmers to sell their livestock to cope, instead of keeping them and waiting until they die. Moreover, most often, rural households share information about risks with their relatives and this enables them to borrow from their relatives in order to cope. Additionally, farmer-to-farmer extension also enables farmers to learn of the existence of and benefits from engaging in food-for-work, and thus enhance their ability to cope.

The fact that farmers living in the *kola* agroecology zone sell livestock to cope could be because *kola* is lowland, and the majority of lowland farmers in Ethiopia rear livestock as their main source of livelihood. Climate factors such as temperature and rainfall also influence how farmers choose to cope. Results show that increasing temperature aggravates drier conditions and forces farmers to sell livestock, to borrow and engage in food-for-work, to depend on food aid and to liquidate other assets. Conversely, increasing precipitation reduces the likelihood of selling livestock, eating less and engagement in food-for-work programmes by easing moisture constraints on crop growth. These results are in line with the fact that Ethiopia is a drought-prone country, where increasing temperatures are stressful to farmers, while precipitation relieves farmers from the constraints of drought.

Although different precipitation prediction models give conflicting results of both increasing and decreasing precipitation, all agree that the temperature in Ethiopia will increase in the coming years (Strzepek and McCluskey, 2006). Strzepek and McCluskey (2006) used three climate prediction models based on two scenarios from the IPCC Special Report on Emission Scenarios (SRES) (Table 17.5). These models are: the Coupled Global Climate Model (CGCM2) (Flato and Boer, 2001); the Hadley Centre Coupled Model (HadCM3) (Senior and Mitchell, 2000); and the Parallel Climate Model (PCM) (Washington et al., 2000). The two SRES scenarios used in the study are the A2 and B2 scenarios. The A2 scenario describes a world in which population growth, per capita economic growth and technological changes are heterogeneous across regions. The B2 scenario describes a world in which the population increases continuously across the globe at a rate less than A2. This scenario is an intermediate level of economic development that is oriented towards environmental protection and social equity, with a focus on local and regional levels (IPCC, 2007). Additionally, forecasts by NMS (2007) indicate that the mean annual temperature of Ethiopia will increase in the range of 1.7–2.1°C by 2050 and 2.7–3.4°C by 2080. Moreover, according to NMS (2007), over the past 60 years, some years in Ethiopia have been characterized by dry rainfall conditions, resulting in droughts and famine, whereas others are characterized by wet conditions; these climatic conditions are expected to be sustained in the future. The fact that the models agree that climate has changed in an unfavourable direction in the past and will continue to persist in the future needs major policy attention to enhance the coping capacity of farmers in Ethiopia.

Thus, different policies should be implemented by the Ethiopian government to enhance the coping capacity of farmers. First, policy that emphasizes building the capacity of female-headed households, such as access to credit and education, can enhance the coping capacity of female-headed households. Teaching farmers about savings and asset holding and investments, such as livestock, can insulate farmers from harsh climatic conditions. Investment in extension can also facilitate the flow of information on coping mechanisms by availing farmers of information on different coping strategies within the reach of farming communities. Encouraging informal social networks can

Table 17.5. Climate predictions for 2050 and 2100 (changes from a 1961–1990 base for SRES A2 and B2). (Source: Strzepek and McCluskey, 2006.)

Model	Temperature change (°C)		Precipitation change (%)	
	2050	2100	2050	2100
CGCM2				
A2	3.30	8.00	−13	−28
B2	2.90	5.10	−13	−28
HadCM3				
A2	3.80	9.40	9	22
B2	3.80	6.70	9	22
PCM				
A2	2.30	5.50	5	12
B2	2.30	4.00	5	12

also strengthen social ties and enhance risk sharing through the exchange of resources in times of unfavourable climatic conditions. Policies that also target the creation of off-farm employment opportunities can prepare Ethiopian farmers to cope better with future extreme climate events.

17.5 Conclusions

Earlier studies in Ethiopia have attempted to analyse the factors that affect the choice of adaptation methods and how farmers perceive long-term climate change at the local level. Although informative, these earlier studies did not include knowledge about how farmers coped with short-term and unexpected extreme climate events and how they chose among short-term coping strategies. This study has contributed to the current literature by identifying the factors that affect farmers' choice of coping strategies to extreme climate events (mainly drought, flood and hailstorm) in the Nile Basin of Ethiopia. The multinomial logit (MNL) model was used to analyse the determinants of farmers' choice of coping strategies based on data obtained from a household survey of farmers during the 2004/05 production year in the country's Nile Basin region. Results from the MNL model show that different socio-economic and environmental factors affect the mechanisms used to cope with extreme climate events. Factors that influence coping positively include education of the head of the household, gender of the household head being male, farm income, livestock ownership, access to extension on crop and livestock production, farmer-to-farmer extension, temperature, ownership of a radio and better-quality homes.

Policies should encourage income generation and asset holding, especially livestock, both of which will enable consumption smoothing during, and immediately after, harsh climatic events. Moreover, government policies should focus on the provision of agroecology-based technology packages and the strengthening of productive safety net programmes to support coping.

Government policies and investment strategies that support the provision of, and access to, education, extension services for crop and livestock production and information on climate and coping measures are necessary to adapt better to climate change. In addition, policy interventions that encourage informal social networks (financially or materially) can promote group discussions and better information flows, thus enhancing the ability to cope with climate change.

Acknowledgement

Support and funding for this chapter was received from the ACCFP Fellowship. The author also would like to thank the START Secretariat for support and administration of the fellowship and IFPRI for hosting him for one year.

References

Belay, K., Beyene, F. and Manig, W. (2005) Coping with drought among pastoral and agro-pastoral communities in eastern Ethiopia. *Journal of Rural Development* 28, 185–210.

Block, S. and Webb, P. (2001) The dynamics of livelihood diversification in post-famine Ethiopia. *Food Policy* 26, 333–350.

Cutter, S.L., Mitchel, J.T. and Scott, M.S. (2000) Revealing the vulnerability of people and places: a case study of Georgetown county, South Carolina. *Annals of the Association of American Geographers* 90, 713–737.

Deressa, T.T., Hassan, R.M., Ringler, C., Alemu, T. and Yesuf, M. (2009) Determinants of farmers' choice of adaptation methods to climate change in the Nile Basin of Ethiopia. *Global Environmental Change* 19, 248–255.

Deressea, T., Hassan, R. and Ringler, C. (2011) Perception of and adaptation to climate change by farmers in the Nile basin of Ethiopia. *Journal of Agricultural Science* 149, 23–31.

Devereux, S. and Guenther, B. (2007) Social protection and agriculture in Ethiopia. Country case study paper prepared for a review commissioned by the Food and Agriculture Organization on Social Protection and Support to Small Farmer Development (http://www.fao.

org/fileadmin/templates/tc/spfs/pdf/FAO_2008_-_Social_Protection__Study_Case_Ethiopia.pdf, accessed 16 December 2009).

Ding, S., Sushil, P., Chuanbo, C. and Humnath, B. (2005) Drought and farmers' coping strategies in poverty-afflicted rural China (http://www.eldis.org/assets/Docs/18238.html, accessed 12 December 2013).

Erkossa, T. and Awulachew, S. (2009) Agricultural productivity in Ethiopian Nile and interventions. Poster presented at the CPWF Nile Basin Focal Project Final Workshop, Dar es Salaam, Tanzania.

Flato, G.M. and Boer, G.J. (2001) Warming asymmetry in climate change simulations. *Geophysical Research Letters* 28, 195–198.

Gbetibouo, G.A., Claudia, R. and Hassan, R. (2010) Vulnerability of the South African farming sector to climate change and variability: an indicator approach. *Natural Resources Forum* 34, 175–187.

Hassan, R. and Nhemachena, C. (2008) Determinants of African farmers' strategies for adapting to climate change: multinomial choice analysis. *African Journal of Agricultural and Resource Economics* 2, 83–104.

IPCC (2007) *Climate Change 2007 – The Physical Science Basis*. Contribution of Working Group I to the Fourth Assessment Report of the Intergovernmental Panel on Climate Change. Solomon, S., Qin, D., Manning, M., Chen, Z., Marquis, M., Averyt, K.B., et al. (eds). Cambridge University Press, Cambridge, UK, and New York.

Kurukulasuriya, P. and Mendelsohn, R. (2008) Crop switching as a strategy for adapting to climate change. *African Journal of Agricultural and Resource Economics* 2, 105–125.

Lautze, S., Aklilu, Y., Raven-Roberts, A., Young, H., Kebede, G. and Learning, J. (2003) Risk and vulnerability in Ethiopia: Learning from the past, responding to the present, preparing for the future. Report for the US Agency for International Development. Addis Ababa.

Maddison, D. (2006) The perception of and adaptation to climate change in Africa. CEEPA. Discussion Paper No 10. Centre for Environmental Economics and Policy in Africa. University of Pretoria, Pretoria, South Africa.

MoFED (Ministry of Finance and Economic Development) (2007) Ethiopia: Building on progress: a plan for accelerated and sustained development to end poverty (PASDEP). Annual Progress Report. MoFED, Addis Ababa.

MoWR (Ministry of Water Resources) (1998) Tekeze River Basin Integrated Development Master Plan Project. Executive Summary, FDRE (Federal Democratic Republic of Ethiopia). MoWR, Addis Ababa.

NMS (National Meteorological Services) (2007) *Climate Change National Adaptation Program of Action (NAPA) of Ethiopia*. NMS, Addis Ababa.

Senior, C.A. and Mitchell, J.F. (2000) The time dependence of climate sensitivity. *Geophysical Research Letters* 27(17), 2685–2688.

Seo, N. and Mendelsohn, R. (2008) Animal husbandry in Africa: climate change impacts and adaptations. *African Journal of Agricultural and Resource Economics* 2, 65–82.

Strzepek, K. and McCluskey, A. (2006) District level hydro-climatic time series and scenario analysis to assess the impacts of climate change on regional water resources and agriculture in Africa. CEEPA Discussion Paper 13. Center for Environmental Economics and Policy in Africa, University of Pretoria, South Africa.

Sutcliffe, J.V. and Parks, Y.P. (1999) The hydrology of the Nile. IAHS Special Publication No 5. IAHS Press, Institute of Hydrology, Wallingford, Oxfordshire, UK.

Tenge De Graaff, J. and Hella, J.P. (2004) Social and economic factors affecting the adoption of soil and water conservation in West Usambara highlands, Tanzania. *Land Degradation and Development* 15, 99–114.

Vyas, S. and Kumaranayake, L. (2006) Constructing socioeconomic status indices: how to use principal component analysis. *Health Policy and Planning* 21, 459–468.

Washington, W.M., Weatherly, J.W., Meehl, G.A., Semtner, J., Bettge, T.W., Craig, A.P., et al. (2000) Parallel Climate Model (PCM): Control and transient scenarios. *Climate Dynamics* 16, 755–774.

Webb, P. and von Braun, J. (1994) *Famine and Food Security in Ethiopia: Lessons for Africa*. John Wiley and Sons, Chichester, UK.

World Bank (2003) Ethiopia: Risk and Vulnerability Assessment (https://openknowledge.worldbank.org/bitstream/handle/10986/8693/262750ET.txt?sequence=2, accessed 12 December 2013).

World Bank (2006) *Ethiopia: Managing Water Resources to Maximize Sustainable Growth*. World Bank Agriculture and Rural Development Department. Washington, DC.

Yirga, C.T. (2007) The dynamics of soil degradation and incentives for optimal management in central highlands of Ethiopia. PhD thesis, University of Pretoria, Pretoria, South Africa.

Index

Agricultural adaptation
    Agricultural Knowledge System (AKS), 253–254
        adaptive capacity, 264–265
        agricultural extension services, 265
        challenges and transformations, 256–257
        climate change adaptation, 257–258
        climate change adaptation options, 260–261
        climate information, 259–260
        collective learning, 261–262
        communication and engagement process, 266–267
        development, 267
        farmer-to-farmer learning, 262
        individual experiential learning, 262
        local experiences and knowledge, 262–264
        participatory engagement, 261–262
        policy-based adaptation, 258
        public extension services, 258–259
        strategies and mechanisms, 259
    farmers' adaptation responses, 265–266
    knowledge needs
        adaptive capacity, 254
        behavioural adaptation, 254–255
        'bolt-on' technologies, 255
        communicating information, 256–257
        multidimensional and multiscale process, 254
        production maximization, 255–256
        stresses and changes, 255
        technological innovations, 255
    policy-based adaptation, 258–259
Agricultural and Grazing Industries Survey (AAGIS), 59
Agricultural Knowledge System (AKS), 253–254
Agricultural Model Intercomparison and Improvement Project (AgMIP), 210–212
Agroecology, 71
Agroforestry
    agricultural land use
        climatic effects of trees, 218
        conservation agriculture, 218
        superficial soil tillage, 217–218
        tree-soil-crop interactions, 218
    buffering functions in landscape, 227
        active and flexible shade management, 219
        buffering factor, 219
        economic portfolio effects, 219
        microclimatic effects, 218
        qualitative initial survey, 219
        tree-soil-crop interactions, 218–220
    global circulation models, 217
    microclimatic differences, 217
    mitigation and adaptation, 216–217
    tree planting, 216
    trees and progressive climate change, 225–226
    trees buffering temperature, 220–222
    urban heat island effect, 217
    water balance modification
        fertility islands, 223–224
        hydraulic equilibration, 224–225
        hydraulic redistribution, 223
        water vapour transport, 223
    wind speed modification, 220
Australia's Commonwealth Scientific and Industrial Research Organisation (CSIRO), 189

Broadacre rainfed farming systems, Australia
  AAGIS survey, 59
  adaptation strategies
    animal genetics, 57
    crop management options, 58
    crop species changes, 57
    erosion management, 58
    grazing management, 56–57
    pest management, 58–59
    planting time, 58
    systemic and transformational adaptation, 53–56
    tactical/incremental adaptations, 53–55
  agricultural communities, 60
  climate change impacts
    biotic stress, 53
    carbon dioxide, 52–53
    extreme climatic events, 53
    precipitation, 51–52
    temperature, 50–51
  climatic variability, 49–50
  drylands syndrome, 61
  farming communities, 60–61
  national importance, 48–49
  rural household, 59

Central West Asia and North Africa (CWANA), 189
Climate projections
  circulation patterns, 13
  event sequences and phenomena, 12
  fossil fuels and land-use change, 8
  global climate models, 8–9
  human activities, 7
  internal variability, 10
  precipitation, 11–12
  regional patterns, 12
  sensitivity, 8–9
  temperature change, 10–11
  uncertainty, 7–8
Climate resilience, cropping systems
  cereal production, 167
  framing adaptation
    adaptive resource management, 168
    context-specific and stakeholder-defined development, 169–170
    ecological stability theory, 168
    external changes, 169
    generic characteristics, 170
    incremental adaptation, 169
    societal objectives, 169
    system property and adaptive capacity, 169
    transformational adaptation, 169
  highly intensive and productive systems, 167
  scale dependency, 177
  soil and crop management options

biophysical parameters, 177
climate forecasts, 176
contour stone bunds, 174
cropping intensity adjustment, 175–176
crop selection, 172
cultivar selection, 171–172
current trends, 178
farmer experimentation, 176–177
intercropping, 172–173
irrigation, 175
landscape diversity, 174–175
pest, weed and disease dynamics, 174
reduced tillage techniques, 173
soil residue retention, 173–174
sowing date changes, 171
CMIP3 coordinated study, 9
Concentrated animal feeding operations (CAFOs), 129
Crop-livestock-woodlands system, 127–128
Cropping system resilience
  complex system behaviour, 204–205
  global warming, 204
  imperfect impact models, 204
  modelling studies
    biophysical variables, 207–208
    challenges, 210–212
    conceptual assessment framework, 208–209
    erosion and soil loss, 209
    generic approach, 209–210
    holistic capacity approaches, 205
    hydrological and livestock models, 207
    impact approaches, 205
    limitations, 206
    midwest USA, 209
    process-based crop simulation, 206
    semi-subsistence systems, 205
    statistical analyses, 206
    weather variability and extremes, 209
  uncertainty, 204

Decision Support System for Agrotechnology Transfer, 188–189
Drylands syndrome, 61
Dynamical forecasts, 236–237

Earth System Models, 9–10
El Niño Southern Oscillation (ENSO), 10, 237
Ethiopia, farm production
  climate coping strategies, 275–278
  data collection and analysis, 273, 274
  drought-coping strategies
    arid and semi-arid pastoralists, 271
    country-level study, 271
    food consumption, 274
    income diversification, 274–275

livestock ownership, 273–274
  response strategies, 273, 275
  shock effects, 275
 drought history, 271
 government policies and investment strategies, 278–279
 long-term climate change, 272
 micro-level adaptation strategies, 271–272
 multinomial logit model, 273
 precipitation prediction models, 278
 study area, 272

Fruit production, UK
 elevated CO_2 and ozone, 89
 everbearing strawberry, 89
 invertebrate pests
  CO_2 levels changes, 100
  precipitation changes, 99–100
  temperature changes, 98–99
 irrigation
  improvements, 96–97
  requirements, 95
  schedule irrigation, 96
  water availability, 95–96
 new pests and diseases, 101–102
 perennial fruit
  chill accumulation, 91
  chilling and heat sum requirements, 89–90
  dormancy and bud break, 89–91
  floral bud development, 90
  low chill cultivars, 91
  orchard management practices, 91
 pests control, 102
 plant pathogens
  climate change, 101
  CO_2 level changes, 101
  precipitation changes, 100–101
  temperature changes, 100
 soft fruit production, 88
 soil flooding
  biochemical changes, 94
  branches, 94
  crop sensitivities, 94
  lowered root hydraulic conductance, 93–94
  prevention, 94–95
 vertebrate pests, 97

Greenhouse gases (GHG) emission
 livestock management, 163–164
 manure management, 163
 nitrogen fertilizer application, 163
 rangeland management, 161–163
 sown pasture management, 164
Global circulation models (GCMs), 217, 237
Global warming, 204

Grazing cover crops, 126–127

Heat stress reduction, dairy cows
 cow cooling
  bottlenecks, 121
  combined soaking and evaporation, 120
  cooler environment, 119–120
  evaporation, 118–119
  fans and sprinkler systems, 121
 dry lots and desert barns, 120
 mechanically ventilated freestalls, 120
 naturally ventilated freestall barn
  advantage, 116–117
  east-west orientation, 116, 117
  four-row, 121
  north-south orientation, 115–117
 temperature humidity index
  controlled environment facilities, 113
  management time interval, 112–113
  minimum THI, 113–114
  thermal load, 113

Indo-gangetic plains (IGP)
 climatic parameters, 33
 intensive rice-wheat system see Intensive rice-wheat system, Asia
 rice-wheat cropping systems, 33
 subregions, 32
Integrated crop-livestock systems, North and South America
 characteristics, 125
 climate change, 124
  adaptation, 134
  forage and grazing lands, 134–135
  landscape features, 135
  sustainable system design, 135
  temperate-humid region, 135–136
 designing and implementation, 124
 greenhouse gas emissions, 132–133
 soil organic C sequestration
  in Brazil, 131–132
  continuous cropping systems, 130
  forages, 129
  maize grain yield, 129
  pasture-crop rotations, 129–130
  in Typic Kanhapludults, 130
  in Uruguay, 130
 temperate-humid climates, 125
 types of
  crop-livestock-woodlands, 127–128
  dual-purpose use of small grains, 127
  grazing cover crops, 126–127
  grazing crop residues, 127
  pasture cropping, 128
  regional integration, 129
  sod-based rotations, 126

Intensive livestock systems
    advantage, 110
    animal productivity, 110
    climate change, 110
    construction and management, 110
    heat stress
        conductive cooling, 114
        evaporative cooling, 114
        hyperthermia, 111
        metabolic pathway, 111
        negative impacts, 110–111
        reduction of see Heat stress reduction, dairy cows
        shading, 115
        variables, 111
        water availability, 114–115
    seasonal patterns, milk yield
        climatic effects, 112
        environment effects, 111
        factors affecting, 111
        heat stress, 112
        protein yield pattern, 111
        temperature and relative humidity, 112
Intensive rice-wheat system, Asia, 31–32
    adaptation strategies
        climate change impacts, 40
        crop improvement, 40–42
        crop management, 42–43
        diversification, 43
        farmers indigenous knowledge, 43–44
        pest management, 43
        weather forecasts and crop insurance, 43
    climate change impacts
        CO_2 enrichment, 35–36
        drought, 36
        flooding, 37
        sea level rise, 36–37
        temperature, 34–35
    photosynthetic processes stimulation, 32
    risk estimation, 37
    vulnerability
        in IGP transects, 39, 41
        spikelet sterility, 40
Intercropping, 172–173

Land management
    data and methods, 144
    discussion, 147–151
    economic value, 146–147
    ecosystem services, 140, 142–144
    energy input and output, 147
    human-environment system, 142
    land use and land-use change, 140–141
    provisioning services, 144–146
Low-input farming systems, 47 see also Broadacre rainfed farming systems, Australia

Millennium Ecosystem Assessment (MEA), 73
MIROC General Circulation Model (GCM), 189
Mixed crop-livestock systems in Asia
    agrosilvopastoral systems, 158
    climate change and adaptation
        adaptive farming system, 161
        ecosystem scale impacts, 159–160
        high stress tolerance, 160–161
        species scale impacts, 159
    crop/pasture rotation, 157–158
    grain crops, 157
    greenhouse gases (GHG) emission
        livestock management, 163–164
        manure management, 163
        nitrogen fertilizer application, 163
        rangeland management, 161–163
        sown pasture management, 164
    ponds, 158–159
    rangeland, 155–157
Model for Interdisciplinary Research on Climate (MIROC), 189
Moderate resolution imaging spectroradiometer (MODIS), 222
Multinomial logit model (MNL), 273

Organic and agroecological farming systems (OAFS), Latin America
    certified organic systems, 71
    climate change projection and assessment
        annual mean warming, 72
        cold spells and frosts, 72–73
        dominant current frameworks, 73
        downscaling experiments, 72
        Millennium Ecosystem Assessment, 73
        regional circulation model, 72
        scaling down and up, 73
        in SESA and SAMS regions, 72
    farmers indigenous knowledge, 71
    organic export products, 71
    soil fertility, 71

Pasture cropping, 128
Projections see Climate projections

Rainfed intensive crop systems
    agroecosystems
        carbon dioxide, 21–22
        crop protection, 25
        soils, 24–25
        temperature and rainfall, 22–24
    climate change, projected impacts of, 25–26
    food and feed, supply of, 17
    modern farm management, 17
    pests and diseases, 17
    temperature and rainfall, 17
    temperature variability, 18
    in warmer and/or drier areas, 18

wheat and grain maize yields
  in North and South America, 20
  in north-west Europe, 18–19
  yield benefits, 20
Regional climate models (RCMs), 12
Representative concentration pathway (RCP)
    scenarios, 8

Seasonal climate forecasts (SCFs)
  adaptation boundary object, 235
  benefits of
    capacity-building programmes, 239
    climate change adaptation, 239
    climate risk management, 237–238
    climate variability, 239–240
    decision support systems, 238–239
    developmental role, 240
    socio-psychological reasons, 239
  climate change adaptation
    adaptive capacity, 234, 244–245
    experiential knowledge and climate
       information, 242–243
    insufficiency critique, 241–242
    misleading concepts, 243
    pros and cons of SCF, 245–246
    robust decision making idea, 244
    skilful SCFs, 240–241
    uncertainty SCFs, 240–241
  climatic variables, 233–234
  coping capacity, 235
  process-based/dynamical modelling, 236–237
  statistical seasonal forecasting, 236
  transformational capacity, 235
  utility, 235–236
Sod-based rotations, 126
South American Monsoon System (SAMS), 72
Southern Oscillation Index (SOI)), 236
Special Report on Emissions Scenarios (SRES), 8
Spikelet sterility, 40
Statistical seasonal forecasting, 236
Sustainable intensive production systems
  adaptation capacity enhancement, 196–197
  breeding
    maize, 191–192
    wheat, 192–193
  climate change impacts
    abiotic stress, 188–190
    biotic stress, 189–191
    high temperatures, 187–188
    rainfall variability, 188
  economic analysis, 199

food and livelihood security, 186–187
food supply chains, 187
food waste, 187
germplasm, 198
improved agronomy
  arid, irrigated, wheat-based system, 193
  cropping system resilience, 194
  fertilizer, seeds and pesticides, 194
  irrigation systems, 193–194
  precision agriculture tools, 194–195
  rainfall infiltration, 193
  residue retention, 194
  semiarid, rainfed, maize-based system,
      193
  temperature changes, 194
  water-use efficiency, 193
  zero-tillage systems, 193, 194
land degradation, 187
maize and wheat production, 186
multidisciplinary approaches, 197–198
postharvest storage technologies, 195–196

Techno-rationalist framing, 243
Tree-soil-crop interactions, 218–220

Urban heat island effect, 217

Vegetable production, UK
  Brassica and *Allium* crops, 89
  elevated CO_2 and ozone, 89
  growth and quality
    broccoli, 92
    cauliflower, 92
    temperature effects, 91–92
    warmer winters and summers, 92–93
  invertebrate pests
    CO_2 levels changes, 100
    precipitation changes, 99–100
    temperature changes, 98–99
  limited soil water availability, 97
  plant pathogens
    climate change, 101
    CO_2 level changes, 101
    precipitation changes, 100–101
    temperature changes, 100
  potato crops, 89
  wet soils, 97

Water Erosion Prediction Project-Carbon
    Dioxide (WEPP-CO2) model, 209
Water, nutrient and light capture in agroforestry
    systems (WaNuLCAS model), 224–225